Maintenance and Operations 2018: Parts 1 and 2

Transportation Research Record: Journal of the Transportation Research Board Volume 2672, Issue 12

ISBN: 978-1-5108-8180-8

Printed from e-media with permission by:

Curran Associates, Inc.
57 Morehouse Lane
Red Hook, NY 12571

Some format issues inherent in the e-media version may also appear in this print version.

Published by Sage Publications USA

ISBN (Print) 978-1-5108-8180-8
ISBN 2018 Printed Set (All Issues) 978-1-5108-7735-1

Printed by Curran Associates, Inc. (2019)

Additional copies of this publication are available from:

Curran Associates, Inc.
57 Morehouse Lane
Red Hook, NY 12571 USA
Phone: 845-758-0400
Fax: 845-758-2633
Email: curran@proceedings.com
Web: www.proceedings.com

Volume 2672 Number 12 December 2018

TRR

JOURNAL OF THE TRANSPORTATION RESEARCH BOARD

Contents

Maintenance and Operations: Parts 1 and 2

Part 1: Maintenance and Preservation

Articles

Field Performance of Chip Seals for Pavement Preservation 1
Adriana Vargas-Nordcbeck

Economic Analysis of Pavement Preservation Techniques 10
Natalia Zuniga-Garcia, Wilfrido Martinez-Alonso, Andre de Fortier Smit, Feng Hong, and Jorge A. Prozzi

Performance-Graded Specifications for Asphalt Emulsions Used in Chip Seal Preservation Treatments 20
Javon Adams, Mohammad Ilias, Cassie Castorena, and Y. Richard Kim

Soft Computing Tools to Predict Progression of Percent Embedment of Aggregates in Chip Seals 32
Aksel Seitllari and M. Emin Kutay

Universal Method to Measure Water Infiltration into Asphalt Pavements Independent of Pavement Preservation Surface Treatment 40
Ben C. Cox, Isaac L. Howard, and Chase A. Hopkins

Field Evaluation and Cost Effectiveness of Crack Sealing in Flexible and Composite Pavements 51
Momen Mousa, Mostafa A. Elseifi, Mohammad Bashar, Zhongjie Zhang, and Kevin Gaspard

Laboratory Investigation of Pavement Backfilling Materials for Micro-Trenching in Cold Regions 62
Leila Hashemian, Vinicius Afonso Velasco Rios, and Alireza Bayat

Field Application of UAS-Based Bridge Inspection 72
Junwon Seo, Luis Duque, and James P. Wacker

Life Cycle Assessment Framework for the U.S. Bridge Inventory 82
Saeed Babanajad, Yun Bai, Helmut Wenzel, Moritz Wenzel, Hooman Parvardeh, Ali Rezvani, Robert Zobel, Franklin Moon, and Ali Maher

Updating Bridge Deck Condition Transition Probabilities as New Inspection Data Are Collected: Methodology and Empirical Evaluation 93
Rabi G. Mishalani, Abdollah Shafieezadeh, and Zequn Li

Senior Community Resilience with a Focus on Critical Transportation Infrastructures: An Accessibility-Based Approach to Healthcare 103
Richard Twumasi-Boakye, John O. Sobanjo, Sylvester Kwame Inkoom, and Eren Erman Ozguven

Systematic Procedures for the Analysis of Agency and User Costs of Bridge Repair Actions 116
Ehsan Fereshtehnejad, Jieun Hur, Abdollah Shafieezadeh, Mike Brokaw, Brad Noll, Jared Backs, and Amjad Waheed

Sunlight Degradation of Polymeric Detectable Warning Surface Products 127
Sukjoon Na, Siavash Vahidi, Hieu Nguyen, Adel ElSafety, and Grace Hsuan

Part 2: Maintenance and Operations

Articles

Parametric Ordinal Logistic Regression and Non-Parametric Decision Tree Approaches for Assessing the Impact of Weather Conditions on Driver Speed Selection Using Naturalistic Driving Data 137
Ali Ghasemzadeh, Britton E. Hammit, Mohamed M. Ahmed, and Rhonda Kae Young

Development of a Pavement Marking Life Cycle Cost Tool 148
Adam M. Pike and Bharadwaj Bommanayakanahalli

The Effect of Additives on the Low Temperature Ice-Melting Capacity of NaCl 158
Kine Nilssen, Alex Klein-Paste, and Johan Wåhlin

2018 D. GRANT MICKLE AWARD: Maintenance and Operations

Bundling Bridge and Other Highway Projects: Patterns and Policies 167
Yu Julie Qiao, Jon D. Fricker, Samuel Labi, and Trevor Mills

Articles

Highway Asset Deterioration Rates 179
William Rasdorf and Ali Almalki

Developing a Pre-Task Safety Briefing Tool for Kentucky Maintenance Personnel 187
Zamaan Al-Shabbani, Roy Sturgill, and Gabriel B. Dadi

Accommodation of Saltwater Temporary Pipelines on the Roadside 198
William Holik, Cesar Quiroga, Ioannis Tsapakis, and Jing Li

Development and Implementation of a Pavement Marking Management System in Florida 209
Bouzid Choubane, Joshua Sevearance, Charles Holzschuher, James Fletcher, and Chieh (Ross) Wang

Modeling Surface Temperatures for Snow-Covered Roads: A Case Study for a Heated Pavement in Bærum, Norway 220
Anne D. W. Nuijten, Inge Hoff, and Knut V. Høyland

Role of Route Optimization in Benefiting Winter Maintenance Operations 232
Tyler Miller, Ben Gleichert, Hiram Crabtree, Jamie Hendershot, Rod Nuveman, and William Schneider

Transportation Officials, Washington, D.C.

K. Jane Williams, *Acting Administrator, Federal Transit Administration, U.S. Department of Transportation*

Paul Zukunft *(Admiral, U.S. Coast Guard), Commandant, U.S. Coast Guard, Washington, D.C.*

---

* Membership as of February 2019.

TRB

TRANSPORTATION RESEARCH BOARD

CORE SPONSORS

Transportation Departments of the 50 States and the District of Columbia

Federal Government

U.S. Department of Transportation
- Federal Aviation Administration (FAA)
- Federal Highway Administration
- Federal Motor Carrier Safety Administration (FMCSA)
- Federal Railroad Administration (FRA)
- Federal Transit Administration
- National Highway Traffic Safety Administration (NHTSA)
- Office of the Assistant Secretary of Transportation for Research and Technology (OST-R)

U.S. Air Force Civil Engineer Center (AFCEC)
U.S. Coast Guard (USCG)
U.S. Department of Energy
U.S. Department of the Interior (DOI)
U.S. Environmental Protection Agency (EPA)

Nongovernmental Organizations

American Public Transportation Association (APTA)
Association of American Railroads (AAR)
California Air Resources Board

The National Academies of
SCIENCES · ENGINEERING · MEDICINE

The **National Academy of Sciences** was established in 1863 by an Act of Congress, signed by President Lincoln, as a private, nongovernmental institution to advise the nation on issues related to science and technology. Members are elected by their peers for outstanding contributions to research. Dr. Marcia McNutt is president.

The **National Academy of Engineering** was established in 1964 under the charter of the National Academy of Sciences to bring the practices of engineering to advising the nation. Members are elected by their peers for extraordinary contributions to engineering. Dr. C. D. Mote, Jr., is president.

The **National Academy of Medicine** (formerly the Institute of Medicine) was established in 1970 under the charter of the National Academy of Sciences to advise the nation on medical and health issues. Members are elected by their peers for distinguished contributions to medicine and health. Dr. Victor J. Dzau is president.

The three Academies work together as the National Academies of Sciences, Engineering, and Medicine to provide independent, objective analysis and advice to the nation and conduct other activities to solve complex problems and inform public policy decisions. The Academies also encourage education and research, recognize outstanding contributions to knowledge, and increase public understanding in matters of science, engineering, and medicine.

Learn more about the National Academies of Sciences, Engineering, and Medicine at **www.national-academies.org.**

The **Transportation Research Board** is one of seven major programs of the National Academies of Sciences, Engineering, and Medicine. The mission of the Transportation Research Board is to increase the benefits that transportation contributes to society by providing leadership in transportation innovation and progress through research and information exchange, conducted within a setting that is objective, interdisciplinary, and multimodal. The Board's varied activities annually engage about 7,000 engineers, scientists, and other transportation researchers and practitioners from the public and private sectors and academia, all of whom contribute their expertise in the public interest. The program is supported by state transportation departments, federal agencies including the component administrations of the U.S. Department of Transportation, and other organizations and individuals interested in the development of transportation.

Learn more about the Transportation Research Board at **www.TRB.org**.

Article

TRR
JOURNAL OF THE TRANSPORTATION RESEARCH BOARD

Transportation Research Record
2018, Vol. 2672(12) 1–9

DOI: 10.1177/0361198118768531
journals.sagepub.com/home/trr
SAGE

Field Performance of Chip Seals for Pavement Preservation

Adriana Vargas-Nordcbeck[1]

Abstract

Over time, new pavements deteriorate due to the effect of traffic loads and the environment. If appropriate treatments are applied during the early stages of deterioration, it is possible to extend the service life of the pavement without incurring in costly rehabilitation or reconstruction activities. Chip seals are preservation treatments that can help protect the pavement structure, reduce the rate of pavement deterioration, improve skid resistance, and address minor surface problems. As part of the National Center for Asphalt Technology Pavement Preservation Study, chip seal test sections were placed in a low traffic volume road (Lee County Road 159) in Auburn, Alabama. The location consists of a two-lane county road that provides dead end access to a quarry and an asphalt plant, resulting in a high percentage of heavy loads. At the time of treatment, the existing pavement was 14 years old and consisted of a 5.5 in. hot-mix asphalt layer over a 6.0 in. granular base. Treatments were applied in the summer of 2012 and have been in service for approximately 4.5 years. During this time, cracking, roughness, rutting, and macrotexture data were collected weekly to evaluate pavement performance. The results determined that the performance of the treated sections is highly dependent on the initial condition of the pavement, particularly the percentage of area cracked. Pavements that are treated while still in good condition tend to remain in that category for a longer time. Macrotexture may also be used to evaluate the functional performance of the chip seals.

Over time, new pavements deteriorate due to the effect of traffic loads and the environment. If appropriate treatments are applied during the early stages of deterioration, it is possible to extend the service life of the pavement without incurring in costly rehabilitation or reconstruction activities. Chip seals consist of an application of asphalt (commonly an emulsion) onto the existing pavement surface followed by an application of aggregate chips, which are immediately rolled (*1, 2*). Double or triple chip seals are constructed by placing two or three applications of a chip seal operation over the same roadway, each subsequent layer being placed after the previous layer has cured (*3*). These preservation treatments can help protect the pavement structure, reduce the rate of pavement deterioration, improve skid resistance, and address minor surface problems.

Chip seal performance depends on many variables, such as traffic, materials, construction practices, and initial condition of the pavement. Typically, performance lives range from 3 to 7 years for single course applications, while multiple applications can increase pavement life by up to 10 years (*3–5*). Eltahan et al. used data from the LTPP SPS-3 experiment to estimate the probability of failure of several preservation treatments—including chip seals—with respect to the original condition at application time (*6*). The results indicated that placing the treatment while the pavement is in good condition results in better performance.

Objectives and Scope of Work

The objective of this study was to evaluate the field performance of chip seal test sections placed on a low traffic volume road. To accomplish this objective, full-scale test sections were treated with chip seal applications and their performance monitored periodically over a 4.5-year period.

Methodology

National Center for Asphalt Technology Pavement Preservation Study

The National Center for Asphalt Technology (NCAT) Pavement Preservation Study was initiated in 2012 in response to the growing need for agencies to obtain reliable performance data for different pavement preservation treatments. The main objective of this study is to develop independent life-extending benefit curves for various preservation treatments.

[1]National Center for Asphalt Technology, Auburn University, Auburn, AL

Corresponding Author:
Address correspondence to Adriana Vargas-Nordcbeck: vargaad@auburn.edu

Table 1. Aggregate Gradation for Chip Seals

Sieve size	Percent passing		
	#7 chip seal	#89 chip seal	W10 chip seal
3/4″	100	100	100
1/2″	97	100	100
3/8″	57	100	100
1/4″	6	74	100
No. 4	2	43	100
No. 8	2	9	88
No. 16	2	4	35
No. 30	2	3	30
No. 50	2	2	24
No. 100	NA	1	NA
No. 200	1.5	0.9	6.8
Design application rate (lb/yd^2)	27.5	16.0	10.0

In Phase I, a low traffic volume road (Lee County Road 159) in Auburn, Alabama, was selected as the test site. The location consists of a two-lane county road that provides dead end access to a quarry and an asphalt plant, resulting in a high percentage of heavy loads. A Lee County report from 2012 indicated that the AADT on Lee County Road 159 was 563 vehicles per day, with a truck volume of 60% (*7*).

At the time of treatment, the existing pavement was 14 years old and consisted of a 5.5 in. hot-mix asphalt layer over a 6.0 in. granular base. The half-mile road segment was in overall sound structural condition, making it an adequate candidate for pavement preservation. The surface condition varied throughout the length of the road, providing the possibility of quantifying performance in terms of initial pavement condition. The road was split into 25, 100-ft long sections, of which 23 sections received a treatment or combination of treatments, and two were left as untreated control sections with varying amounts of cracking.

Treatment Description

This paper focuses on a subset of six treatments that can be classified as a variation of a chip seal, either as a stand-alone, as a multi-layer application or a combination with other treatments. Overall, three chip seal designs were performed using the McLeod Method (*8*), for three different aggregate sizes. The aggregate gradations are shown in Table 1, and Table 2 presents the design properties of each treatment.

Four of the test sections received a single layer application using the same aggregate gradation. These included a stand-alone chip seal and a chip seal placed after sealing the existing cracks with a hot-applied asphalt-based product. A single layer chip seal was also placed over a specially formulated, polymer modified, crack resistant membrane. Finally, a scrub seal—a more advanced and aggressive multi-step chip seal process that uses a specialized emulsion as the chip binder and rejuvenator in conjunction with a mechanized scrub broom that forces the optimum amount of emulsion into the cracks—was also applied using the same aggregate gradation. In addition, sections were also treated with double and triple layer chip seals.

Treatments were applied in the summer of 2012 and have been in service for approximately 4.5 years. During this time, pavement performance has been monitored periodically to determine the life-extending benefit of the various pavement preservation treatments.

Data Collection

Performance data were obtained weekly using a data collection vehicle equipped with an inertial profiler, an INO laser rut measurement system (LRMS), and high resolution cameras. The inertial profiler collects the longitudinal profile of the pavement surface for both wheel paths. These profiles are used to measure the roughness of the pavement in terms of the international roughness index (IRI). The LRMS system can obtain the transverse profile and measure rut depths, while multiple laser sensors can also be used to calculate macrotexture data. Finally, the images obtained with the high resolution cameras are processed using special software to assess the extent of detected cracking.

Test sections were further subdivided into 5 × 10 ft subsections that could be monitored in order to quantify the life-extending benefit of the pavement preservation treatments as a function of initial condition (*9*). Subsections were classified as good, fair, or poor depending on the pre-treatment amount of cracking, as shown in Table 3. These category ranges were selected to be consistent with the performance criteria established by the Federal Highway Administration (*10*). Since the treatments included in this paper are not expected to have a significant impact on performance in terms of roughness and rutting, these measures were not included in the assignment of the condition categories.

Traffic

One particular characteristic on Lee County Road 159 is that although both lanes are subjected to the same traffic volume, traffic loads vary significantly as trucks travel unloaded in the inbound direction (to the quarry and asphalt plant) and exit loaded in the outbound direction. Data-sharing agreements between the quarry and asphalt plant and NCAT provide load history throughout the life of the project. Equivalent single-axle loads (ESALs) are calculated based on truck weight and axle configuration. Figure 1 shows the cumulative ESALs in both lanes through 4.5 years of service.

Table 2. Treatment Properties

Section	Aggregate gradation	Emulsion type	Design emulsion rate (gal/yd^2)
Single layer chip seal	#89	CRS-2HP	0.26
Single layer chip seal + crack seal	#89	CRS-2HP	0.26
Double layer chip seal	#89 (top)	CRS-2HP	0.26
	#7 (bottom)	CRS-2HP	0.30
Triple layer chip seal	W10 (top)	CRS-2HP	0.26
	#89 (middle)	CRS-2HP	0.30
	#7 (bottom)	CRS-2HP	0.15
Single layer chip seal on fiber membrane	#89	CRS-2L	NA
Scrub seal	#89	CMS-1P(CR)	0.35

Table 3. Pavement Condition Categories

Initial condition	% of area cracked
Good	<5
Fair	5–20
Poor	>20

Existing Pavement Structural Condition

Prior to treatment application, the condition of the pavement structure was evaluated by means of the falling weight deflectometer (FWD) in accordance with ASTM D 4694. FWD data were collected at two random locations per test section. At each location, testing was performed in the inside, outside, and between wheel paths, in the inbound and outbound directions. A Dynatest 8000 FWD was used with nine sensors spaced at 0, 8, 12, 18, 24, 36, 48, 60, and 72 in. from load center, and a load plate with a radius of 5.91 in. At the time of testing, three repetitions at four load levels (approximately 6, 9, 12, and 16 kips) were completed at each location, and surface temperatures were recorded.

FWD deflections recorded were used to calculate deflection basin parameters (DBPs) that characterize the structural condition of the various pavement layers. In this study, three DBPs were considered: the area under the pavement profile (AUPP), the base damage index (BDI) and the base curvature index (BCI), as shown in Table 4.

The AUPP parameter correlates with the horizontal tensile strain at the bottom of the asphalt concrete layer. Higher AUPP values indicate higher strains, and therefore a higher potential for bottom-up fatigue cracking. The BDI and BCI indices represent the structural condition of the base and subgrade layers, respectively. Higher values indicate more damage in the structure.

Figure 2 shows the pretreatment values for all three DBPs in each of the test sections, and the descriptive statistics are summarized in Table 5. It can be observed that the structural condition of the pavement was fairly uniform among the different test sections.

Results and Discussion

Cracking

For each treatment, three data sets were generated based on the initial condition of the pavement. Although there is a significant difference in traffic loads between both lanes, the level of cracking observed in the test sections after 4.5 years is primarily influenced by the type of treatment applied and the initial condition of the surface. An analysis of variance was performed to study the effect of three factors on cracking after a 4.5 year period: treatment type, travel lane (inbound lane or outbound lane), and initial condition category (good, fair, or poor). As shown in Table 6, the traffic level resulting from the difference in travel lanes had no effect on the amount of cracking in year 4.5 at the 0.05 significance level (p-value = 0.067). Therefore, the results presented hereafter are not separated by travel lane.

Table 7 summarizes the amount of cracking and the corresponding condition category after 4.5 years, as well as the number of years the pavement remained in good condition after treatment. It can be seen from Figure 3, that the amount of cracking present after 4.5 years is far greater in the control (untreated) sections than in the pavements that received treatments. In addition, the percentage of cracking is generally greater for pavements in which the initial condition category was worse.

While the control sections have deteriorated to a "poor" condition regardless of initial condition category, treated sections remain in "good" to "fair" condition, with the exception of the single layer chip seal applied on a "poor" surface. It is evident that by treating pavements while they are still in "good" condition, they will remain in that category for longer than if the treatments are deferred until the pavement has reached a higher level of deterioration.

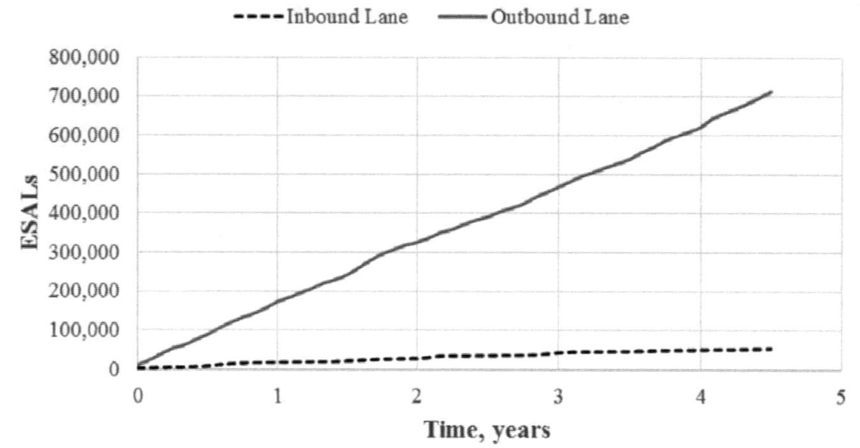

Figure 1. Cumulative ESALs over time.

Table 4. Deflection Basin Parameters

Parameter	Equation	Layer characterized
Area under pavement profile	$AUPP = \frac{5D_0 + 2D_{12} + 2D_{24} + D_{36}}{D_0}$	Upper layers
Base Damage Index	$BDI = D_{12} - D_{24}$	Base
Base Curvature Index	$BCI = D_{24} - D_{36}$	Subbase or subgrade

Note: D_0, D_{12}, D_{24}, and D_{36} are the measured deflections at a distance of 0, 12, 24 and 36 in. from the center of the loading plate, respectively.

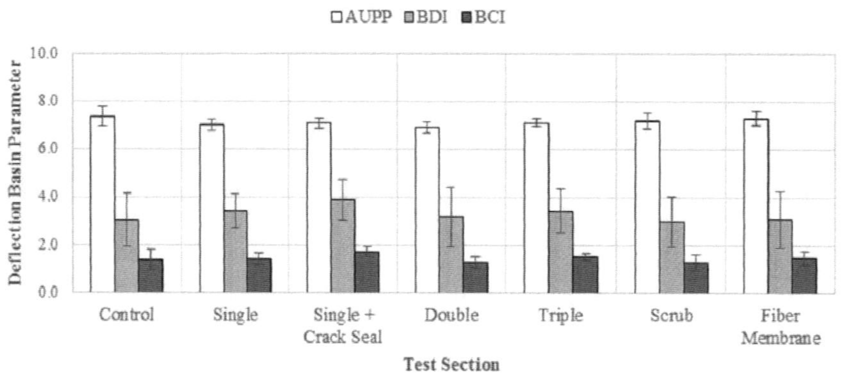

Figure 2. Pretreatment DBP values for all test sections.

Table 5. DBP Descriptive Statistics

Parameter	Mean	Min.	Max.	SD
AUPP	7.2	6.9	7.4	0.3
BDI	3.3	3.0	3.9	1.1
BCI	1.4	1.3	1.7	0.4

Note: Min. = minimum; Max. = maximum; SD = standard deviation.

Compared with the control sections, pavements that are treated while in "good" condition will remain that way for at least 1.5 years longer, and in some cases are projected to outlast the untreated pavements by more than 3.3 years. NCAT's Pavement Preservation Study is still ongoing, and a full assessment of the life-extending benefit of the treatments will be performed when all sections deteriorate beyond the "good" category.

Roughness

Roughness was measured in terms of the IRI for the entire length of each section. As mentioned earlier, the preservation

Table 6. Results from Analysis of Variance

Source	df	Adj. SS	Adj. MS	*F*-value	*p*-value
Treatment type	7	92,642	13,234.6	75.46	0.000
Travel lane	1	595	594.8	3.39	0.067
Initial condition	2	10,306	5,153.1	29.38	0.000
Error	301	52,791	175.4		
Total	311	164,384			

Note: Adj. SS = adjusted sum of squares; Adj. MS = adjusted mean squares.

Table 7. Effect of Initial Condition on Treatment Performance

Treatment	Initial condition	% cracking at year 4.5	Condition at year 4.5	Years in good condition
Single layer chip seal	Good	10.0	Fair	2.7
	Fair	5.6	Fair	4.5
	Poor	34.6	Poor	2.7
Single layer chip seal + crack seal	Good	2.3	Good	4.5+
	Fair	11.4	Fair	2.7
	Poor	16.3	Fair	3.3
Double layer chip seal	Good	5.2	Fair	4.5
	Fair	8.6	Fair	3.4
	Poor	19.3	Fair	3.4
Triple layer chip seal	Good	3.5	Good	4.5+
	Fair	4.0	Good	4.5+
	Poor	9.9	Fair	2.7
Single layer chip seal on fiber membrane	Good	4.3	Good	4.5+
	Fair	7.2	Fair	2.1
	Poor	10.5	Fair	2.7
Scrub seal	Good	5.7	Fair	4.5
	Fair	4.1	Good	4.5+
	Poor	6.6	Fair	1.3
Control	Good	29.6	Poor	1.2
	Fair	43.9	Poor	NA
	Poor	93.3	Poor	NA

Note: NA = Pavement was deteriorated beyond good condition at the beginning of the study.
[+]Pavement remains in good condition after 4.5 years.

treatments included in this study are not expected to have a significant effect on pavement roughness, nonetheless, this performance measure is monitored periodically to identify any potential failures. As seen on Figure 4, the IRI remains relatively constant over time, and therefore the life-extending benefit of the treatments is not expected to be driven by changes in roughness.

Rutting

As with IRI, changes in rutting performance have not been significant enough over time to influence the determination of the life-extending benefit of the treatments. Figure 5 shows the average rut depths of the test sections prior to treatment application and after 4.5 years. Although rutting appears to have decreased as a result of the treatments, it is important to note that the results are generally low. Because chip seals are not intended to improve rutting, the differences observed may be actually attributed to random measurement error and seasonal variations. It can be seen that all sections remain under 5 mm of rutting after 4.5 years, which can be categorized as "good" (*10*).

Macrotexture

Average mean texture depth (MTD) values were obtained for each of the test sections. Although macrotexture affects pavement surface friction, MTD alone is not sufficient to characterize skid resistance, and therefore should not be viewed as a pavement performance indicator, but rather a pavement property. Previous studies have proposed the use of macrotexture in chip seals as a performance indicator for the treatment itself (*1*).

Figure 6 presents the MTD values over time along with the initial MTD for each of the treatments. The data compiled to date suggest that MTD decreases exponentially

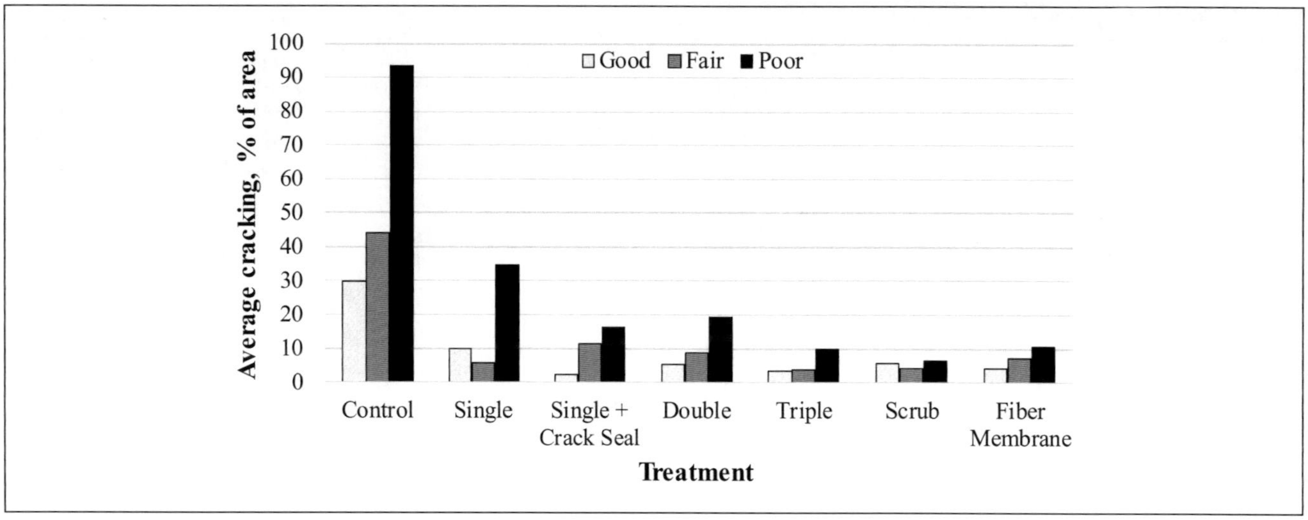

Figure 3. Cracking area after 4.5 years.

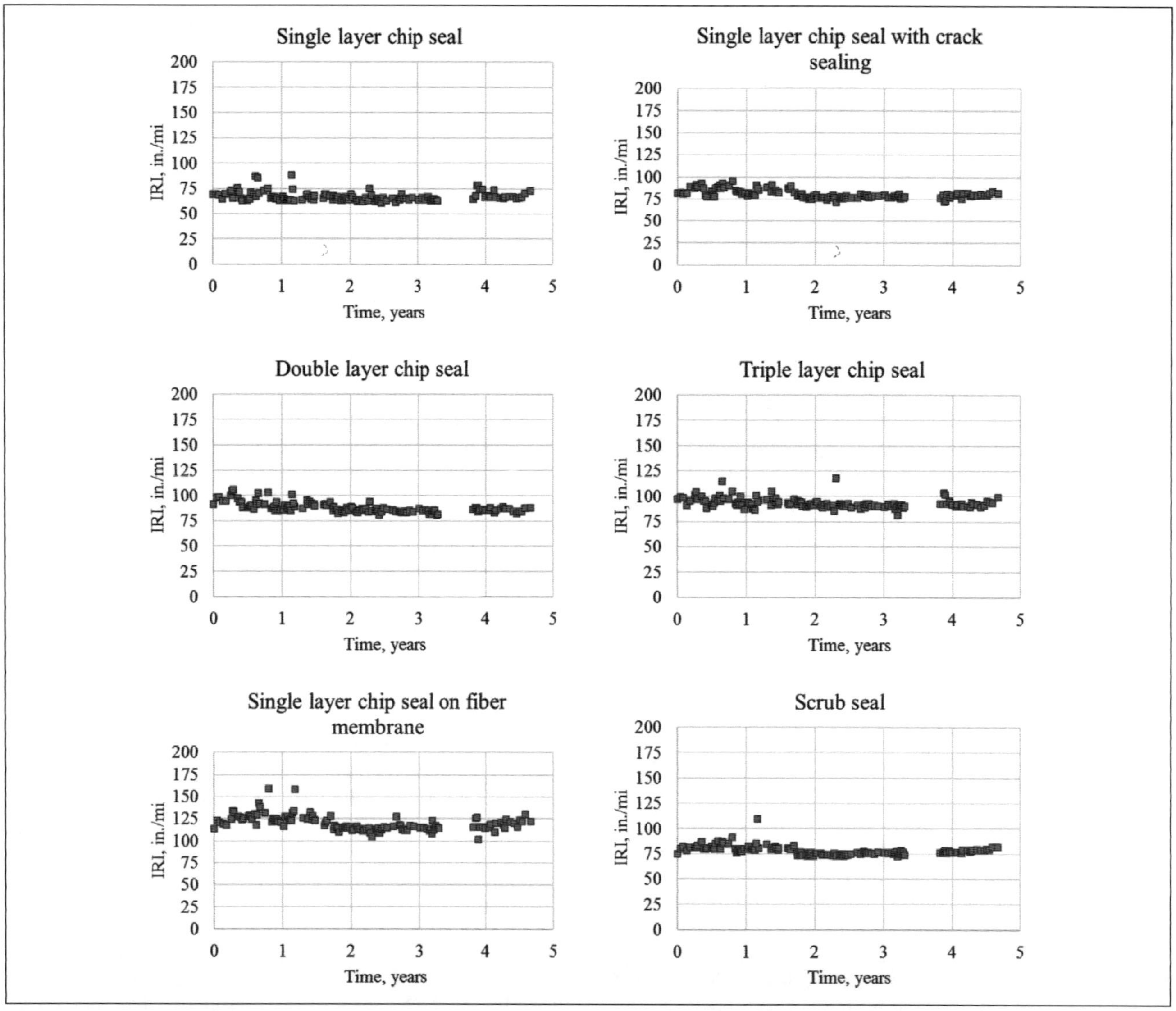

Figure 4. IRI over time for chip seal treatments.

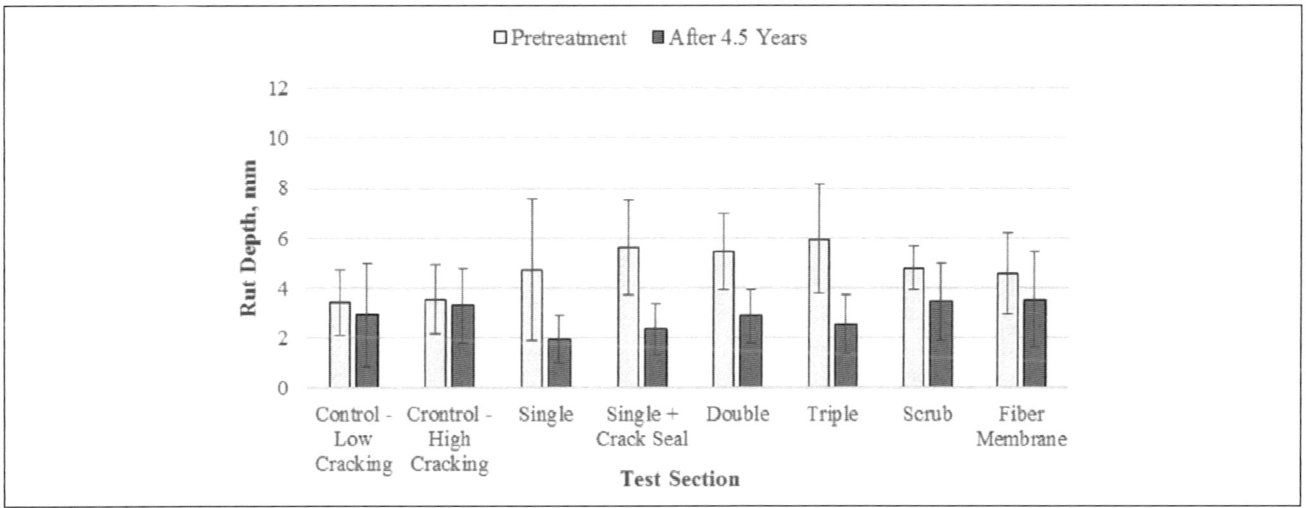

Figure 5. Average rut depths.

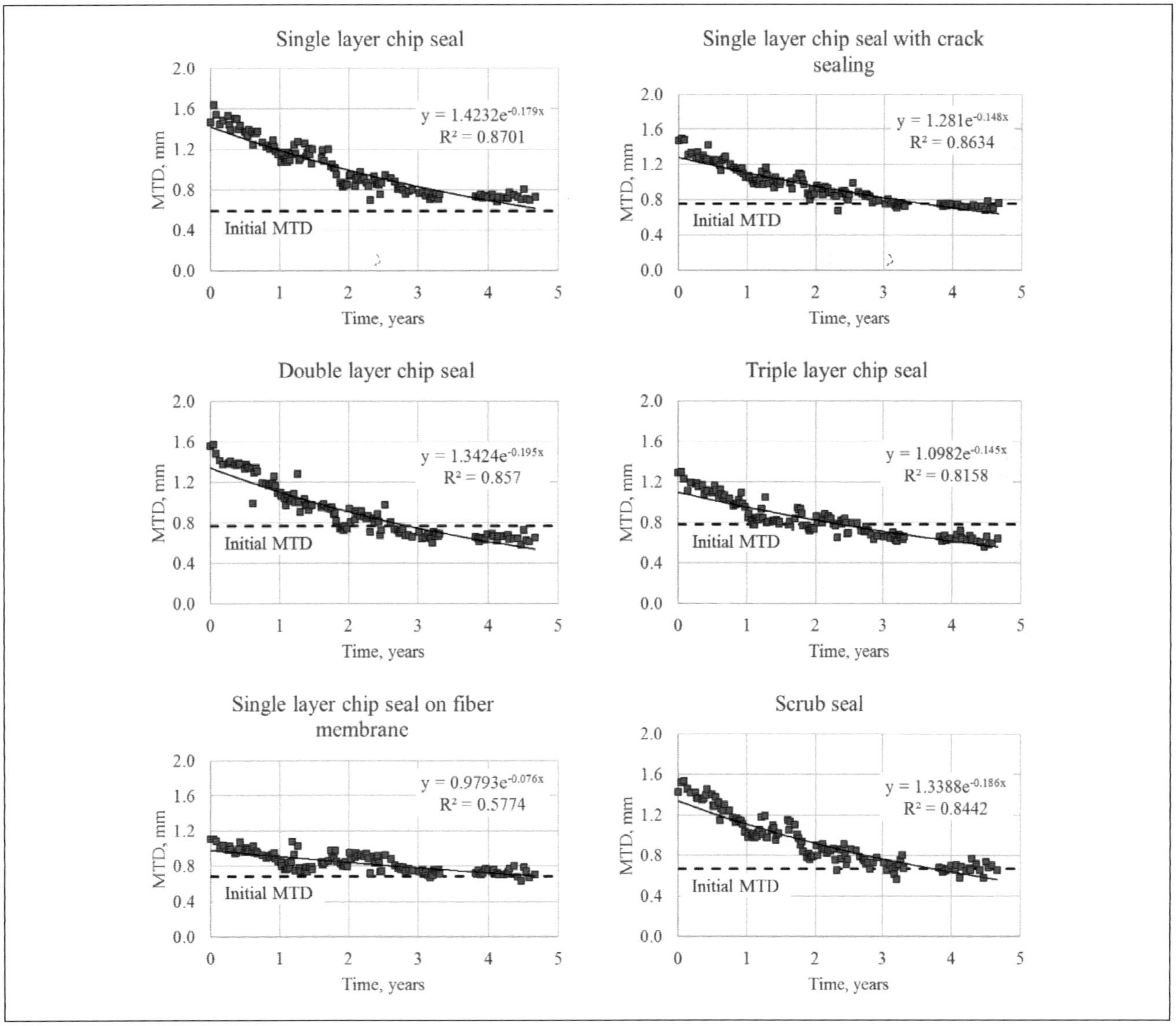

Figure 6. MTD over time for chip seal treatments.

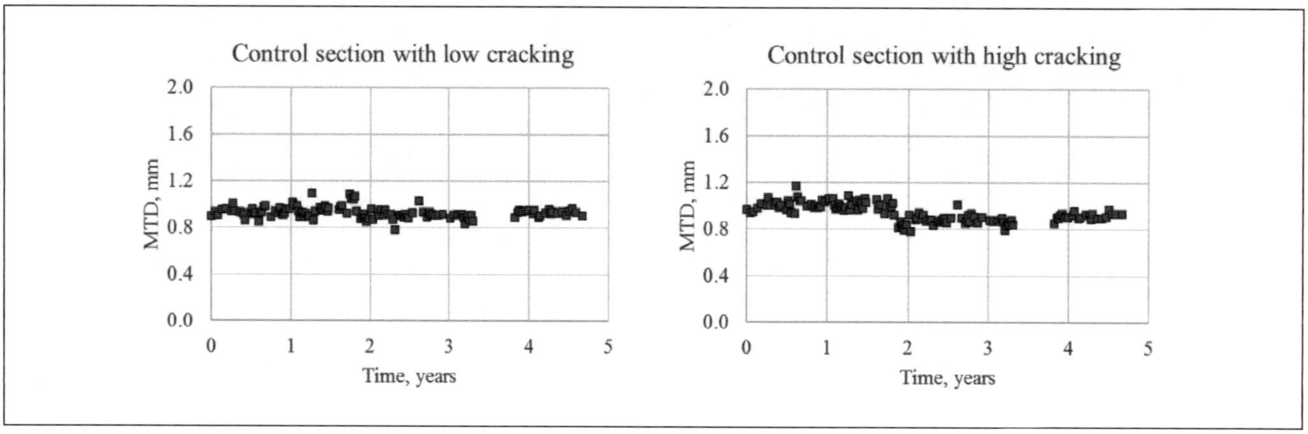

Figure 7. MTD over time for untreated sections.

with the age of the chip seal applications. In contrast, the untreated sections (shown in Figure 7) exhibited little variation in macrotexture during the same period. If these treatments are applied to address low friction, a way to determine the functional life of the treatment would be to estimate the time needed to return to the pretreatment MTD value. For the treatments considered in this study, that time varies between 2.5 and 4.5 years, depending on the type of treatment and initial MTD. Another approach could be to use the curves shown in Figure 6 to estimate the time required to reach a threshold MTD value, defined based on experience.

Conclusions

This study presented the preliminary findings related to field performance of a group of pavement preservation treatments included in NCAT's Pavement Preservation Study. Due to the long-term nature of the study, more time is needed to monitor performance and draw definitive conclusions. However, during the first 4.5 years of service of the chip seal treatments applied on Lee County Road 159 in Auburn, Alabama, the following observations can be made:

- Performance of the treated pavements is mainly affected by the percentage of area cracked. Pavement deterioration in terms of roughness and rutting occurs at a slower rate and has not represented a major factor during the first 4.5 years.
- The amount of cracking developed after treatment application depends on the initial condition of the pavement. This emphasizes the importance of timely intervention, as pavements that are treated while still in good condition tend to remain in that category for a longer time.
- Compared with the untreated sections, the percentage of cracking after 4.5 years was significantly reduced for all treated pavements, regardless of initial condition category.
- Macrotexture may be used to evaluate the functional performance of the treatments, especially if they are applied with the objective of improving pavement friction. For the available data, the decrease in MTD over time follows an exponential relationship.

The pavement preservation treatments included in this study are part of a broader research project that includes multiple treatment types and combinations of treatments. This study aims at evaluating the effect of additional factors such as traffic and climate on treatment performance. Data collection efforts continue and additional test sections have been constructed in new locations since Phase I of the experiment was implemented in 2012. These newer sections have been in service for less than 2 years, but will be studied in the long-term and are expected to provide valuable information that will help agencies estimate the performance of their treated pavement more accurately according to their local conditions.

Acknowledgments

The author thanks the Departments of Transportation of Alabama, Colorado, Georgia, Illinois, Kentucky, Michigan, Minnesota, Mississippi, Missouri, New York, South Carolina, Oklahoma, Tennessee, and Wisconsin, and FP^2 Inc. for their sponsorship of this project.

References

1. Gransberg, D., and D. James. *NCHRP Synthesis 342: Chip Seal Best Practices*. HRB, National Research Council, Washington, D.C., 2005.
2. Shuler, S., A. Lord, A. Epps-Martin, and D. Hoyt. *NCHRP Report 680: Manual for Emulsion-Based Chip Seals for Pavement Preservation*. HRB, National Research Council, Washington, D.C., 2011.
3. Wade, M., R. DeSombre, and D. Peshkin. *High Volume/High Speed Asphalt Roadway Preventive Maintenance Surface*

Treatments. Report SD99-09. South Dakota Department of Transportation, Pierre, SD, 2001.

4. Peshkin, D., K. L. Smith, A. Wolters, J. Krstulovich, J. Moulthrop, and C. Alvarado. *SHRP 2 Report S2-R26-RR-2: Guidelines for the Preservation of High Traffic Volume Roadways*. Transportation Research Board of the National Academies, Washington, D.C., 2011.
5. Wilde, W. J., L. Thompson, and T. J. Wood. *Cost-Effective Pavement Preservation Solutions for the Real World*. Final Report No. MN/RC 2014-33. Minnesota Department of Transportation, Paul, MN, 2014.
6. Eltahan, A. A., J. F. Daleiden, and A. L. Simpson. Effectiveness of Maintenance Treatments of Flexible Pavements. *Transportation Research Record: Journal of the Transportation Research Board*, 1999. 1680: 18–25.
7. *Lee County Highway System Index*. Lee County, Florida. http://www.leeco.us/business/highway/docs/FY_11_12_Highway_System_Index_for_Website.pdf. Accessed March 14, 2017.
8. McLeod, N. W. A General Method of Design for Seal Coats and Surface Treatments. *Proceedings of the Association of Asphalt Paving Technologists*, Vol. 38, 1969, pp. 537–628.
9. Powell, B. Development and Validation of a Nondestructive Methodology to Measure Subgrade Moisture Contents at the NCAT Pavement Test Track. In *The Roles of Accelerated Pavement Testing in Pavement Sustainability* (J. Aguiar-Moya, A. Vargas-Nordcbeck, F. Leiva-Villacorta, and L. Loria-Salazar, eds.), Springer International Publishing, New York City, 2016, pp. 621–632.
10. Federal Highway Administration. National Performance Management Measures; Assessing Pavement Condition for the National Highway Performance Program and Bridge Condition for the National Highway Performance Program. *Federal Register*, Vol. 82, No. 11, 2017, pp. 5963–5968.

The Standing Committee on Pavement Preservation (AHD18) peer-reviewed this paper (18-02655).

Article

TRR
JOURNAL OF THE TRANSPORTATION RESEARCH BOARD

Transportation Research Record
2018, Vol. 2672(12) 10–19

DOI: 10.1177/0361198118768515
journals.sagepub.com/home/trr
SAGE

Economic Analysis of Pavement Preservation Techniques

Natalia Zuniga-Garcia[1], Wilfrido Martinez-Alonso[1], Andre de Fortier Smit[1], Feng Hong[2], and Jorge A. Prozzi[1]

Abstract

This paper summarizes the research study conducted to develop and implement a methodological framework, using an economic analysis technique, to evaluate the cost effectiveness of the three different preventive maintenance treatments applied to roadways in Texas: chip seals, microsurfacing, and thin overlays. The analysis is based on a stochastic evaluation of the effective life and cost of more than 14,000 maintenance and rehabilitation projects undertaken from 1994 to 2015. The effect of traffic loads, traffic volume, and roadway type was also evaluated. The life-cycle cost of the preventive maintenance techniques was obtained using a Monte Carlo simulation. Among the principal results, it was found that chip seals are the most cost-effective treatment and present the lowest life-cycle cost variability. The effective life of all three treatments was found to be quite similar. Additionally, it was found that the chip seals and microsurfacing tend to present comparable life-cycle costs when used on heavy traffic roadways.

Highways and road networks represent a critical infrastructure asset for governments and citizens. However, keeping the pavements in good condition has always been a challenge for transportation agencies. The American Society of Civil Engineers (ASCE) has rated the current condition of roads in America with a D+ (at risk), placed in the lowest possible tier (*1*). The Federal Highway Administration (FHWA) estimates $170 billion dollars would be needed on an annual basis to improve the conditions and performance of existing roads (*1*).

A significant share of the highway cost is assigned to the implementation of pavement preservation techniques. These techniques are applied to extend the life of a pavement and in some cases to increase its structural capacity. Despite the improvements in data collection for estimating the service life of maintenance and rehabilitation (M&R) practices, these data have not been extensively used (*2*). Further, there is a lack of a sound methodology to quantify objectively the benefits of applying minor rehabilitation work, preventative maintenance or routine maintenance. Empirical evidence has shown timely maintenance to be the best approach to delay the deterioration rate of a given pavement surface, thus extending its service life.

Economic analysis is regularly used in transportation research, mainly as an evaluation tool to compare the life-cycle cost (LCC) of alternative projects (*3–5*). Many research efforts have produced knowledge directed to improve the life-cycle cost analysis (LCCA) of pavements. Some studies have contributed to the field by exploring how governmental agencies can benefit from it, studying management strategies and developing sensitivity analysis for factors affecting LCCAs (*3–6*). However, no significant research has been done to quantify the service-life cost of M&R treatments due to the short service life and the lack of availability of the actual cost and lifespan information. Access to data on service life of M&R treatments is limited (*6, 7*) and it is subjected to intense data processing (*2*). Some researchers have based their analysis on field tests (*8*), previous research findings (*9*), or example data sets from pavement condition surveys (*10*).

The main objective of this research study was to develop and implement a methodological framework, using economic analysis, to evaluate the cost effectiveness of the three primary preventive maintenance (PM) treatments applied in Texas: chip seals, microsurfacing and thin overlays. A stochastic LCCA framework was developed to accomplish the goal, using historical information from more than 14,000 M&R projects undertaken between 1994 and 2015, obtained from the Texas Department of Transportation's (TxDOT) databases.

The main contributions of this study can be summarized as follows: (*a*) it entails the use of actual cost and life data of PM treatments, obtained after processing M&R projects databases; (*b*) it uses a stochastic approach that allows the inclusion of uncertainty into the analysis; (*c*) it provides insights into the cost effectiveness of chip seals, microsurfacing, and thin overlays; and (*d*) it evaluates the effect of traffic volume, traffic load, and facility type on the LCC of the treatments.

[1]The University of Texas at Austin, Austin, TX
[2]Texas Department of Transportation, Austin, TX

Corresponding Author:
Address correspondence to Natalia Zuniga-Garcia: nzuniga@utexas.edu

The subsequent sections of this paper are organized as follows. "Preventive Maintenance Treatments" provides a general description of the main M&R concepts and describes the treatments analyzed. "Life Cycle Cost Analysis" presents an overview of the principal characteristics of the LCCA. "Case Study" provides a detailed explanation of the data processing and considerations regarding the case study. "LCCA of PM Treatments" presents the methodology framework, results, and discussion. The final section, "Summary and Conclusions," summarizes the main findings.

Preventive Maintenance Treatments

FHWA defines pavement preservation as a program employing a network-level, long-term strategy, to enhance the condition of the pavement network by implementing an integrated and cost-effective set of practices that extend the life of a pavement, improve safety, and meet motorists' expectations (*11*). Pavement preservation practices have three main components: minor or non-structural rehabilitation, preventative maintenance (PM), and routine maintenance (*11*). The PM process is the systematic application of a series of maintenance actions over the service life of a pavement, targeted to maintain a good condition, extend its lifespan, and minimize the LCC (*11*).

The application of PM treatments is a critical part of the pavement preservation program. Using PM treatments decreases the rate of pavement deterioration in order for it to meet performance standards. PM treatments are applied while the roadway is still in a good condition and shows only minimal distresses, before the pavement falls into a condition where placing structural overlays, major milling or reclaiming, or replacement is necessary (*12*).

Specific PM treatments exist for bituminous-surfaced and concrete-surfaced pavements, and may also include the maintenance of drainage features. This study examines three bituminous-surfaced treatments commonly used in Texas: chip seals, microsurfacing and thin overlays.

Chip Seal

Chip seal (or seal coat) is a surface treatment in which asphaltic material (typically asphalt cement or emulsified asphalt) is sprayed over the pavement surface followed by a uniform graded aggregate cover (*13*). It is employed to correct minor deficiencies in the surface such as cracking, raveling, bleeding, and lack of skid resistance (*14*). Although a chip seal can be applied to high-traffic volume roadways, it is generally limited to low-traffic volume roadways. Facilities with average daily traffic in excess of 10,000 vehicles per day will be considered as high traffic (*14*). In general, chip seals are more effective in preventing reappearance of cracking than microsurfacing, with the asphalt rubber chip seal performing best overall (*13*).

Microsurfacing

Microsurfacing comprises a mixture of cationic polymer-modified asphalt emulsion, mineral aggregate, mineral filler, water, and other additives (*15*). It is applied to prevent raveling and oxidation (*13*). In addition to a lane or road-width treatment, microsurfacing has been used to fill minor rutting and it proved to be more effective than chip seals in preventing the reappearance of bleeding (*13*).

Thin Overlay

Thin overlays consist of an overlay of hot-mix asphalt (HMA) less than one inch thick. It is composed of a compacted mixture of aggregate and asphalt binder mixed hot in a mixing plant (*15*). Thin overlays are placed to improve friction, correct surface irregularities, and reduce surface permeability. This treatment is not recommended when the surface presents existing rutting (*14*).

Life Cycle Cost Analysis

The LCCA is a decision-support tool frequently employed by transportation agencies to compare total user and agency costs for different project alternatives. LCCA is considered a type of benefit–cost analysis (BCA), which is an economic analytical tool that compares benefits and costs for the different alternatives, allowing decision makers to select the optimal option (*16*).

Types of LCCA

The FWHA defines two approaches to prepare a LCCA: deterministic and stochastic (probabilistic). The methods differ in the manner they address the variability and uncertainty associated with the LCCA input parameters including activity cost, activity timing, and discount rate (*16*).

Deterministic LCCA Approach. The deterministic LCCA involves the use of fixed input values that result in deterministic output values. The value for each input parameter is usually estimated based on either historical evidence or engineering judgment (*16*). Sensitivity conducted to test input assumptions by varying one input and holding other inputs constant should be conducted as a minimum requirement in deterministic LCCAs. This helps to determine the effect of the variation of parameters in the outputs. Some flaws in the deterministic approach include its failure to address simultaneous variation in multiple-input cases as well as the inability to convey the degree of uncertainty associated with the LCC estimates (*16*).

Stochastic LCCA Approach. The stochastic or probabilistic LCCA allows the value of individual input parameters to be defined by a frequency (probability) distribution (*16*). The probabilistic LCCA is more robust than the deterministic one, and involves the modeling of uncertainty as it takes probabilities into account (*9*). To characterize these uncertainties, a stochastic LCCA approach combines probability descriptions of random variables and computer simulation techniques, commonly known as Monte Carlo simulation (MCS) (*17*).

LCCA Methodology

FHWA has developed a methodology to estimate LCCAs for different alternatives (*16*). It can be synthesized in five steps: (*a*) establish design alternatives; (*b*) determine activity timing, (*c*) estimate costs; (*d*) compute, life-cycle costs; and (*e*) analyze the results. This methodology was used for conducting the LCCAs in this study.

The LCCA process is initiated after an asset has been selected to be improved and a range of possible alternatives have been identified to help accomplish that improvement (*16*). Each design alternative will have an expected initial design life, periodic maintenance treatments, and often a series of rehabilitation activities (*17*). At least two mutually exclusive options ought to be considered. The economic difference between alternatives is then assumed to be attributable to the total cost that each of them represents (*16*). The identification of maintenance and rehabilitation activities is often based on historical practice, research, and agency policies (*16*).

The second step consists of the activity timing. The service life of the initial pavement design and subsequent rehabilitation and maintenance activities have an impact on the LCCA outcomes as they directly affect the frequency of agency intervention. These will, in turn, affect agency costs along with user costs during the periods when the pavement is subjected to construction and maintenance activities (*17*). The timing of activities should be based on existing performance records, such as those available from pavement management systems (*16*).

The third step is to estimate costs. The LCCA considers costs accrued to highway agencies and to users of the highway system, as a result of the agency construction and maintenance activities (*16*). LCCAs do not require all costs associated with each alternative to be calculated. Only costs demonstrating differences between alternatives need be considered (*16*). Costs common to all alternatives cancel out and these cost factors are consequently excluded from the LCCA calculations (*17*).

After that, the next step is to compute the LCC. Projected activity costs for alternatives need to consider the value of money over time (*16*). Methods from the field of economics are applied to transform anticipated future costs to present value, so that the lifetime costs of different alternatives can be compared in a direct manner (*16*).

Several economic indicators can be considered during a LCCA. The most common ones include benefit/cost (B/C) ratios, internal rate of return (IRR), net present value (NPV), and equivalent uniform annual costs (EUAC). The B/C analysis represents the net discounted benefits divided by net discounted costs for a given alternative. This methodology is not recommended for pavement analysis because of the difficulty in sorting out reliable benefit and cost estimates (*12*). The IRR represents the discount rate necessary to make discounted cost and benefits equal. This index provides valuable information when budgets are constrained, or if the accuracy of the adopted discount rate is doubtful (*17*). NPV and EUAC are typically used to convert cost streams into a single economic value by using a discount rate that resembles reality in a reliable manner (*5*).

The NPV or net present worth (NPW) is the discounted monetary value of the expected net benefits. Present-worth costs of the strategies provide a fair comparison basis (*5*). There is a strong agreement in the literature that NPV should be the economic efficiency indicator of choice (*5, 17*). The NPV considered those costs accruing to highway agencies and to users of the highway system as a result of agency construction and M&R activities (*16*). The salvage value includes the remaining service life values and it is subtracted from the NPV. The NPV for the lifespan of a pavement section can be estimated using the following equation:

$$\text{NPV} = \text{Initial Cost} + \sum_{k=1}^{N} \text{M \& R Cost}_k \left[\frac{1}{(1+i)^{n_k}} \right] - \text{Salvage Value} \left[\frac{11}{(1+i)^{n_k}} \right] \quad (1)$$

where i is the discount rate, and n is the years of expenditure.

When the time interval is not defined in round years, a continuous compounding should be implemented using:

$$\text{NPV} = \text{Initial Cost} + \sum_{k=1}^{N} \text{M \& R Cost}_k \left[\frac{1}{\exp(i * n_k)} \right] - \text{Salvage Valuue} \left[\frac{1}{\exp(i * n_k)} \right] \quad (2)$$

The EUAC combines every NPV obtained for all discounted costs for a studied option and the benefits of an alternative to that option into equal annual payments over the analysis period. The EUAC indicator is particularly meaningful when budgets are established on an annual basis (*9*). It can be calculated by estimating the NPV in the first place and implementing the following equation afterwards:

$$\text{EUAC} = \text{NPV} \left[\frac{i(1+i)^n}{(1+i)^n - 1} \right] \quad (3)$$

Finally, once the deterministic or probabilistic LCCAs have been computed, the present values of the differential costs may be compared across competing alternatives.

Case Study

In the present research study, the analysis used pavement duration data containing historical information about M&R projects in Texas. The available data was compiled from different databases of the TxDOT. The final database required processing and merging information from a design and construction information system (DCIS), pavement management information system (PMIS), maintenance management information system (MMIS), site manager (SM) and Compass.

TxDOT databases containing M&R project-related information can be divided into two groups regarding data formatting and content (*18*). The first group includes DCIS and SM and contains data from contracted projects. The second group includes MMIS and Compass and contains data from internal or in-house projects performed by TxDOT personnel. The first group is the primary source of the data used in this study because it provides more detailed and precise information. The objective of processing the information is to obtain information about the M&R projects such as the location, effective life, cost, and other traffic information.

The final compiled database contains historical information on M&R projects constructed between 1994 and 2015. The present analysis only involves data from PM projects, including a total of 14,372 projects. The PM work only includes sections with non-structural pavement damage.

Effective Life

The effective life of a treatment is a fundamental input in the LCCA. Therefore, obtaining this information from the databases is critical. This section explains the process used to obtain the effective life of the PM treatments analyzed.

Data Processing. Every contracted M&R job uses an identification number called Control Section Job (CSJ). The CSJ contains nine digits, where the first six numbers refer to the control section (location of the roadway segment in the Texas highway network) and the last three numbers identify the job. The first step of the data processing consisted of obtaining initial and final dates for each CSJ. The initial date of a PM treatment corresponds to the day the work was completed and opened to traffic, in a particular control section. The final date corresponds to the day another treatment was applied in the same section. This information is obtained from the SM. The final date was reviewed and carefully corrected using information from the MMIS database.

In the present research, the difference between the initial and final date is defined as "effective life." The elapsed time between applying the treatment and applying another surface over it is an indicator of the treatment's effectiveness. This duration time does not discriminate between reactive and scheduled work; however, it provides an estimate of the treatment's "effective life" based on realistic field pavement data.

The databases were also used to verify that the final dates correspond to actual improvements in the pavement condition. This process was possible using performance curves from the PMIS database, which contains performance information regarding the condition score (CS). This indicator combines the ride quality and surface distress severity into an index that ranges from 0 to 100, where 100 indicates best conditions. Every time a new PM treatment is applied, the performance curve shows an increase. All pavement performance curves were inspected visually to verify that the final dates correspond to an increment in the CS value.

The service life of a surface depends mainly on wear and tear caused by traffic. Traffic information was extracted primarily from the PMIS database. Three indicators describe the traffic information: annual average daily traffic (AADT), equivalent single axle load (ESAL) and highway designation or roadbed type. The processing of this information consisted of obtaining average values for the PMIS sections located within the analyzed section. The traffic volume is represented by the AADT, which measures how busy a road is. AASHTO defines the AADT as the total amount of traffic on a highway segment for one year, divided by 365 days (*19*). The traffic loads are characterized by ESAL, which is a concept developed for the AASHTO Road Test to establish a damage relationship that compares the effects of different axles carrying different loads. The reference axle load is an 80 kN (18,000 lb) single axle with dual tires (*19*).

The type of highway corresponds to TxDOT designations. TxDOT assigns a specific designation to highways located in Texas depending on their construction standards, design requirements, and purpose (*20*). A hierarchy of four roadway categories was defined for the analysis. Primary routes include two categories: Interstate highways (IH) and US highways (US), which form part of a system of expressways that go through more than one State (*20*). US were implemented before IH. State highways (SH) corresponds to a network connecting internal, state-maintained roads. They can belong to both primary and secondary routes (*20*). Farm to market highways (FM) are roadways that connect rural or agricultural areas to market towns and are part of a system of secondary routes.

Survival Analysis. The effective life of the evaluated PM treatment is highly variable. Thus, the most realistic approach to the LCCA is the stochastic method, including a probabilistic effective life. Survival analysis was applied to the PM treatments. This type of analysis allows the incorporation of both observed and censored data. Censored data corresponds to the cases where the treatment is still in use and its service life is not yet consumed. Including censored information in the

LCCA allows an unbiased and more robust estimation. The survival analysis applied used the approach as suggested by Serigos et al. (*18*).

The survival probabilities of the three PM treatments were jointly estimated to reduce the impact of potential confounding factors on the comparison of the different PM treatment's effectiveness. Additionally, the development of an accelerated lifetime model (ALM) adopting a Weibull distribution allows influence factors to be accounted for. This survival model was specified using chip seal as the base treatment and three covariates: traffic volume (in 10^3 ESAL), traffic load (AADT), and roadway type. This ALM model was estimated using the R statistical programming language (R Core Team 2014) employing the SURVIVAL package. By using this model, it was possible to obtain the scale (α) and shape (γ) parameters of the Weibull distribution:

$$f\left(x|\alpha,\gamma\right)=\frac{\gamma}{\alpha}\left(\frac{x}{\alpha}\right)^{\gamma-1}e^{-\left(x/\alpha\right)^{\gamma}} \quad (4)$$

where x is a random variable, $\alpha > 0$ is the scale parameter, and $\gamma > 0$ is the shape parameter.

The Weibull probability distribution was also estimated for each treatment based on traffic volume, traffic load, and roadway type. This modeling was applied to analyze the effect of the traffic variables in the LCC. The traffic volume consisted of three AADT ranges: lower than 5,000, between 5,000 and 10,000, and more than 10,000. The traffic load has three categories: the first corresponds to loads lower than 10^6 ESALS; the second group includes loads between 10^6 and 10^7 ESALs, and the last category incorporates loads greater than 10^7 ESALs. The roadway type categories were based on the four TxDOT designations explained previously.

Cost

The cost for each PM treatment was obtained from the final cost of each project, estimated once the treatment was applied. The information was obtained from the PMIS and the SM databases. Information was also extracted on the total length of the section and the number of lanes. These data allow the estimation of the cost per lane-mile. The cost of each pavement segment is converted into its 2016 value using the inflation index suggested by the United States Bureau of Statistics (*21*). The variation of this cost is a consequence of the variable duration of the applied PM treatments, changes in the value of the materials used, the location of the project, and the workforce employed, among other variables. The project cost was also modeled using a probabilistic approach to take these variabilities into consideration in the LCCA.

The cost was modeled using a Lognormal probabilistic distribution, using the cost obtained from the databases. The Lognormal probability distribution avoids simulating negative costs and better represents the extreme values found in the analyzed data. This distribution also presents a mathematical advantage as it makes it possible to use transformations to estimate costs parameters linearly. The Lognormal distribution requires the location (μ) and scale (σ) parameters; these parameters were obtained using the information from the final PM projects' database. The distribution was also estimated for each treatment based on the traffic volume, traffic load, and roadway type using the categories defined previously. The Lognormal distribution is presented in the following equation:

$$f\left(x|\mu,\sigma\right)=\frac{1}{x\sigma\sqrt{2\pi}}e^{-\frac{\left(lnx-\mu\right)^{b}}{2\sigma^{2}}} \quad (5)$$

where $x \geq 0$ is a random variable, μ is the location parameter, and σ is the scale parameter.

The parameters μ and σ are estimated as follow:

$$\mu=\log\left(m^{2}/\sqrt{v+m^{2}}\right) \quad (6)$$

$$\sigma=\sqrt{\log\left(v/m^{2}+1\right)} \quad (7)$$

where m is the mean, and v is the variance of the log-normal distribution.

LCCA of PM Treatments

A stochastic approach was utilized in this study because it makes it possible to account for the uncertainty of the input parameters and to measure the variability of the LCCA outputs, which can be used for reliability analyses.

Methodology Framework

The analysis period was 25 years. An important assumption is that once a PM treatment was implemented in each section, the same type of treatment would be applied throughout time, as in common practice (*22*). It is important to mention, however, that in some cases a pavement with 25 years of successive maintenance may need a major repair to maintain good condition. Thus, this period is established mainly for the purpose of analysis.

Figure 1 describes the methodological framework used for the analysis. The time between each PM treatment application corresponds to the effective life, which presents a Weibull probability distribution obtained from the survival analysis. When the treatment reached this effective life, another treatment was applied. Note that the terminal condition criteria may vary as the replacement of the treatment is based only on the effective life, which is considered an indicator of the treatment's effectiveness.

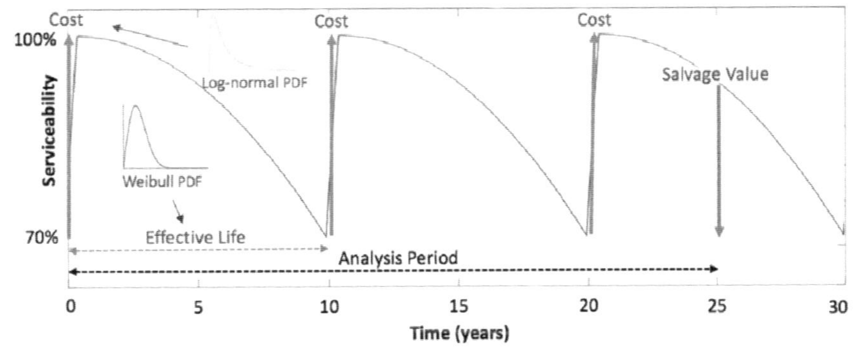

Figure 1. Life cycle cost analysis methodological framework.

The cost of the treatment follows a Lognormal probability distribution. The total cost of maintenance of a pavement segment was computed as the sum of the cost for all the PM treatments applied during the analysis period minus the salvage value at the end of the analysis period. The salvage was the estimated monetary value a pavement section would have at the end of its lifespan, as illustrated in Figure 1. Therefore, this value was calculated as the cost of the last PM cycle in the section times the percentage of the remaining life of that treatment:

$$S_{val} = \frac{m_{res}}{m}(\text{Cost}) \tag{8}$$

where m is the effective life, and m_{res} is the residual life.

The analysis consisted of Monte Carlo simulation (MCS) using 100,000 repetitions. This type of evaluation allows the simulation of both the effective life and the cost using the respective probability distributions. It is possible to obtain a NPV for each of the simulations and thus a probability distribution for the NPV of each PM treatments. These results allowed analysis of the variability of the outputs. The interest rate used was 4%, as suggested in the literature for estimating the NPV for highway projects in Texas (*22*).

The MCS was applied using MATLAB software. The first step was the simulation of the effective life using a random number generator in MATLAB. It was possible to obtain a total of 100,000 values that follow the Weibull distribution described by the scale (α) and shape (γ) parameters obtained from the survival analysis. Similarly, the simulation of the cost consisted of generating 100,000 random values that follow the Lognormal distribution described by the location (μ) and scale (σ) parameters. Finally, the LCC is estimated using the NPV for each of the 100,000 outputs using the simulated life and cost. Thus, a total of 100,000 NPVs are estimated which allows their variability to be observed. This process was applied for the three PM treatments using the complete database, but also for each of the three categories of AADT, three classes of ESAL, and four roadway types, segregated by the PM treatment. Thus, a total of 33 LCC distributions were estimated. The NPV was obtained by implementing the following equation in MATLAB:

$$NPV_{jk} = C_{jk} + \sum_{x=1}^{z_1-1} \frac{C_{jk}}{\exp\left[i \cdot \left(x \cdot m_{jk}\right)\right]} - \frac{S_{val}}{\exp(25 \cdot i)} \tag{9}$$

where,

j = PM treatment (chip seal, microsurfacing, thin overlay)

k = Monte Carlo iteration (1 to 100,000)

NPV_{jk} = Net Present Value of PM treatment j during iteration k

C_{jk} = cost of PM treatment j during iteration k

m_{jk} = effective life of PM treatment j during iteration k

$z = 25 / m_{jk}$ (applications of the PM treatment j during the 25 years' analysis period)

z_1 = round down $\{z\}$

$S_{val} = \left[(z - z_1) \cdot C_{jk}\right] / m_{jk}$ (salvage value)

i = 0.04 (interest rate)

Results and Discussion

This section presents the results obtained from the analysis of the databases and the application of the LCCA framework. It also provides an examination and discussion of the main findings.

Effective Life and Cost. The effective life and cost data obtained from the analysis of the databases are shown in Figure 2, *a* and *b*, respectively. The three PM treatments presented similar effective life distribution, with a mean of 7.6 (chip seals), 8.6 (thin overlays), and 8.4 (microsurfacing) years. It is important to mention that thin overlays present a very similar probability distribution, with a higher variability compared with chip seals.

The cost of the PM treatments shows distinctive distributions. The chip seal showed a mean of approximately $14,500 per lane-mile with estimates located on a tight cost interval and maximum amounts of around $50,000. Microsurfacing

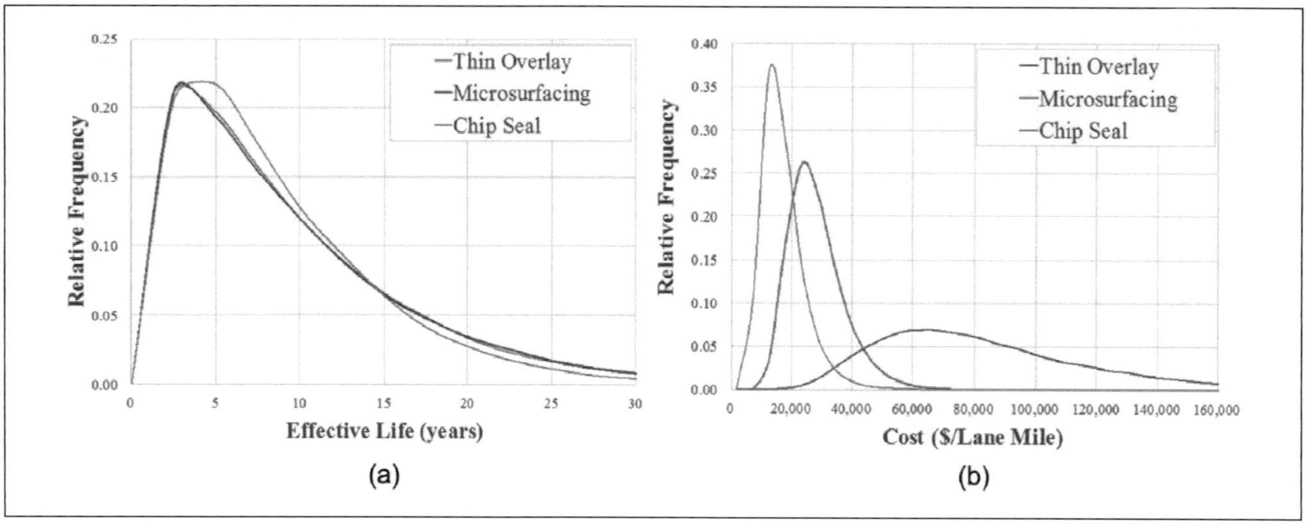

Figure 2. Effective life and cost results: (*a*) effective life per treatment and (*b*) cost per treatment.

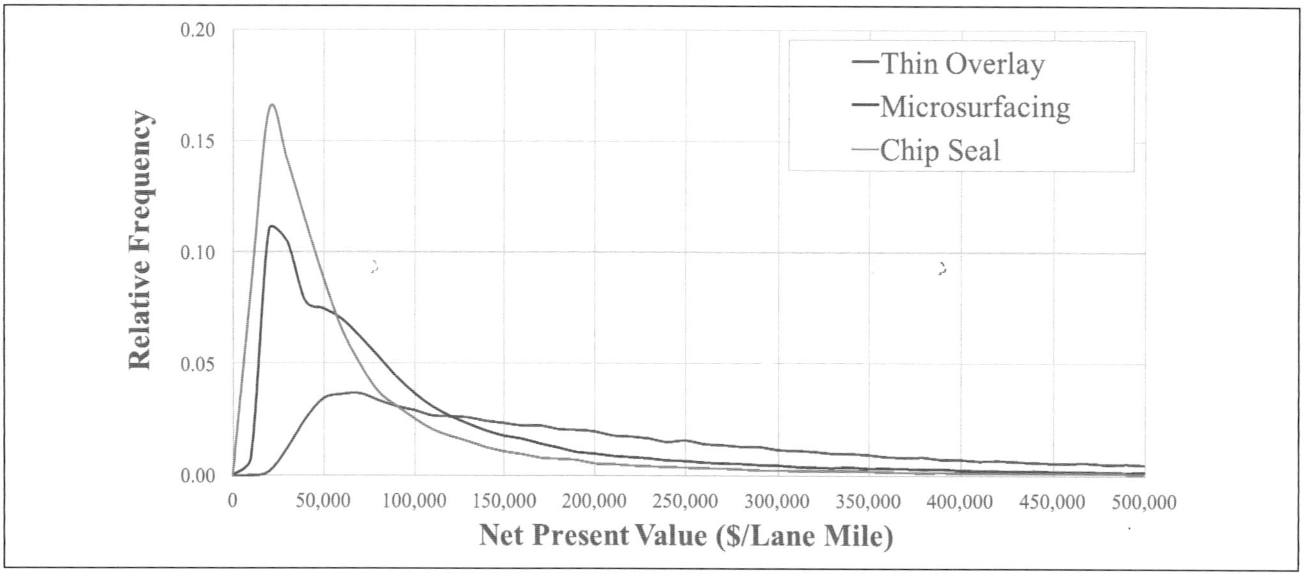

Figure 3. Life cycle cost analysis results.

showed higher variability than chip seal, with an average of $25,600 per lane-mile. Thin overlays presented the greatest cost and variability, with an average of $83,400 per lane-mile, approximately three times as expensive as microsurfacing and almost six times as chip seals.

LCCA Results. The results obtained from the LCCA (using MC) are shown in Figure 3. The probability distribution of the NPV, which represent the LCC, shows a marked difference between the cost of the treatments. The median was the preferred statistical indicator to describe the results because it is less sensitive to the extreme values, as compared with the mean value. Chip seals have the lowest life-cycle cost with a median of $39,000. This treatment also presents the least variability. Microsurfacing presents a LCC 70% higher, with a median of $66,000. Furthermore, thin overlay shows the highest LCC with a median of approximately $190,000 and a high variability. This cost is about five times greater than chip seals and about three times greater than microsurfacing.

Although the LCCA appears to indicate chip seals as the overall best PM treatment option, the selection of the treatment depends on different factors, based on the characteristics of each treatment type. These factors include the environmental conditions of the pavement section because climate and soil has a lasting impact on pavements. In addition, rural roads are generally cheaper to maintain than urban roads. The location within the road is also relevant as segments where acceleration/deceleration occurs need to be more resistant, for

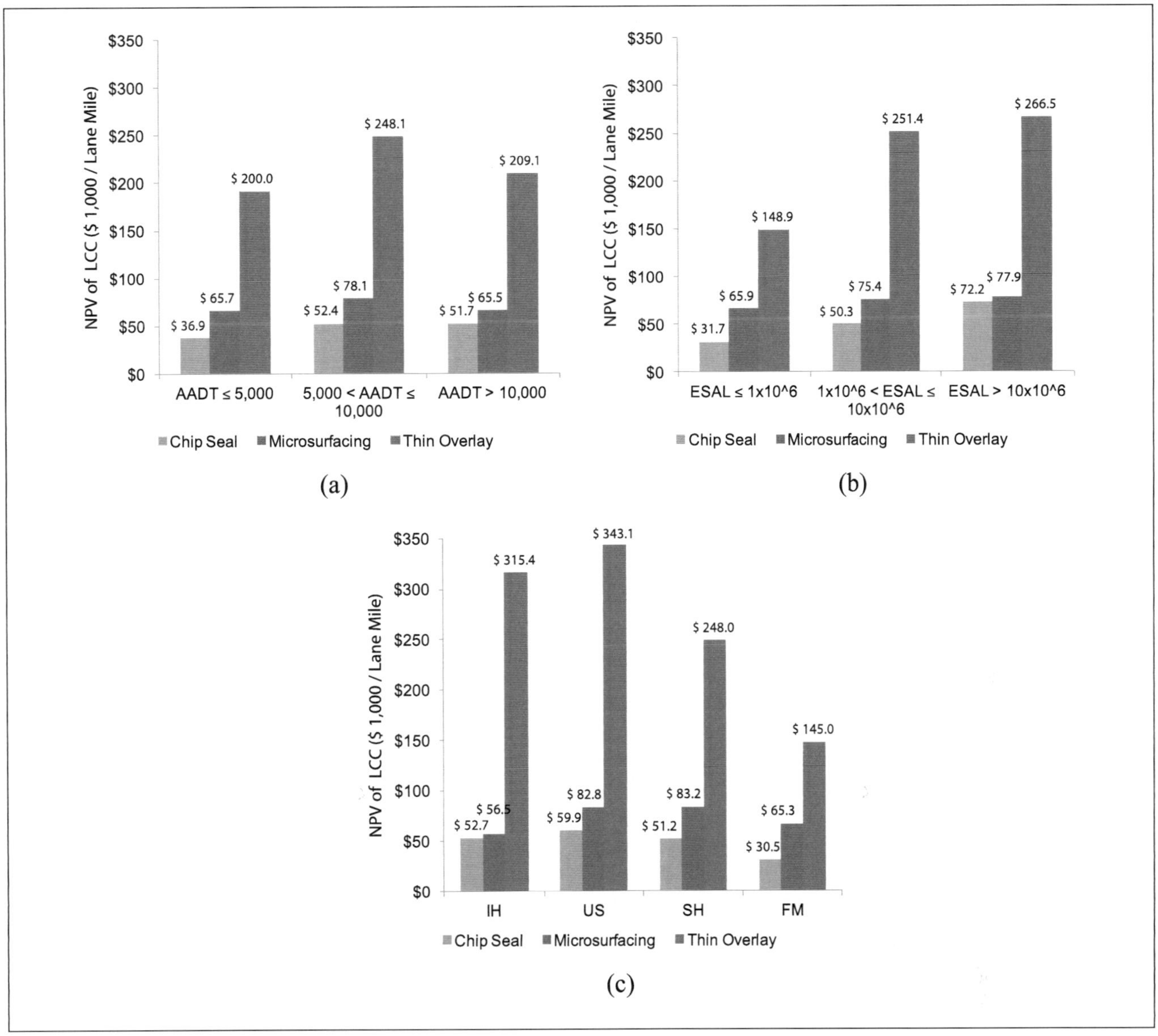

Figure 4. Effect of traffic volume and loads on the LCCA results (median values): (*a*) effect of AADT; (*b*) effect of ESALs; and (*c*) effect of highway type.

example, thin overlays are applied at intersections. Other factors include the availability of required materials as well as design traffic volumes and loads, and previous experience of the transportation agency.

Effect of Traffic Volume and Loads. Traffic volume and load information were also included to evaluate their influence on the effective life, cost, and LCC. The results of the median LCC values based on the AADT are shown in Figure 4*a*. The figure indicates that the second category presents the highest cost. However, the variation of the LCC is not significant among the three classes selected. This suggests that the PM treatment cost does not vary significantly for different ranges of AADT.

The effect of the traffic loads was analyzed using the ESALs. The results are shown in Figure 4*b*. In contrast to the AADT effect, the weight of the vehicles on a pavement section has a significant impact on the lifespan of the pavement, particularly when the loads are noticeably heavy. The cost per PM treatment increases as the weight of loads increases. The LCCA suggests that chip seals are the most efficient option, and thin overlays are the least efficient alternative, but microsurfacing performed better than seal coats under heavier traffic.

The roadway type was also included in the analysis. The results of the median value of the LCC is shown in Figure 4*c*. In this case, the thin overlay presented the highest variability among the evaluated roadway types. FM presented the loweest LCC, followed by SH. IH and US, corresponding to the principal roads of the country, presented the highest LCC.

Cost Probability. The probability that a given PM treatment is more cost-effective (lower LCC) than another treatment option is an important indicator as it can be used for sensitivity analyses and to indicate which option to select in the case of uncertainty. Results of a sensitivity analysis indicating the likelihood of treatments being the most effective option were obtained using 1,000,000 MCS iterations.

The probability of the LCC of chip seals being less than microsurfacing is 70%, and likely 85% compared with thin overlays. This suggests that overall seal coats are effective preventative maintenance options. The LCC of microsurfacing, in turn, is likely to be 75% less than thin overlays. These results provide an approximate comparison of the cost of PM treatments and can be used by practitioners as a first step in the process of decision making when the information is limited.

Summary and Conclusions

This study developed a LCCA framework aimed at evaluating and comparing the three primary PM treatments used in Texas: chip seals, microsurfacing, and thin overlays. The case study was developed based on 14,372 projects implemented between 1994 and 2015. The data collection included processing and merging information from different TxDOT databases. Information about effective life and cost of the project was gathered from the data available. A stochastic approach was used to take into account the data variability. Information on traffic volumes, loads and roadway type of the pavement sections was taken into consideration to gauge the impact of the variation on the effective life and costs affecting the LCCA. The LCC was estimated using the MCS technique to simulate the effective life, cost, and NPV using different probability density functions.

The results of this study are based on data from real projects. This study thus provides an accurate estimation of the effective life, costs, and LCC of the PM treatments. The major findings and conclusions from this study are summarized as follows:

- The effective life of the PM treatments analyzed is similar.
- The LCC of the chip seal is significantly lower than the thin overlay and microsurfacing, and it presents the lower variability.
- Chip seals present the most cost-effective PM treatment, and thin overlays are the least cost-effective. Microsurfacing is, in general, more expensive than chip seals but less costly than thin overlays.
- LLCs are not overly influenced by traffic volume expressed in terms of AADT.
- The LCCA based on ESAL suggests the cost of implementing chip seals notably increases as loads increase from low to medium and heavy. On the other hand, the LCCs for microsurfacing are not significantly changed as traffic loads increase.
- The LCC ratio between chip seals and microsurfacing decreases as loads increase. This suggests that for heavily trafficked pavement sections applying chip seals may be at least as effective as applying microsurfacing.

As the study only considered PM treatments in pavements that were structurally sound, it is recommended that the analysis be extended to investigate the relative benefits of chip seals when considering damaged pavement sections. Additionally, it is recommended to include into future studies the effect of other variables that were not considered and may affect LCC, such as the climate and type of material.

References

1. American Society of Civil Engineers (ASCE). *America's Structures Score*. 2013 Infrastructure Report Card. 2013. http://www.infrastructurereportcard.org/
2. Pour, S. A., and D. H. S. Jeong. Realistic Life-Cycle Cost Analysis with Typical Sequential Patterns of Pavement Treatment Through Association Analysis. *Transportation Research Record: Journal of the Transportation Research Board*, 2012. 2304: pp. 104–111.
3. Federal Highway Administration (FHWA). *Life Cycle Cost Analysis in Pavement Design – In Search of Better Investment Decisions*. Report FHWA-SA-98-079. U.S. Department of Transportation, Washington, D.C., 1998.
4. Ferreira, A., and J. Santos Life-Cycle Cost Analysis System for Pavement Management at Project Level: Sensitivity Analysis to the Discount Rate. *International Journal of Pavement Engineering*, Vol. 14, No. 7, 2013, pp. 655–673.
5. Beg, M. A., Z. Zhang, and W. R. Hudson. *A Rational Pavement Type Selection Procedure*. Center for Transportation Research, University of Texas at Austin, 1998.
6. Peshkin, D. G., T. E. Hoerner, and K. A. Zimmerman. *NCHRP Report 523: Optimal Timing of Pavement Preventive Maintenance Treatment Applications*. Transportation Research Board of the National Academies, Washington, D.C., 2004.
7. Gransberg, D. D., and E. Scheepbouwer. Infrastructure Asset Life Cycle Cost Analysis Issues. *AACE International Transactions*, 2010, pCSC.03.1.
8. Pittenger, D. M. *Stochastic Life Cycle Cost Analysis Model for Sustainable Pavement Preservation Treatments*. Doctoral dissertation. University of Oklahoma, 2012.
9. Pittenger, D. M., D. D. Gransberg, M. Zaman, and C. Riemer. Stochastic Life-Cycle Cost Analysis for Pavement Preservation Treatments. *Transportation Research Record: Journal of the Transportation Research Board*, 2012. 2292: 45–51.
10. Harvey, J., A. Rezaei, and C. Lee. Probabilistic Approach to Life-Cycle Cost Analysis of Preventive Maintenance Strategies on Flexible Pavements. *Transportation Research Record: Journal of the Transportation Research Board*, 2012. 2292: 61–72.

11. Gieger, D. R. *Memorandum: Pavement Preservation Definitions*. FHWA, U.S. Department of Transportation, 2016. https://www.fhwa.dot.gov/pavement/preservation/091205.cfm.
12. Chang, J. R., D. H. Chen, and C. T. Hung. Selecting Preventing Maintenance Treatments in Texas: Using the Technique for Order Preference by Similarity to the Ideal Solution for Specific Pavement Study 3-Sites. *Transportation Research Record: Journal of the Transportation Research Board*, 2008. 1933: 62–71.
13. Texas Department of Transportation (TxDOT). *Pavement Manual*, Texas Department of Transportation, 2017. http://onlinemanuals.txdot.gov/txdotmanuals/pdm/pdm.pdf
14. Texas Department of Transportation (TxDOT). *Seal Coat and Surface Treatment Manual*, Texas Department of Transportation, 2017. http://onlinemanuals.txdot.gov/txdotmanuals/scm/scm.pdf
15. Texas Department of Transportation (TxDOT). *Standard Specifications for Construction and Maintenance of Highways, Streets, and Bridges*, Texas Department of Transportation, 2014. http://ftp.dot.state.tx.us/pub/txdot-info/des/spec-book-1114.pdf
16. Federal Highway Administration (FHWA). *Life-Cycle Cost Analysis Primer*. Federal Highway Administration, Office of Asset Management, U.S. Department of Transportation, Washington, D.C., 2012.
17. Walls, J., and M. R. Smith. *Life-Cycle Cost Analysis in Pavement Design - Interim Technical Bulletin*. Publication No. FHWA-SA-98-079. Federal Highway Administration, U.S. Department of Transportation, Pavement Division, Washington, D.C., 1998.
18. Serigos, P., A. Smit, and J. Prozzi. *Performance of Preventive Maintenance Treatments for Flexible Pavements in Texas*. TxDOT Technical Report No: 0-6878-2. Center for Transportation Research, University of Texas at Austin, 2017.
19. Huang, Y. H. *Pavement Analysis and Design*. Pearson Education, Inc., Hoboken, N.J., 2004.
20. Texas Department of Transportation (TxDOT). *Transportation Planning and Programming Division, Highway Designation Files*, 2016. https://www.txdot.gov/inside-txdot/division/transportation-planning/highway-designation/glossary.html
21. Bureau of Labor Statistics. *Commodity Data Database*. U.S. Department of Labor, Washington, D.C., 2016. https://www.bls.gov/data/
22. Wimsatt, A., C. Chang, P. Krugler, T. Scullion, T. Freeman, and M. Valdovinos. *Considerations for Rigid vs. Flexible Pavement Designs When Allowed as Alternate Bids*. Technical Report No: FHWA/TX-09/0-6085-1. Texas Transportation Institute, College Station, 2005.

The Standing Committee on Pavement Preservation (AHD18) peer-reviewed this paper (18-00843).

Article

TRR
JOURNAL OF THE TRANSPORTATION RESEARCH BOARD

Performance-Graded Specifications for Asphalt Emulsions Used in Chip Seal Preservation Treatments

Transportation Research Record
2018, Vol. 2672(12) 20–31

DOI: 10.1177/0361198118770169
journals.sagepub.com/home/trr
SAGE

Javon Adams[1], Mohammad Ilias[2], Cassie Castorena[3], and Y. Richard Kim[3]

Abstract

This paper details the development of a framework for emulsion performance-grade (EPG) specifications for chip seal treatments. Chip seals are preservation surface treatments that are designed to improve the condition of the pavement surface while mitigating deterioration of the overall pavement structure. Asphalt emulsions used in chip seals often are selected based on factors that are not necessarily related to performance. Aggregate loss and bleeding have been identified as the most critical chip seal distresses that are related to binder performance. Storage stability, sprayability, and drain-out have been determined to be the most critical constructability concerns. For this study, binder and mixture test methods were identified to reflect the failure mechanisms for each critical distress type. The emulsion residue test methods that were identified to capture chip seal performance are the multiple stress creep and recovery test for bleeding and the dynamic shear rheometer frequency sweep test for low-temperature aggregate loss. The fresh emulsion test methods that were identified to capture chip seal constructability are the three-step shear test and storage stability test. The proposed EPG specifications for the fresh emulsion properties that are related to constructability were developed using statistical analysis of the binder test results. The proposed EPG specifications for the residual binder properties were developed by defining the temperature-independent relationships between the emulsion residue properties and mixture performance that correspond to each critical distress. Preliminary specification limits were then established based on the values of the binder properties that correspond to the critical mixture performance thresholds.

Chip seals are pavement preservation surface treatments that are applied in a thin layer to an existing roadway surface to slow deterioration and to maintain or improve the roadway's surface condition without increasing its structural capacity. One of the primary constituents of a chip seal treatment is asphalt emulsion. Although it is well known that the properties of the emulsion are critical to the performance of the chip seal, current specifications are empirical in nature and are not based on emulsion properties that relate directly to performance.

Emulsion performance-grade (EPG) specifications that specify the asphalt emulsion properties that relate directly to performance will improve the selection process of binders for a given chip seal to prevent premature pavement failure. The urgent need for EPG specifications for emulsions used in chip seals is clear in the responses to surveys conducted by Johnston and King (*1*) in which pavement practitioners in the United States were asked about the need for such specifications. The survey results strongly suggest the need to develop a climate-driven, performance-related grading system for asphalt emulsions used in chip seals. New EPG specifications should incorporate the residual binder material properties that correlate to critical chip seal mixture performance measures. The EPG specifications should also account for the climatic and traffic loading conditions that a chip seal will experience during its anticipated service life. The EPG specifications should address both fresh emulsion and the emulsion residue properties that relate to constructability and in-service conditions, respectively.

Objective

The objective of this study is to develop proposed EPG specifications for chip seal emulsions.

Distresses

A literature review determined that the following performance characteristics (distresses) are critical to the assessment of chip seal materials (*2–4*): aggregate loss, bleeding, and rutting (in multilayered chip seals).

[1]College of Engineering, Office of Academic Affairs, North Carolina State University, Raleigh, NC
[2]Fugro Roadware, Inc., Austin, TX
[3]Department of Civil, Construction, and Environmental Engineering, North Carolina State University, Raleigh, NC

Corresponding Author:
Address correspondence to Y. Richard Kim: kim@ncsu.edu

Excessive bleeding in the wheel path can lead to loss of friction and can become a safety hazard for drivers. Although the terms *bleeding* and *flushing* often are used interchangeably in the literature, the mechanism behind flushing is construction/design-related, whereas bleeding is related directly to the viscoplasticity of the binder at high temperatures. Therefore, bleeding, not flushing, was considered as a critical distress in the development of the proposed EPG specifications.

In multilayer chip seal surface treatments, rutting can occur in the wheel path under repeated loading. During rainy conditions, these ruts fill with water, which can lead to hydroplaning. Rutting resistance is related to the viscoplastic nature of the asphalt binder.

Aggregate loss (also referred to in the literature as raveling) is a critical performance characteristic because the loss of aggregate particles can cause damage to vehicles and is a safety concern. Significant raveling can also lead to bleeding. Aggregate loss at an intermediate temperature is driven by the lack of compatibility between the aggregate and emulsion and therefore cannot be accounted for in an emulsion purchase specification by itself (*5*). Aggregate loss at a low temperature is driven by cohesive failure of the binder. Consequently, aggregate loss at a low temperature and bleeding at a high temperature have been identified as the most critical chip seal distresses that are related to binder performance.

Material Properties and Proposed Test Methods

In developing the proposed EPG specifications, test methods were identified that address 1) the fresh emulsion properties that relate to storage and constructability and 2) the residual binder properties that relate to chip seal performance.

Fresh Emulsion Testing

The properties of an emulsion during its storage, transportation, and construction significantly affect the performance of chip seals. Asphalt emulsions used in the construction of chip seals are multiphase systems that are composed of water, asphalt binder, emulsifier, and in some cases, modifiers. *Storage stability* is defined as the ability of an emulsion to resist significant changes in properties over time (*6*). In the proposed EPG specifications, storage stability is measured using a modified version of ASTM D6930 (*7*) in which the storage conditions are consistent with ASTM D6930. However, instead of comparing the asphalt content of the emulsion that is extracted from the top and bottom of a sample following storage, the viscosity of the top and bottom portions of the stored specimen is evaluated using a rotational viscometer (RV), as viscosity is related directly to the performance of fresh emulsions. In addition, the middle portion of the sample is remixed, and the viscosity is then measured and compared to an unconditioned reference sample to assess stability. Two parameters are reported: (1) the separation ratio (R_s), which is the ratio of the viscosity of the top portion to the bottom portion of the specimen and is used to evaluate the resistance of the emulsion to sedimentation and creaming (*6*), and (2) the stability ratio (R_d), which is used to compare the viscosity of the remixed emulsion and original emulsion samples after storage and to evaluate the emulsion's ability to be remixed following storage.

The viscosity of the emulsion is also critical during chip seal construction and is characterized by the sprayability and drain-out parameters. *Sprayability* is defined as the ability of an emulsion to be sprayed in a uniform thickness across the surface of an existing pavement (*8*). An emulsion that is too viscous will result in streaking, spot bleeding, and partial loss of the cover aggregate in the chip seal. *Drain-out* is defined as the ability of an emulsion to resist draining off the pavement surface via gravity after spraying (*9*). Large amounts of drain-out lead to premature aggregate loss and can reduce the amount of binder that is available for proper aggregate embedment.

Sprayability and drain-out are measured using a three-step shear test that employs an RV. The test subjects an emulsion to three successive shear rates to quantify its thixotropic and shear thinning behavior. An initial low shear rate of 5 rpm simulates the circulation of the emulsion in a tank, a second step at a high shear rate of 150 rpm simulates spraying through a nozzle, and a third step at a low shear rate of 5 rpm simulates the flow under gravity once placed. Sprayability is assessed by the viscosity value at the high shear rate. Drain-out is assessed by the viscosity value at the last (third step) low shear rate (*9*).

Existing Emulsion Tests Retained in Proposed EPG Specifications

The existing AASHTO emulsion test methods for fresh emulsion and recovered residue that are retained in the proposed EPG specifications are:

Particle Charge and Sieve – AASHTO T 59 (*10*);
Residue Recovery and Percentage – AASHTO R 78 Method B (*11*);
Solubility – AASHTO T 44 (*12*);
Float – AASHTO T 50 (*13*).

Residual Binder Testing

As discussed earlier, the most critical distresses to be considered in developing the proposed EPG specifications for chip seals are bleeding and aggregate loss. Various test methods have been investigated for use in the proposed EPG specifications (*14*). The selected binder test methods allow for the failure properties of asphalt emulsions at high and low temperatures to be evaluated under conditions that simulate various climatic and traffic conditions.

For bleeding at the high-temperature EPG, the multiple stress creep and recovery (MSCR) test was used to measure the non-recoverable creep compliance (J_{nr}). To address aggregate loss at a low temperature, the dynamic shear rheometer (DSR) frequency sweep test was used to measure the dynamic shear modulus ($|G^*|$) values at critical phase angle (δ_c) values. AASHTO R 78 Method B (*11*) is the residue recovery method that is used here and is recommended for residue testing at both low and high temperatures in the chip seal EPG specifications.

Materials

To develop the proposed EPG specifications for asphalt binders used in chip seals, a set of asphalt emulsions was acquired from seven suppliers located in various parts of the United States. Each supplier was required to provide standard emulsion formulations used for chip seal construction projects.

A single granite aggregate source was utilized for all the chip seal specimens to allow isolation of the binder effects in mixture performance. The chip seal mixture specimens were fabricated using a performance-based mix design and a sample fabrication procedure that closely simulates the field construction process (*15*). Each chip seal specimen was designed such that the residual asphalt content was 67% after curing. Granite 78M aggregate was used for all the fabricated specimens with an aggregate application rate of 15.5 lb/yd^2 that was found to yield a single stone coverage.

The emulsions were acquired in two phases. The following Phase 1 emulsions were used for all the fresh emulsion testing and both the residual binder and mixture high-temperature testing:

- CRS-2 (A);
- CRS-2P (A);
- CRS-2L (A);
- CRS-2 (C);
- CRS-2L (C);
- CRS-2P-HP (E);
- CRS-2 (E);
- CRS-2P (E);
- CRS-2 (F);
- CRS-2L (F);
- CRS-2 (NC);
- CRS-2L (NC);
- PP-CRS-2 (A);
- CRS-1-(B).

The following Phase 2 emulsions were used for residual binder and mixture low-temperature characterization:

- CRS-2P (A);
- CRS-2L (C);
- CRS-2 (F);
- CRS-2L (F);
- CRS-2 (NC);
- CRS-2L (NC).

Note that in the lists of the Phase 1 and Phase 2 emulsions acquired, the letter in parentheses following the emulsion name identifies (anonymously) the emulsion company that supplied the material for the tests in this project. The 'PP' designation before the emulsion name indicates that the emulsion was produced to be purposely "poor performing" while still meeting the current emulsion specifications.

EPG Specification Framework

Intended Use of the EPG Specifications

The proposed EPG specifications are intended to provide a performance grading system for emulsions for use in chip seals, similar to the current Superpave performance grading system for hot-mix asphalt (HMA). The EPG specifications assume that no pre-existing structural distresses are present within the pavement structure on which the chip seal is to be applied.

EPG Specification Design Temperature Grade Concepts

High Temperatures. The high-temperature climatic HMA performance grade requirements for a given location were determined using a model that is based on a depth of 20 mm, yearly degree days above 10°C, and the assumed threshold for the allowable rut depth of 12.7 mm (*16*). Because the proposed EPG specifications use surface temperature instead of temperature measured at a 20-mm depth, as recommended in the existing HMA specifications, an adjustment to the HMA high-temperature performance grade requirements was necessary. Based on the Long-Term Pavement Performance (LTPP) High Pavement Temperature Model (*17*), the difference between the temperature of a pavement at its surface and at a depth of 20 mm is 3.9°C. Therefore, the proposed EPG specifications require a high-temperature grade that is 3°C higher than the corresponding HMA performance grade temperature for a given location. Although a shift of 4°C would be closer to expected temperature gradients, 3°C is more practical because the existing HMA performance grade specifications are based on 6°C increments. For example, if the existing HMA specification high-temperature performance grade is PG 58, the EPG specification high-temperature grade would be EPG 61.

It is important to note that this change in specification grade does not mean that the base asphalt performance grade used to fabricate the emulsion would necessarily be higher than the performance grade of asphalt binders currently used successfully in the field. In fact, all of the emulsions tested in this research were made from an original base asphalt with a high-temperature performance grade of either 58°C or 64°C, yet every emulsion tested had a higher temperature grade in the EPG specifications than the original base asphalt performance grade used to fabricate the emulsion.

Low Temperatures. The low-temperature grade in the HMA performance grade specifications is defined based on the annual one-day minimum pavement surface temperature. Therefore, the depth correction required for high-temperature emulsion performance grades does not apply to low-temperature grade determination. However, it would make practical sense to shift both the high- and low-temperature grades from the HMA performance grade specifications to the EPG specifications for simplicity. Therefore, temperature data from the LTPP database were analyzed to determine the grade requirements for a given location if the grade thresholds were increased by 3°C. When determining the grade requirement for a given location, the continuous grade was rounded (conservatively) to the nearest available grade. For example, if the low-temperature continuous grade was determined to be −18°C, then the HMA low-temperature grade requirement would be −22°C. The corresponding EPG low-temperature grade requirement would be −19°C if the available low-temperature performance grades were increased by 3°C. As a result of this rounding, most locations would have emulsion low-temperature performance grades that are 3°C higher than their corresponding HMA low-temperature grade.

Traffic Designations for the EPG Specifications

Because the performance of chip seals depends on the traffic conditions that the pavement surface treatment undergoes, the developed EPG specifications provide material test limits for grading binders according to the expected traffic. A survey of emulsion professionals, a literature review (*18*), and the authors' experience have led to the following recommendations for categorizing the low, medium, and high traffic classes that are proposed in the EPG specifications based on the average annual daily traffic (AADT) volume (for vehicles):

- Low Traffic: 0–500 AADT;
- Medium Traffic: 501–2,500 AADT;
- High Traffic: 2,500–20,000 AADT.

An upper limit of 20,000 vehicles is recommended for high traffic based on a study of high traffic chip seal practices across the United States as well as the authors' experience.

Binder Grade Naming Designations for EPG Specifications

The graded binders in the developed EPG specifications consist of high- and low-temperature grade designations that relate to climate and traffic. The EPG specifications retain the currently used designations for the emulsifier charge and set rate. For example, a sample emulsion grade may be CRS-EPG 67–19*M*, which is defined as follows:

> CRS = cationic rapid-setting; the binder type designation denotes the charge, special float type, if any (i.e., high float), and setting rate for the binder.
> EPG = emulsion performance grade.
> 67 = average annual seven-day maximum pavement surface temperature (Celsius).
> −19 = minimum pavement surface temperature (Celsius).
> *M* = medium traffic volume; traffic volume categories are low (L), medium (M), or high (H) in the proposed EPG specifications.

Underlying Concepts and Outline for the Development of the Proposed EPG Specifications

Fresh Emulsion Concept. Statistical analysis was conducted to develop the specification test limits for fresh emulsion properties under the assumption that the standard emulsions tested in this study represent a broad range of materials that perform acceptably in service. Because the poor-performing emulsions were fabricated by the emulsion suppliers intentionally to exhibit inferior performance, test data for the poor-performing emulsions were not used to determine the fresh emulsion specification limits but were used only for validation.

A box-and-whisker statistical analysis procedure was used to derive the fresh emulsion property limits. The box-and-whisker plot method is a standardized method that is employed to display the distribution of data and to identify outliers statistically within a normally distributed dataset (*19*). Identified outliers were removed from the dataset and the remaining data were used to determine the specification limits based on the 98% reliability concept coupled with engineering judgment.

Residual Binder Concept. The relationship between each binder property evaluated was compared against the mixture performance to which the binder property relates to develop the specifications for residual binders. Based on this relationship, the corresponding binder test limit could be determined to ensure that the target mixture performance was achieved. Figure 1 shows an example of the underlying concept used in the development of the EPG specifications for residual binders. The MSCR test results are plotted against mixture bleeding for both modified and unmodified binders. A temperature-independent relationship was found between the binder property, that is, the non-recoverable creep compliance, J_{nr}, and bleeding performance of the chip seal mixtures. Here, the upper limit for bleeding performance was used to define the maximum J_{nr} value for the MSCR tests. The underlying concept is that if a binder meets this MSCR J_{nr} limit, the bleeding of the mixture should not exceed the

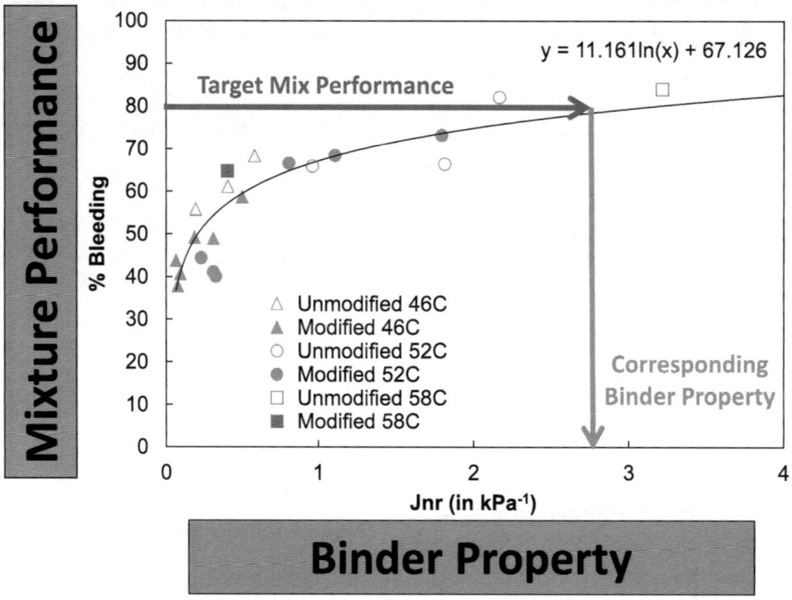

Figure 1. Example of underlying concept used in EPG specifications for residual binder.

maximum bleeding threshold because of inadequate binder performance.

Experimental Plan

Fresh Emulsion

Fresh emulsion three-step shear rate tests were conducted at 60°C. Storage stability conditioning followed ASTM D6930. After conditioning, samples were partitioned and subjected to RV testing at 60°C.

High Temperatures

To develop criteria for the EPG specifications that address high-temperature performance, the J_{nr} value of the binders at a stress level of 3.2 kPa was evaluated with regard to chip seal performance. MSCR binder testing was conducted at temperatures ranging from 46°C to 70°C. To verify that the binder J_{nr} values related to bleeding performance and that these J_{nr} values could be used subsequently to establish specification limits, the binder J_{nr} test results were correlated with chip seal performance that was measured using the one-third scale model mobile load simulator (MMLS3). The MMLS3 simulates traffic loading on chip seals at an accelerated rate of 990 wheel applications every 10 minutes inside a temperature-controlled chamber (*2*). For direct comparison to the high-temperature binder properties measured from the MSCR tests, the mixture specimens were fabricated and subjected to high temperatures of 46°C, 52°C, and 58°C. The mixture test temperature of 58°C represents the highest temperature that the MMLS3 chamber is capable of maintaining during mixture testing. Extrapolation of the relationships between the binder and mixture test results allowed for consideration of the binder test results at typical emulsion performance grade temperatures of 61°C and 67°C. The percentage of bleeding for a chip seal specimen was obtained by dividing the area of bleeding on the chip seal specimen that is because of MMLS3 loading by the total area of the specimen.

To establish traffic-dependent specification criteria, MMLS3 traffic loads were converted to equivalent field wheel loads that represent low, medium, and high traffic conditions; this conversion process is fully detailed in the report for the proposed EPG specifications (*14*). Specification limits were derived based on the J_{nr} value that corresponds to the maximum allowable chip seal mixture performance threshold at each traffic level, which was determined to be 80% bleeding for chip seals. The 80% maximum bleeding limit for chip seals defines the amount of bleeding under MMLS3 loading that, when exceeded, is unacceptable for chip seal performance (*2*). Lee's data revealed that significant bleeding under MMLS3 loading is 80% or above.

Low Temperatures

Chip seal mixture low-temperature aggregate loss was measured using Vialit tests conducted at temperatures ranging from 0°C to −28°C. Fracture properties of the residue binders were obtained via single-edge notched beam (SENB) tests (*14*) at −16°C, −22°C, and −28°C. The residual binder rheological properties were obtained via DSR frequency sweep testing at temperatures ranging from 5°C to 15°C. Note that DSR testing at low temperatures is not possible in most DSRs. Although both rheology-based and fracture mechanics-based properties were found to relate to chip seals at low temperatures, the rheological properties, which demonstrated the strongest relationship to mixture performance, were ultimately selected for the specifications on the basis of efficiency and cost-effectiveness. DSR testing also is required for high-temperature emulsion performance grade determination, and therefore, the DSR, as opposed to other low-temperature test set-ups such as the SENB test set-up, was employed to minimize equipment requirements. In addition, DSR tests require less residual binder, involve simpler specimen preparation, and take less time than SENB tests. Details about all the residual binder properties evaluated for the low-temperature specifications for chip seal emulsions are provided in (*14*).

Although numerous rheological properties were evaluated in this study, the property that demonstrated the strongest relationship to chip seal aggregate loss is the $|G^*|$ at the critical phase angle, δ_c. The δ_c varies as a function of the low-temperature performance grade of interest. The use of $|G^*|$ values at δ_c for the low-temperature specifications was motivated by the discovery of a strong relationship between the crossover modulus ($|G^*|_c$) and low-temperature mixture performance. The $|G^*|_c$ is defined as the $|G^*|$ that corresponds to the reduced frequency where the phase angle equals 45°.

Note that the finding that the $|G^*|$ value at the δ_c is related to fracture resistance is corroborated by previous investigations into the relationship between binder and mixture fracture resistance (*20*). Rowe et al. (*20*) demonstrated that a higher R-value correlates with better strain-controlled fatigue cracking resistance in asphalt mixtures, thus suggesting that aggregate loss in chip seals may be a strain-controlled phenomenon. The R-value is the difference between the glassy modulus value and the $|G^*|$ value at the reduced frequency where the phase angle equals 45°. For most binders, the glassy modulus value is approximately 10^9 Pa (*21*). Thus, the R-value is essentially an indicator of the asphalt binder $|G^*|_c$, that is, the $|G^*|$ value at the δ_c.

Because the $|G^*|_c$ is a temperature-independent parameter whereas the mixture performance changes as a function of temperature, the $|G^*|_c$ cannot be used in the framework of the EPG specifications development. Therefore, to adapt the concept of the $|G^*|_c$ to allow its incorporation into the EPG specification framework, δ_c values were determined as a function of mixture test temperature to produce a temperature-independent relationship between the mixture's low-temperature performance and the corresponding $|G^*|$ value. To determine the $|G^*|$ value at the δ_c of interest, DSR frequency sweep tests were conducted at both 5°C and 15°C regardless of the performance grade. That is, the δ_c values for which the $|G^*|$ values were determined at varied as a function of the low-temperature performance grade rather than the test temperature. However, consistent with other performance grading specification criteria, the specification limits for the $|G^*|$ values are independent of climate.

To determine the specification limits, the residual binder property values that correspond to critical mixture performance thresholds were determined. The mixture performance thresholds were determined based on the results obtained in this study. Chip seal aggregate loss at low temperatures is dependent on traffic level; therefore, the specification limits were defined for high, medium, and low traffic levels.

Performance-Graded Specification Limits

Fresh Asphalt Emulsion EPG Specification Limits

Figure 2 depicts how the fresh emulsion specification test limits were determined using the test data obtained from the storage stability three-step shear rate tests. Any outliers identified in Figure 2, *a, c, e*, and *g* were removed based on the box-and-whisker plot method (*19*). Figure 2, *b, d, f* and *h* show that the specification limits were determined based on the 98% reliability concept and engineering judgment. For example, the stability ratio of all emulsions exceed one, and therefore, a maximum limit was established as 2 using the 98% reliability threshold, as shown in Figure 2*d*.

Fresh Emulsion Limit Summary. The specification test limits derived for the fresh emulsions used in the proposed EPG specifications are as follows:

- Separation Ratio – 0.5–1.5;
- Stability Ratio – maximum 2;
- Sprayability – maximum 400 cP;
- Drain-out – minimum 50 cP.

Residual Binder Test Limits

This section presents the methodology used to derive the binder specification criteria that relate to the performance of chip seals with respect to critical distresses.

High Temperatures. Figure 3 presents the correlations between the binder J_{nr} test results and the amount of bleeding measured from the MMLS3 tests of chip seal mixtures at low, medium, and high traffic volumes. Note that the binder J_{nr} values do not change with traffic volume. Each data point represents the average of three MMLS3 bleeding test replicates and four MSCR test replicates. The results show a strong temperature-independent relationship between the binder J_{nr} values and the mixtures' bleeding performance. The results indicate that the binder J_{nr} value captures the binder's contribution to the bleeding in the chip seal and matches expected trends, as a lower J_{nr} value indicates more permanent strain resistance and thus less susceptibility to

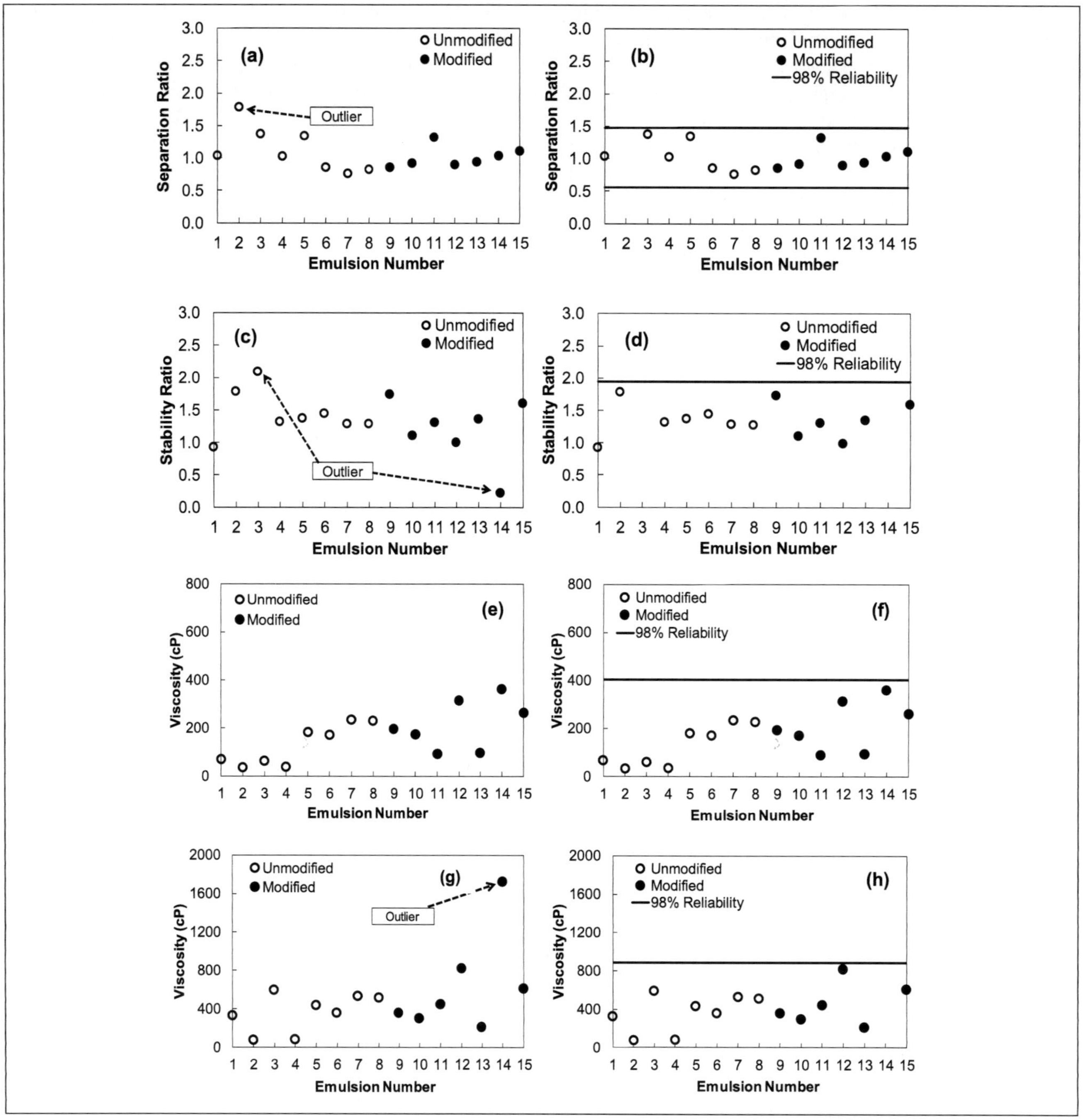

Figure 2. Examples of: (*a*) outlier identification for separation ratio data, (*b*) separation ratio specification limit determination, (*c*) outlier identification for stability ratio data, (*d*) stability ratio specification limit determination, (e) outlier identification for sprayability data, (*f*) sprayability specification limit determination, (*g*) outlier identification for drain-out data, and (*h*) drain-out specification limit determination.

bleeding. Figure 3*d* shows the relationship between the J_{nr} values and '% bleeding' plotted as a function of traffic level. As expected, the results shown in Figure 3*d* indicate that the severity of the bleeding increases with an increase in MMLS3 traffic loading.

Curves were fitted to the bleeding versus J_{nr} relationships shown in Figure 3 to allow extrapolation to higher J_{nr} values measured at 61°C and 67°C, which are the temperatures at which the corresponding mixture bleeding performance could not be measured because of the limitation of the MMLS3 temperature control system. Figure 3 presents the results for the low, medium, and high traffic volumes that were used to establish preliminary specification limits for the EPG specifications. To establish limits for the binder J_{nr} value at each traffic

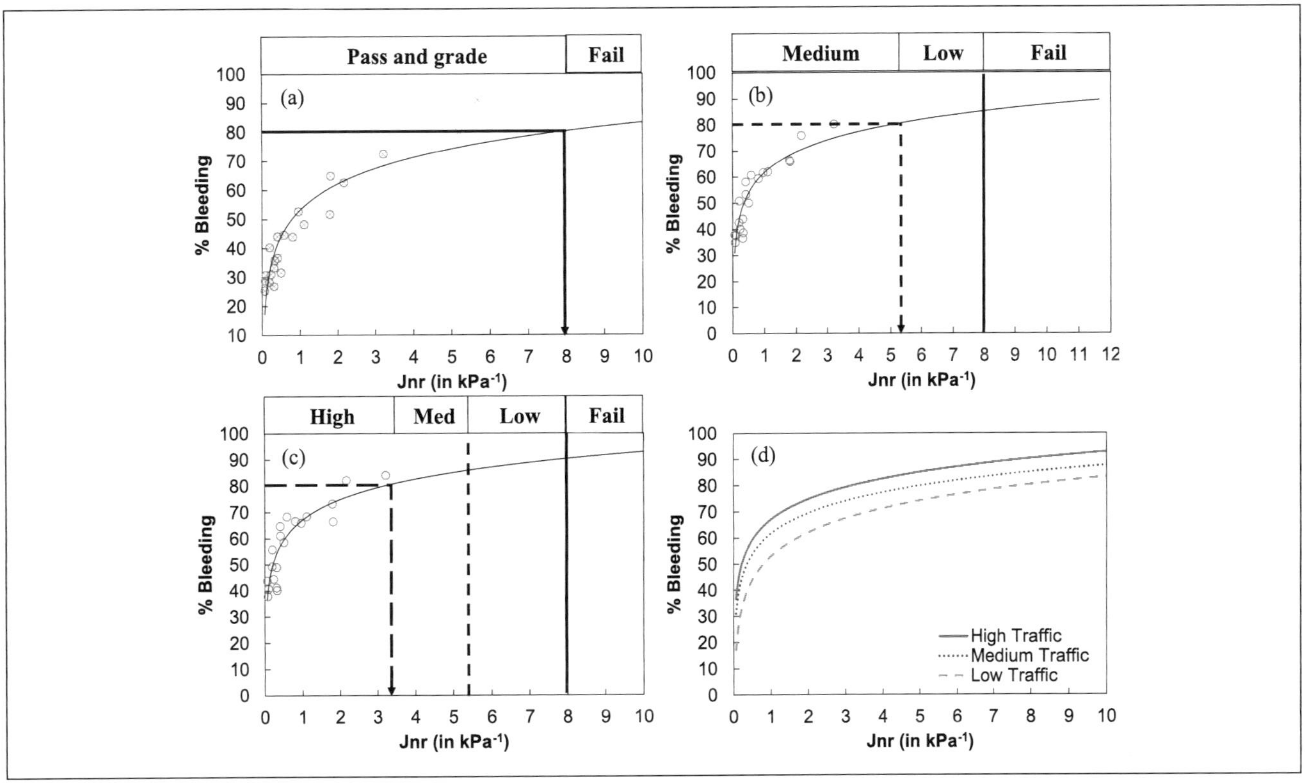

Figure 3. Developed maximum MSCR J_{nr} limits established for high-temperature grading in the EPG specifications based on (*a*) low traffic, (*b*) medium traffic, (*c*) high traffic, and (*d*) relationship shown as a function of traffic level.

level, a maximum allowable bleeding threshold of 80% was utilized for the laboratory-tested chip seal mixture specimens. The maximum limit for the J_{nr} value for each traffic volume is defined as the corresponding J_{nr} value at the point where the J_{nr} versus bleeding curve intersects the 80% bleeding severity threshold, as depicted in Figure 3.

The proposed MSCR test limits for the high-temperature grades of binders used in EPG chip seal specifications are as follows:

- Low Traffic: maximum J_{nr} @ 3.2 kPa of 8.0 kPa^{-1};
- Medium Traffic: maximum J_{nr} @ 3.2 kPa of 5.5 kPa^{-1};
- High Traffic: maximum J_{nr} @ 3.2 kPa of 3.5 kPa^{-1}.

Low Temperatures. Similar to the derivation of the high-temperature specification limits, specifications to address aggregate loss at low temperatures were derived using the relationship between binder properties and mixture performance combined with a critical mixture performance threshold. Note that aggregate loss typically occurs during the first year following construction. Because low winter temperatures occur only a few months after the initial construction, it is assumed that the low-temperature distresses are the most critical distresses before the residue ages significantly. Therefore, it is recommended that unaged residue be used for low-temperature grading.

Trends in the test results combined with a priori knowledge of the emulsions' performance were utilized to establish critical performance thresholds for the Vialit test aggregate loss of the chip seal mixtures at low temperatures, as shown in Figure 4*a*. The aggregate loss threshold for low traffic was set to 35% based on the known success of CRS-2-NC emulsion that has been used for low traffic roadways in a −19°C region. To determine the high traffic threshold, trends in aggregate loss with temperature also were used. As shown in Figure 4*a*, once the temperature is below −19°C, the sensitivity of aggregate loss to temperature for the CRS-2L-NC mixture is much greater than for the other modified mixtures, such as CRS-2P-A and CRS-2L-F. Also, a field study by Im (*22*) demonstrated that CRS-2L-NC emulsion performs satisfactorily under high traffic conditions in North Carolina, which is a −19°C region. It is noted that the aggregate loss of the CRS-2L-NC mixture at −19°C is about 25%. Similarly, the CRS-2L-C emulsion experienced an abrupt increase in the sensitivity of aggregate loss to temperature when the temperature dropped below −25°C. The aggregate loss of the CRS-2L-C mixture at −25°C also is around 25%. Based on these observations, the high traffic aggregate loss threshold was set at 25% in the proposed EPG specifications. In addition, CRS-2P-A emulsion has been used successfully in −25°C regions on medium traffic roadways and passes this criterion, further validating the reasonableness of its usage. The medium traffic aggregate loss threshold was selected simply as the average of the high and low traffic thresholds.

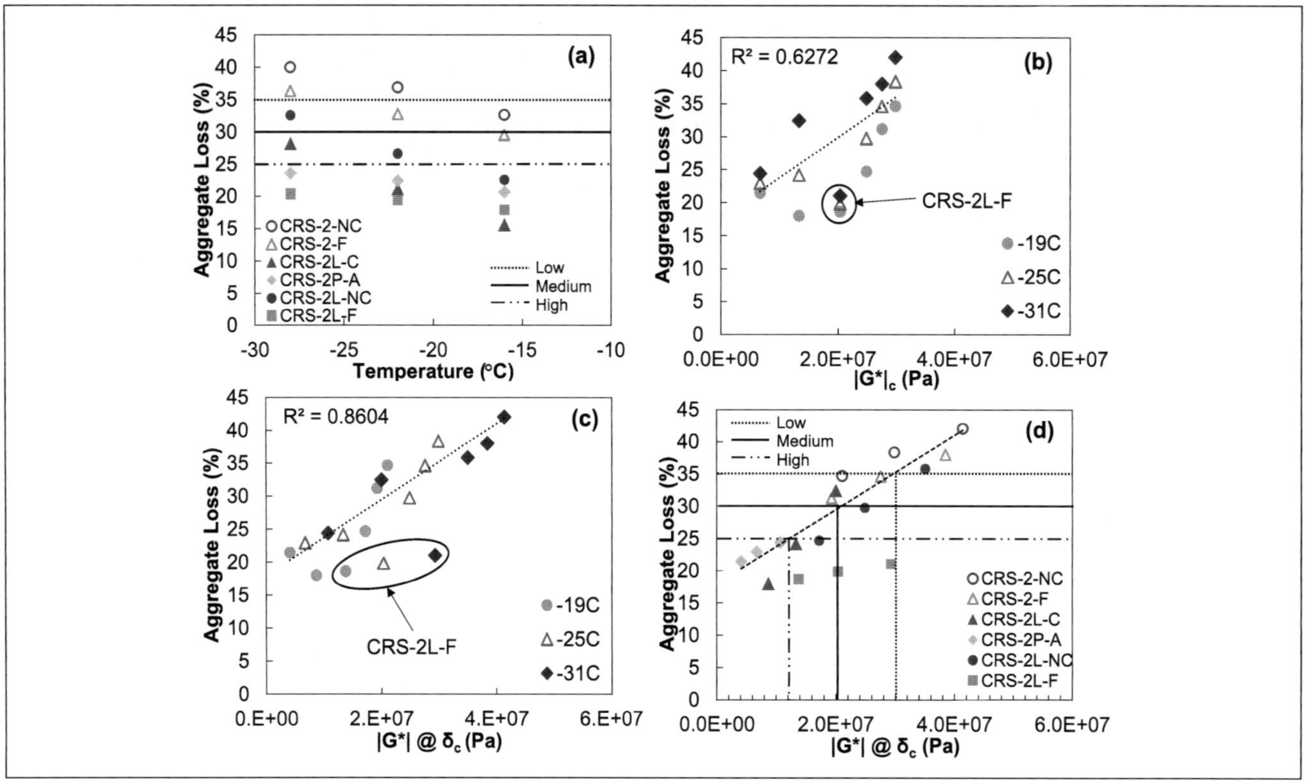

Figure 4. (*a*) Chip seal aggregate loss results as a function of temperature, (*b*) chip seal aggregate loss versus residual binder $|G^*|$ value at δ_c corresponding to $\delta = 45°$, (*c*) chip seal aggregate loss versus residual binder $|G^*|$ value at adjusted phase angle as a function of temperature, and (*d*) maximum critical $|G^*|$ limits for residual binders.

To determine the δ_c values as a function of the low-temperature emulsion performance grade, the chip seal mixtures' aggregate loss was correlated with the binder $|G^*|$ values that correspond to various phase angle values. The phase angle values used for each mixture performance test temperature were adjusted until a temperature-independent relationship between mixture aggregate loss and binder $|G^*|$ was obtained. Figure 4, *b* and *c* present the results. Figure 4*b* shows that the relationship between aggregate loss and the $|G^*|_c$ depends on temperature, because the mixture performance changes as a function of temperature but the $|G^*|_c$ does not. Figure 4*c* shows the relationship between aggregate loss and the $|G^*|$ when the δ_c values for which the $|G^*|$ is determined are adjusted as a function of temperature. The δ_c that corresponds to −25°C remains 45°. However, the δ_c values that correspond to −19°C and −31°C are 48° and 42°, respectively. These results demonstrate good agreement (R^2 = 0.86) between the aggregate loss and $|G^*|$ values regardless of the mixture test temperature, with the exception of CRS-2L-F. Aggregate loss in the CRS-2L-F mixture does not vary appreciably with test temperature, unlike the other emulsions, but its rheology does change with temperature, which explains its outlier behavior.

The relationships between the low-temperature emulsion performance grades and the δ_c values shown in Figure 4*c* were extended to determine the δ_c values that correspond to all low-temperature emulsion performance grades. The δ_c values for each low-temperature emulsion performance grade are:

- −7°C (δ_c = 54°);
- −13°C (δ_c = 51°);
- −19°C (δ_c = 48°);
- −25°C (δ_c = 45°);
- −31°C (δ_c = 42°);
- −37°C (δ_c = 39°).

The residual emulsion specification limits were defined based on the relationships between aggregate loss and the $|G^*|$ values that correspond to the δ_c values and the established allowable aggregate loss thresholds for low, medium, and high traffic volumes. Figure 4*d* presents the results. The corresponding maximum $|G^*|$ limits for low, medium, and high traffic volumes are 30 MPa, 20 MPa, and 12 MPa, respectively. The relationships between the low-temperature emulsion performance grades and δ_c values shown in Figure 4*c* were used to determine the δ_c values that correspond to all low-temperature emulsion performance grades.

Table 1 shows the proposed EPG specifications for chip seal treatments developed in this research.

Table 1. Proposed EPG Specifications for Chip Seals

	Chip seal emulsion performance grade											
	EPG 61				EPG 67				EPG 73			
	−13	−19	−25	−31	−13	−19	−25	−31	−13	−19	−25	−31
Average 7-day maximum pavement surface design temperature, °C[a]	<61				<67				<73			
Minimum pavement surface design temperature, °C[a]	>−13	>−19	>−25	>−31	>−13	>−19	>−25	>−31	>−13	>−19	>−25	>−31
Proposed test methods[b]	Proposed testing temperature (°C)											
Tests on original emulsion												
Storage stability	60											
Modified AASHTO T 59												
Measured responses: rotational viscosity, η,												
A – 24-hour separation ratio (Rs): 0.5–1.5												
B – 24-hour stability ratio (Rd): maximum 2												
Sprayability	60											
Modified AASHTO TP 48												
Measured response: viscosity @ 3 shear rates,												
Maximum 400 cP @ high shear rate (150 rpm)												
Resistance to Drain-out	60											
Modified AASHTO TP 48												
Measured response: viscosity @ 3 shear rates,												
Min. 50 cP @ low shear rate (5 rpm)												
Demulsibility	25											
AASHTO T 59												
Measured response: % demulsibility												
Min. 40% (anionic)												
Min. 60% (cationic)												
Particle charge	25											
AASHTO T 59												
Measured response: particle charge												
Positive (cationic)												
Sieve test	25											
AASHTO T 59												
Measured response: % mass												
Maximum 0.1%												
Solubility	25											
AASHTO T 44												
Measured response: % solubility												
Min. 97.5%												
Float[c]	60											
AASHTO T 50												
Measured response: float time												
Min. 1200 seconds												
Percent residue	25											
AASHTO R 78												
Measured response: % residue												
Min. 65% (cationic)												
Min. 63% (anionic)												
Tests on residue recovered using AASHTO R 78 Method B												
Resistance to bleeding and rutting	61				67				73			
AASHTO T 350												
Measured response: non-recoverable creep compliance, J_{nr}												
Max Jnr @ 3.2 kPa, 8 kPa^{-1} (low traffic)												
Max Jnr @ 3.2 kPa, 5.5 kPa^{-1} (medium traffic)												
Max Jnr @ 3.2 kPa, 3.5 kPa^{-1} (high traffic)[d]												
Resistance to low temperature raveling	5°C and 15°C											
DSR temperature frequency sweep	Critical phase angle, δc (°)											
Measured response: |G*| at critical phase angle, δc												
Maximum |G*| @ δc: 30 MPa (low traffic)	51	48	45	42	51	48	45	42	51	48	45	42
Maximum |G*| @ δc: 20 MPa (medium traffic)												
Maximum |G*| @ δc: 12 MPa (high traffic)												

[a]Pavement surface temperatures are estimated from air temperatures using an algorithm contained in the LTPP Bind program, or may be provided by the specifying agency.
[b]Vialit testing should be performed in accordance with British Standard EN12272-3 to measure resistance to aggregate loss because of compatibility issues between aggregate and emulsion at the intermediate temperature grade, which is the average of the high and low emulsion performance grades, plus 4 degrees.
[c]For high float emulsions only.

Conclusions and Recommendations

This study established proposed EPG specifications. The proposed EPG specifications include fresh emulsion properties that address the stability and constructability of emulsions used in chip seals. The chip seal EPG specifications for residual asphalt binders address bleeding and low-temperature aggregate loss. Preliminary specification limits are proposed in the EPG specifications based on critical mixture performance thresholds. The following conclusions and recommendations are drawn from this study:

- Based on an analysis of the LTPP High Pavement Temperature Model, it is recommended that the EPG specifications define the climatic conditions for which an emulsion can be utilized in chip seals by assigning an emulsion performance grade temperature that is 3°C higher than the performance grade temperatures found in the Superpave specifications for asphalt binders used in equivalent locations.
- Fresh emulsion and residual binder tests in the EPG specifications can effectively capture the materials' failure mechanisms that are related to critical mixture performance and can differentiate binder types.
- A temperature-independent relationship exists between binder non-recoverable creep compliance measured at the 3.2 kPa stress level in the MSCR test and chip seal bleeding and rutting performance. A maximum limit should be placed on the binder non-recoverable creep compliance to preclude chip seal failure because of bleeding and rutting.
- The residual binder crossover modulus is related to chip seal aggregate loss at low temperature. To adapt the crossover modulus for use in EPG specifications, the critical phase angle values at which the dynamic shear modulus is evaluated should be varied as a function of the EPG low temperature. A maximum limit should be placed on dynamic shear modulus values at critical phase angle values to preclude premature failure because of low-temperature aggregate loss.
- It is recommended that long-term field validation be conducted to validate the developed test methods and refine the specification limits for a variety of climatic, aggregate, and traffic conditions.

Acknowledgment

The authors would like to acknowledge the financial support from the National Cooperative Highway Research Program (NCHRP) under project NCHRP 9-50.

References

1. Johnston, J. B., and G. King. *Using Polymer Modified Emulsions in Surface Treatments*. Federal Lands Highway Interim Report, 2008.
2. Lee, J. *Performance-Based Evaluation of Asphalt Surface Treatments Using the Third Scale Model Mobile Loading Simulator*. PhD dissertation. North Carolina State University, Raleigh, N.C., 2007.
3. Lawson, W. D., M. Leaverton, and S. Senadheera. *Maintenance Solutions for Bleeding and Flushed Pavements Surfaced with a Seal Coat or Surface Treatment*. Report 0-5230-1. Texas Department of Transportation Research and Technology Implementation Office, 2007.
4. Walubita, L., A. Epps-Martin, and C. J. Glover. *A Surface Performance Graded (SPG) Specification for Surface Treatment Binders: Development and Initial Validation*. Publication FHWA-TX-05/1710-2. FHWA, US Department of Transportation, 2005, pp. 26–47.
5. Adams, J., C. Castorena, J. H. Im, M. Ilias, and Y. R. Kim. Addressing Raveling Resistance in Chip Seal Specifications. *Transportation Research Record: Journal of the Transportation Research Board,* 2017. 2612: 39–46.
6. Redelius, P., and J. Walter. *Bitumen Emulsions. Emulsions and Emulsions Stability*. CRC Taylor and Francis, 2006, p. 383.
7. American Society for Testing and Materials. *ASTM D 6930: Standard Test Method for Settlement and Storage Stability of Emulsified Asphalts*. ASTM, 2010.
8. Asphalt Institute. *A Basic Asphalt Emulsion*. Manual Series No. 19, Asphalt Institute, Lexington, Ky., 2008.
9. Bahia, H., K. Jenkins, and A. Hanz. Performance Grading of Bitumen Emulsions for Sprayed Seals. *Proc., 1st International Sprayed Sealing Conference*, Adelaide, Australia, Australian Road Research Board, 2008, pp. 1-13.
10. American Association of State Highway and Transportation Officials. *AASHTO T 59: Standard Method of Test for Emulsified Asphalts*. AASHTO, 2015.
11. American Association of State Highway and Transportation Officials. *AASHTO R 78: Standard Practice for Recovering Residue from Emulsified Asphalt Using Low-Temperature Evaporative Techniques*. AASHTO, 2013.
12. American Association of State Highway and Transportation Officials. *AASHTO T 44: Standard Method of Test for Solubility of Bituminous Materials*. AASHTO, 2015.
13. American Association of State Highway and Transportation Officials. *AASHTO T 50: Standard Method of Test for Float Test for Bituminous Materials*. AASHTO, 2015.
14. Kim, Y. R., J. Adams, C. Castorena, M. Illias, J. Im, H. Bahia, and P. Johannes. *NCHRP Report 837: Performance-Related Specifications for Asphalt Binders Used in Preservation Surface Treatments*. National Research Council, Washington, D.C., 2017.
15. Kim, Y. R., and J. Adams. *Development of a New Chip Seal Mix Design Method*. Publication HWY-2008-04. N.C. Department of Transportation, 2011.
16. Mohseni, A., S. H. Carpenter, and J. D'Angelo. Development of Superpave High-Temperature Performance Grade (PG) Based on Rutting Damage (with Discussion and Closure). *Journal of the Association of Asphalt Paving Technologists*, Vol. 74, No. 1, 2005, pp. 197–254.
17. Mohseni, A. *LTPP Seasonal Asphalt Concrete (AC) Pavement Temperature Models*. FHWA-RD-97-103. Pavement Systems (PavSys), FHWA, McLean, Va., 1998.
18. Gransberg, D. D., and D. M. B. James. *NCHRP Synthesis Report 342: Chip Seal Best Practices*. HRB, National Research Council, Washington, D.C., 2005.
19. Sim, C. H., F. F. Gan, and T. C. Chang. Outlier Labeling with Boxplot Procedures. *Journal of the American*

Statistical Association, Vol. 100, No. 470, 2005, pp. 642–652.

20. Rowe, G. M., M. J. Sharrock, and S. Raposo. Cracking and Linear Viscoelastic Binder Properties, *Proc., 8th RILEM International Conference on Mechanisms of Cracking and Debonding in Pavements*, Vol. 13, 2016, pp. 3–8.
21. Anderson, D. A., D. W. Christensen, H. U. Bahia, R. Dongre, M. G. Sharma, C. E. Antle, and J. Button. *SHRP A-369: Binder Characterization and Evaluation Volume 3: Physical Characterization*. Strategic Highway Research Program, National Research Council. Washington, D.C., 1994.
22. Im, J. H. *Performance Evaluation of Chip Seals for High Volume Roads Using Polymer-Modified Emulsions and Optimized Construction Procedures*. PhD dissertation. North Carolina State University, Raleigh, N.C., 2013.

The Standing Committee on Pavement Preservation (AHD18) peer-reviewed this paper (18-04652).

Article

TRR
JOURNAL OF THE TRANSPORTATION RESEARCH BOARD

Transportation Research Record
2018, Vol. 2672(12) 32–39

DOI: 10.1177/0361198118756868
journals.sagepub.com/home/trr
SAGE

Soft Computing Tools to Predict Progression of Percent Embedment of Aggregates in Chip Seals

Aksel Seitllari[1] and M. Emin Kutay[1]

Abstract

In this study, soft computing and multilinear regression techniques were employed to develop models for prediction of progression of chip seal percent embedment depth (P_e). The model uses inputs such as cumulative equivalent traffic volume, Vialit test results, dust content of aggregates, and initial embedment depth. Multilinear regression, adaptive neuro-fuzzy system, and artificial neural network techniques were used to estimate the P_e. The contribution of the variables affecting P_e was evaluated through a sensitivity analysis. The results indicate that while most of the proposed models were able to predict the P_e reasonably, the artificial neural network model performed the best.

The initial cost, ease, and speed of construction make chip seals a popular surface treatment method (*1, 2*). The aim of using chip seals is to extend pavement life by slowing down the aging of asphalt pavement layers, improve the skid resistance, sustain pavement surface against abrasive effects, and reduce water infiltration to the pavement (*3*). The most common distresses in chip seals are bleeding and chip loss. Bleeding (or flushing) occurs when an excessive amount of asphalt emulsion/binder is used during the construction of chip seals (*4*). Raveling, or loss of aggregate, is the failure of bonding between the aggregate and binder as a result of inadequate emulsion application rate, aggregate properties, traffic load, and climate (*5, 6*).

Asphalt pavement performance measures often do not apply to deterioration mechanisms on chip seals. Roque and Anderson pointed out that macrotexture depth (which relates to aggregate percent embedment) is a good indicator of chip seal performance (*6*). Gürer et al. studied the effects of construction parameters on chip seal performance. Five chip seal test roads were monitored and evaluated using nondestructive tests. It was concluded that pavement deterioration was significantly affected by macrotexture loss over time (*7*). New Zealand chip seal specifications recommend a minimum chip seal macrotexture mean texture depth of 0.9 mm (*8*). Chip loss and flushing distresses are thought to relate to the change in macrotexture with time (*9*). Macrotexture reduction is typically a result of change of embedment with time, aggregate wearing, and aggregate rotation/translation (especially flat/elongated aggregates). Adams and Kim stated that aggregate embedment depth is the most crucial factor influencing bleeding performance and aggregate loss (*10*).

An accurate model to predict the change in embedment depth is important to estimate the life of a chip seal project. Praticò et al. developed a macrotexture prediction model based on mix design-related factors (*11*). Aktaş et al. established a multilinear regression (MLR) model to predict chip seal life (*9*).

The main objective of this research was to employ different soft computing approaches to evaluate chip seal performance using percent embedment (P_e) depth as chip seal performance indicator. Several multilayer perceptron (MLP) model combinations were run to predict the P_e. Moreover, neuro-fuzzy models were developed and generalized to predict P_e based on the input data set that includes cumulative equivalent traffic volume (CEV), product of Vialit test results and dust content of aggregates ($V \times D$), and initial ($P_{e\text{-init}}$) embedment depth of aggregates into the binder. In addition, the MLR technique was also evaluated in the study and the results were compared with adaptive neuro-fuzzy system and MLP techniques.

Data Collection and Analysis

The data utilized in this study were extracted from Aktaş et al., which included macrotexture, traffic, volumetric

[1]Department of Civil and Environmental Engineering, Michigan State University, East Lansing, MI

Corresponding Author:
Address correspondence to Aksel Seitllari: seitllar@egr.msu.edu

properties of chip seals (average least dimension, dust content, voids, etc.), pavement temperature, and Vialit test results (*9*).

It is well known that the amount of traffic on a chip-sealed pavement affects the change in macrotexture and embedment depth. The amount of traffic was quantified using the CEV parameter. CEV is the sum of number of equivalent light vehicles per lane per month. The concept of equivalent single-axle load (ESAL) was not used because ESAL does not include light vehicles, which do cause damage to chip seals (in addition to the heavy vehicles). Therefore, CEV, which includes light vehicles, is thought to be more appropriate. A schematic view describing the effect of CEV on embedment depth of chips is illustrated in Figure 1.

One of the methods of calculation of embedment depth of aggregates is the following formula:

$$P_e(\%) = \frac{(ALD - MTD)}{ALD} \quad (1)$$

where P_e is the percent embedment (%), ALD is the average least dimensions of aggregate, MTD is the mean texture depth of the macrotexture of chip seal. The ALD can be obtained from the grain size distribution of the aggregates. The MTD can be obtained using sand patch test or laser-based methods (*12, 13*).

It is worth mentioning that the MTD obtained from sand patch and laser methods assumes perfectly flat substrate and ignores aggregate embedment into substrate. In addition, it is assumed that all the aggregates are lying on their flattest side. These assumptions are not always accurate and the P_e calculated from MTD can be misleading when these assumptions are not accurate. There are more accurate methods of directly measuring P_e. Kutay et al. recently developed a method to compute the P_e of aggregate using image processing techniques (*14*). The method involves coring and vertical slicing of the chip-sealed pavements and imaging the vertical slice. Then, image processing and analysis techniques were used to compute the P_e distribution of aggregates directly. This method is more accurate than the sand patch or laser-based methods since it can consider the embedment of the aggregates into the substrate and the leakage of the binder into the cracks or voids of the substrate.

Finally, the Vialit test results and dust content of aggregates were also employed in the models as a relevant parameter influencing the percent embedment depth (*4*).

Theoretical Background

In this study, MLR, adaptive neuro-fuzzy interface system (ANFIS) and artificial neural network (ANN) techniques were used to develop predictive models for chip seal embedment depth. Brief descriptions of these three models are provided below.

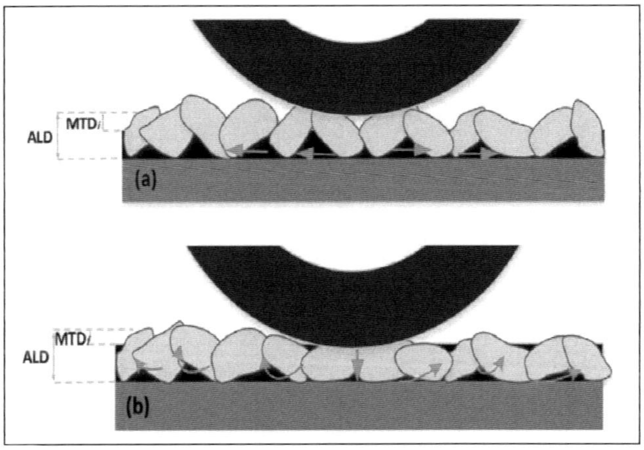

Figure 1. The effect of CEV on chip seal microstructure.
Note: MTD_i = initial mean texture depth; MTD_f = final mean texture depth; ALD = average least dimension.

MLR

The MLR technique is used to model the linear relationship between one dependent variable and multiple independent variables. Assuming y to be the dependent variable and x_1, $x_2 \ldots, x_z$ as independent variables, then the linear equation can be written as

$$y = \alpha_0 + \alpha_1 x_1 + \alpha_2 x_2 + \ldots . + \alpha_z x_z \quad (2)$$

where α_0 represents the intercept regression coefficient, α_1, $\alpha_2, \ldots, \alpha_z$ are the regression coefficients which are computed by minimizing the sum of squares of differences between the predicted and observed values in the data (*15*).

In the MLR analysis, various combinations of the inputs were used to evaluate their relation to P_e. The training data set, which was 4/5 of total data set, was used to determine the MLR model regression coefficients for each combination. The coefficients generated from the training data set were used to evaluate the model using the testing data set. The input data set combinations were selected based on principal component analysis (*16*). First, the influence of individual input parameters on P_e was evaluated. Based on the obtained sensitivity results, the following input combinations showed potential correlation based on P_e data conducted in this study: CEV, $(V \times D)$ $P_{e\text{-init}}$.

ANFIS

ANFIS is a multilayer adaptive network based fuzzy inference system first introduced by Jang et al. (*17*). The capability of using a hybrid learning procedure enables ANFIS to model the linguistic approach of expert knowledge (if–then rules) without the need for exact quantitative analysis. An ANN structure is composed of several nodes connected through

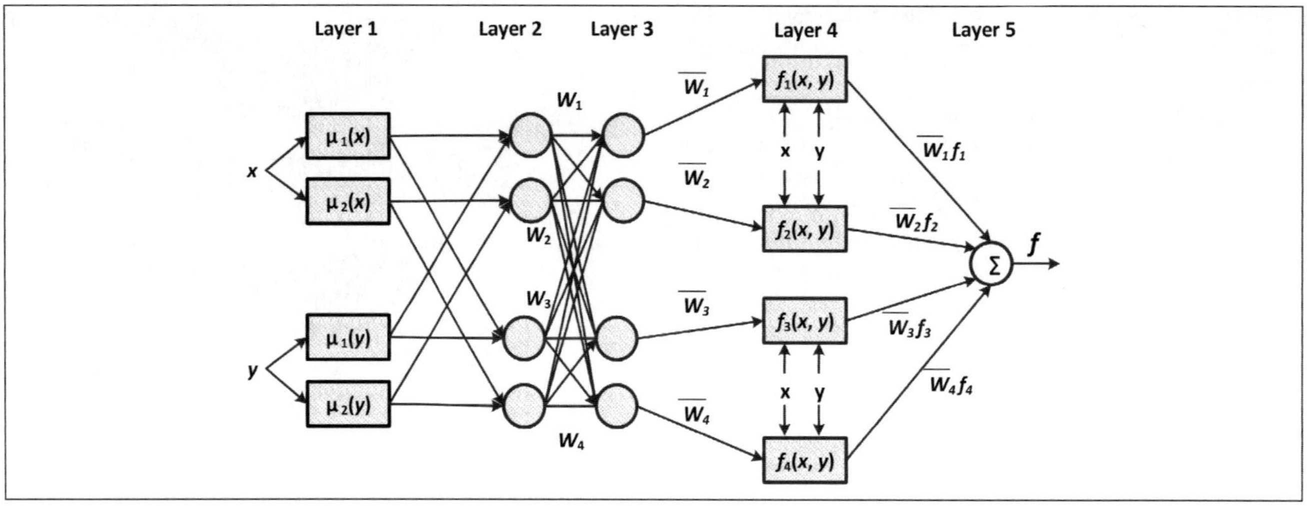

Figure 2. ANFIS architecture based on two inputs (*x* and *y*).

directional links. Every node is distinguished by a node function which is either a fixed or adjustable parameter. The training phase of a fuzzy inference system determines the parameter values while minimizing the error measure. The learning algorithm employed is the back-propagation method, which minimizes sum of squared differences between observed and estimated outputs (*18*).

In general, fuzzy inference systems are classified into three types: Mamdani's system (*19*), Tsukamoto's system (*20*), and Sugeno's system (*21*). As applied to the current research, Sugeno's system is a compact and efficient computation method. The defuzzification operation is time-saving and reduces the output to a single value or crisp variable. Further details regarding the deffuzification (or round off) methods are available in Takagi and Sugeno (*21*).

In addition to numerical variables, the fuzzy system requires some linguistic variables. The if–then rules take place while establishing a relationship between the fuzzy variables. A first-order Sugeno's system with two fuzzy if–then rules can be expressed as follows:

$$\text{Rule 1: If } x \text{ is } A_1 \text{ and } y \text{ is } B_1, \text{then } f_1 = a_1 x + b_1 y + r_1 \tag{3}$$

$$\text{Rule 2: If } x \text{ is } A_2 \text{ and } y \text{ is } B_2, \text{then } f_2 = a_2 x + b_2 y + r_2 \tag{4}$$

where x and y are two input variables, A and B are the linguistic labels, and a, b, and r are the output function parameters.

An example illustration of an ANFIS architecture for two inputs (x and y) is illustrated in Figure 2. As shown in Figure 2, ANFIS is composed of five layers: fuzzification layer (Layer 1), rule inference layer (Layer 2), normalization layer (Layer 3), defuzzification layer (Layer 4), and output layer (Layer 5).

Each node in Layer 1 (fuzzification layer) represents membership functions described by a generalized bell function, which can be defined as follows:

$$\mu_i(x) = \frac{1}{1 + \left|\dfrac{x - c_1}{a_1}\right|^{2b_1}} \tag{5}$$

where $\mu_i(x)$ is the ith output of the first layer, x is the input to the node, and a_1, b_1, c_1 are the variables representing the shape of the bell function. Variables on the first layer are known as the "premise set." The output of this layer is the membership values of the "premise part" of the model.

Each node in Layer 2 (rule inference layer) is based on a product of the incoming signals, and the output represents the firing strength of a rule.

$$W_i = \mu_{ki}(x)\mu_{li}(y), \text{i} = 1, 2, \ldots \tag{6}$$

where W_i is the ith output of the second layer, $\mu_{ki}(x)$ and $\mu_{li}(y)$ are the various combinations of incoming signals from the first layer. The nodes in Layer 3 (normalization layer) calculate the proportion of the firing strength of the ith rule to the sum of all rules firing strength using the following formula:

$$\overline{W}_\text{i} = \frac{W_\text{i}}{W_1 + W_2 + W_3 + W_4}, \text{i} = 1, 2, \ldots \tag{7}$$

In Layer 4 (defuzzification layer), every node is adapted with the node function.

$$\overline{W}_i f_i = \overline{W}_i\left(a_i x + b_i y + r_i\right), \text{i} = 1, 2, \ldots \tag{8}$$

where $\overline{W}_i$ is the normalized output from Layer 3 and $\{a_i, b_i, c_i\}$ are the parameter set.

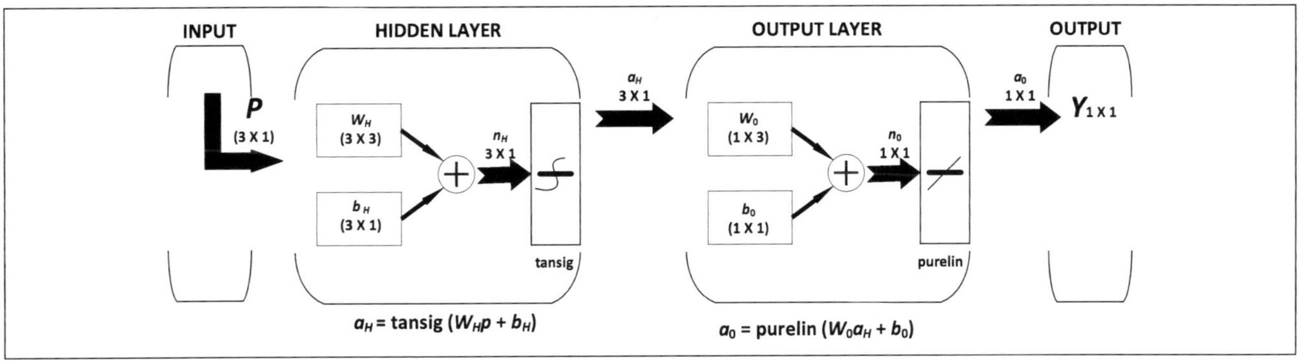

Figure 3. ANN model structure used in this study.

Layer 5: the single node in this layer is a fixed node, which computes the final output as the summation of all incoming signals:

$$f = \sum_{i=1}^{n} W_i f_i \tag{9}$$

Detailed description of ANFIS can be found in Jang et al. (*17*).

Several neuro-fuzzy (NF) structures for the same input combination (i.e., CEV, ($V \times D$), $P_{e\text{-init}}$) were tried in an algorithm developed in MATLAB®. Various NF structures were tested and compared by means of R^2, mean square errors (MSE), and mean absolute relative errors (MARE) statistics during both training and testing phase. The best combination included 2 triangular, 2 Gaussian, and 2 Gaussian membership functions ($\mu_i(x)$; see Equation 5), which are essentially curves defining the mapping of each input value to a degree of membership between 0 and 1.

ANNs

ANN is a powerful computational tool that was developed after inspiration from biological neural networks. It is composed of one or more hidden layers, of which the computation nodes are known as hidden neurons of hidden units. In this study, a multi-layer ANN, illustrated in Figure 3, is used. ANN may have more than one hidden layer; however, theoretical studies have shown that one hidden layer is sufficient for the ANN to anticipate any complex linear function (*22, 23*). The number of hidden neurons in the hidden layer was determined by trial and error. Sigmoid and linear functions were used as activation functions for the hidden and output nodes, respectively. Further details regarding the ANN structure are explained elsewhere (*24, 25*).

The abbreviations W_H and W_0 represent the weights of hidden and output layers. Likewise, b_H and b_0 are the biases for hidden and output layers, respectively. Increasing the number of hidden layers enables the network to achieve high-order statistics; however, this may cause overtraining. Detailed information regarding ANNs can be found elsewhere (*26*). In this study, the Levenberg–Marquard technique is employed in training the ANN. According Cigizoglu and Kişi, this technique is more powerful than the conventional gradient descent technique (*27*). This technique, different from the back-propagation with gradient descent technique (which is the steepest descent algorithm) is an approximation of Newton's method (*28*). The key step in the Levenberg–Marquard algorithm is the calculation of the Jacobean matrix. As for neural network mapping process, Jacobean matrix terms can be computed by a simple modification to the back-propagation algorithm (*29*).

To assess and compare the ability of two above computing techniques relative to each other, ANN models are developed. These models are trained in accordance with the best input combinations selected during the application of MLR. An ANN model, namely MLP, is developed for estimating percent embedment depth. For this purpose, a program code including a neural network toolbox was advanced using MATLAB® for the simulations. Different ANN structures were developed by a trial and error procedure, and the best fitting model structure was determined in terms of R^2, MSE, and MARE statistics for training and testing phases as illustrated in Table 1.

It is worth mentioning that the activation functions used for the input layer to hidden layer then to output layer is Tansig–Pureline. It is observed that the ANN model whose inputs are CEV ($V \times D$) $P_{e\text{-init}}$ has $R^2 = 0.94$, MSE = 14.18, and MARE = 0.05. Based on statistical criteria it is easily recognized that the application of the ANN technique significantly improves the model. The weights and biases for each ANN layer after the training are shown in Table 2.

Results and Discussion

In this study, CEV, product of Vialit test results and dust content of aggregates ($V \times D$), and $P_{e\text{-init}}$ embedment depth of aggregates into the binder were used as inputs to the three models described above to estimate progression of percent embedment (P_e) with time for the chip seal sections (*30*). In total 40 data points obtained from the field were included in the model development database. Randomly, 30 of the points were used for training and 10 points were used for testing. The statistical parameters for all input data values are illustrated in Table 2.

Table 1. Training and Testing Input Data Statistics

Input	$\bar{x}$	s_x	c_{sx}	x_{min}	x_{max}
Training data					
Cumulative equivalent traffic volume (CEV)	3,128,098	3,467,031	0.842	1	9,115,810
Vialit Test × Dust Content ($V \times D$)	13.43	9.88	0.139	0.002	28.03
Initial embedment depth of aggregate, $P_{e\text{-init}}$	29.48	2.93	−0.304	25.30	32.90
Embedment depth of aggregate, P_e	50.71	16.56	0.442	25.30	80.62
Testing data					
Cumulative equivalent traffic volume (CEV)	4,035,947	2,624,371	0.160	1	9,115,810
Vialit Test × Dust Content ($V \times D$)	14.70	8.34	0.056	0.002	28.03
Initial embedment depth of aggregate, $P_{e\text{-init}}$	30.38	2.60	−1.052	25.30	32.90
Embedment depth of aggregate, P_e	59.42	15.25	−0.788	30.74	80.62

Note: $\bar{x}$ = overall mean; s_x = standard deviation; c_{sx} = skewness coefficient; x_{min} = minimum; x_{max} = maximum.

Table 2. Weights and Biases of the ANN Model

	Weight Matrix for the Hidden Layer (W_H)			Bias Vector for the Hidden Layer (b_j)
Element no.	1	2	3	
1	−0.9034	−2.9080	−0.3904	1.4351
2	−1.0332	−9.6239	−1.8480	5.8955
3	−0.3190	2.7417	−5.8115	−0.7013
Element no.	Weight Vector of the Output Layer (W_0)			Bias for the Output Layer (b_0)
1	−2.0681	−0.4781	−1.4089	0.5810

Table 3. MSE, MARE, and R^2 Statistics of the Best MLR, ANFIS, and ANN Models

Method	MSE	MARE	R^2
Training Phase			
MLR	12.64	0.09	0.90
ANFIS	5.20	0.08	0.93
ANN	5.31	0.06	0.98
Testing Phase			
MLR	24.52	0.09	0.86
ANFIS	2.54	0.04	0.89
ANN	14.18	0.05	0.94

MSE, MARE, and coefficient of determination (R^2) statistics are used as evaluation criteria. Coefficient of determination (R^2) measures the degree to which the engaged data are linearly related. MSE and MARE provide different type of information about measure of the error and the accuracy of the model. The R^2, MSE, and MARE are defined as follows:

$$R^2 = \frac{\Sigma(m-p)^2}{\Sigma\left(p-p_{avg}\right)^2} \tag{10}$$

$$\text{MSE}=\frac{1}{n}\sum_{i=1}^{n}(m-p)^2 \tag{11}$$

$$\text{MARE}=\frac{1}{n}\sum_{i=1}^{n}\left|\frac{m-p}{m}\right|\times 100 \tag{12}$$

where:
m = measured output;
p = predicted output;
p_{avg} = the average estimated output;

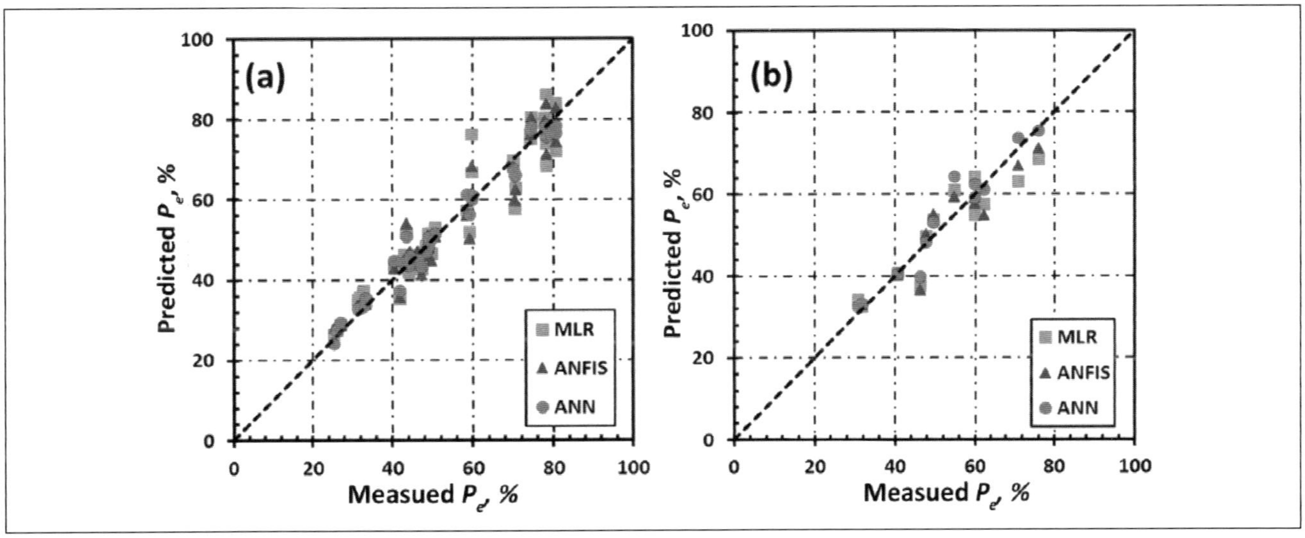

Figure 4. Predicted versus measured values of P_e for (*a*) training and (*b*) testing data sets.

n = total number of data;
MSE = mean square error; and
MARE = mean absolute relative error.

Predicted versus measured values of P_e for MLR, ANFIS and ANN models are given in Figure 4, *a* and *b*, for training and testing phase, respectively. As shown, the data scatter is quite close to the line of equality (45°) for ANN in the both scatterplots combinations compared with that of MLR and ANFIS.

The MSE, MARE, and R^2 statistics for respective MLR model for training and testing period are given in Table 3. The following relationship was developed for MLR model:

$$P_e = -9.4 + 0.496 \times (V \times D) + 0.000004 \times \text{CEV} + 1.42 \times P_{e-init} \quad (13)$$

where P_e is the percent embedment (%), $(V \times D)$ product of Vialit test results and dust content of aggregates, CEV is cumulative equivalent traffic volume and $P_{e\text{-init}}$ is the initial embedment depth of aggregates into the binder.

The results shown in Table 3 indicate that the MLR model performs with good accuracy, whereas ANFIS performs slightly better than the MLR. The ANN seems to be the best predictor among the three models evaluated.

In practice, engineers can use the models presented here to predict the progression of P_e with time, by measuring the initial P_e right after construction and performing Vialit tests using the aggregates and binder/emulsion used in the project. Assuming the dust content of the aggregates is known and the average annual traffic level is known for a given project, one can compute the progression of P_e with time and can estimate the projected design life of the chip seal. This could be useful for planning and cost allocation purposes.

The plots of percent embedment (P_e) data measured for eight different sections against the CEV are shown in Figure 5. As shown, all three models are quite good in predicting the progression of the P_e with traffic. Overall, the ANN model seems to be more accurate than the MLR and ANFIS.

Conclusions

In this study, MLR, ANFIS, and ANN techniques were used to estimate the progression of aggregate P_e in chip seal projects. The inputs included the cumulative equivalent traffic volume (CEV), Vialit test results (*V*), dust content of the aggregates (*D*), and the initial percent embedment ($P_{e\text{-init}}$). These inputs can easily be obtained for a given chip seal project and the projected life of a chip seal can be estimated using the models presented in this paper. The results revealed that even though all three models were quite good in predicting the percent embedment with traffic, in general, the ANN model performed better than the ANFIS model, and the ANFIS performed better than the MLR model. As more data become available, the authors recommend further validation of the models.

References

1. Labi, S., and K. C. Sinha. Effectiveness of Highway Pavement Seal Coating Treatments. *Asphalt*, Vol. 130, 2004, pp. 14–23.
2. Lee, J., and Y. Kim. Understanding the Effects of Aggregate and Emulsion Application Rates on Performance of Asphalt Surface Treatments. *Transportation Research Record: Journal of the Transportation Research Board*, 2008. 2044: 71–78.
3. Gransberg, D. D. Chip Seal Program Excellence in the United States. *Transportation Research Record: Journal of the Transportation Research Board*, 2005. 1933: 72–82.
4. Benson, F. J., and B. M. Gallaway. Retention of Cover Stone by Asphalt Surface Treatments. *Bulletin Texas Engineering Experiment Station System College Station*, Vol. 133 SRC, 1953.

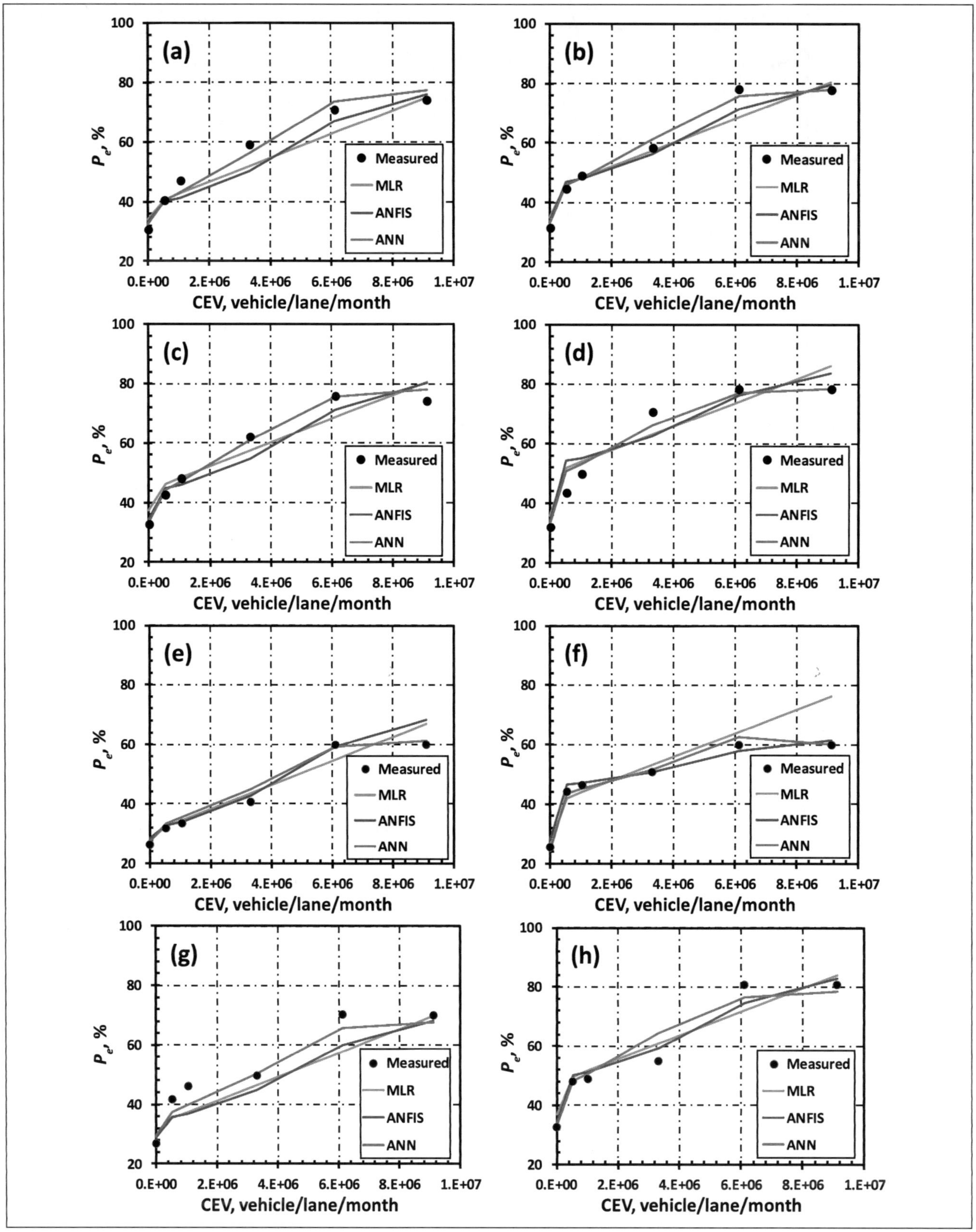

Figure 5. Measured and predicted P_e depth versus CEV for each test section.

5. Shuler, S. Chip Seals for High Traffic Pavements. *Transportation Research Record: Journal of the Transportation Research Board*, 1990. 1259: 24–34.
6. Roque, R., and D. Anderson. Effect of Material, Design, and Construction Variables on Seal-Coat Performance. *Transportation Research Record: Journal of the Transportation Research Board,* 1991. 1300: 108–115.
7. Gürer, C., M. Karaşahin, S. Çetin, and B. Aktaş. Effects of Construction-Related Factors on Chip Seal Performance. *Construction and Building Materials*, Vol. 35, 2012, pp. 605–613, 2012.
8. Transport New Zealand. *Notes for the Specification for Bituminous Reseals (TNZ P17)*. Wellington, New Zealand: Transport New Zealand, 2002.
9. Aktaş, B., M. Karaşahin, and M. Tiğdemir. Developing a Macrotexture Prediction Model for Chip Seals. *Construction and Building Materials*, Vol. 41, 2013, pp. 784–789.
10. Adams, J. M., and Y. R. Kim. Mean Profile Depth Analysis of Field and Laboratory Traffic-Loaded Chip Seal Surface Treatments. *International Journal of Pavement Engineering*, Vol. 15, No. 7, 2013, pp. 645–656.
11. Praticò, F. G., R. Vaiana, and T. Iuele. Surface Performance Characterization of Single-Layer Surface Dressing: A Macrotexture Prediction Model, *8th RILEM International Symposium on Testing and Characterization of Sustainable and Innovative Bituminous Materials*, RILEM Bookseries, Vol. 11, 2016, pp. 459–470.
12. Flintsch, G., E. de León, K. McGhee, and I. AI-Qadi. Pavement Surface Macrotexture Measurement and Applications. *Transportation Research Record: Journal of the Transportation Research Board*, 2003. 1860: 168–177.
13. China, S., and D. E. James. Comparison of Laser-Based and Sand Patch Measurements of Pavement Surface Macrotexture. *Journal of Transportation Engineering*, Vol. 138, No. 2, 2012, pp. 176–181.
14. Kutay, M. E., U. Ozdemir, D. Hibner, Y. Kumbargeri, and M. Lanotte. *Development of an Acceptance Test for Chip Seal Projects*. No. SPR-1649. Michigan State University, 2016.
15. Chapra, S. C., and C. P. Raymond. *Numerical Methods for Engineers*. New York: McGraw-Hill, 2010.
16. Bro, R., and A. Smilde. Principal Component Analysis. *Analytical Methods*, Vol. 6, No. 9, 2014, p. 2812.
17. Jang, J. S. R., C. T. Sun, and E. Mizutani. Neuro-Fuzzy and Soft Computing-A Computational Approach to Learning and Machine Intelligence. *IEEE Transactions on Neural Networks*, Vol. 8, No. 5, 1997, pp. 1219.
18. Daldaban, F., N. Ustkoyuncu, and K. Guney. Phase Inductance Estimation for Switched Reluctance Motor Using Adaptive Neuro-Fuzzy Inference System. *Fuel and Energy Abstracts*, Vol. 47, No 5, 2006, p. 347.
19. Mamdani, E., and S. Assilian. An Experiment in Linguistic Synthesis with a Fuzzy Logic Controller. *Readings in Fuzzy Sets for Intelligent Systems*, 1993, pp. 283–289.
20. Jang, J. ANFIS: Adaptive-Network-Based Fuzzy Inference System. *IEEE Transactions on Systems, Man, and Cybernetics*, Vol. 23, No.3, 1993, pp.665–685.
21. Takagi, T., and M. Sugeno. Fuzzy Identification of Systems and Its Applications to Modeling and Control. *Readings in Fuzzy Sets for Intelligent Systems*, 1993, pp. 387–403.
22. Cybenko, G. Approximation by Superpositions of a Sigmoidal Function. *Mathematics of Control, Signals, and Systems*, Vol. 2, No. 4, 1989, pp. 303–314.
23. Hornik, K., M. Stinchcombe, and H. White. Multilayer Feedforward Networks Are Universal Approximators. *Neural Networks*, Vol. 2, No. 5, 1989, pp. 359–366.
24. Demuth, H., and M. Beale. Neural network toolbox: for use with Matlab: Computation Visualization Programming Neural Network Toolbox: User's Guide, Version 4. The Mathworks.
25. Ozturk, H. I., and M. E. Kutay. An Artificial Neural Network Model for Virtual Superpave Asphalt Mixture Design. *International Journal of Pavement Engineering*, Vol. 15, No. 2, 2014, pp. 151–162.
26. Haykin, S. *NEURAL NETWORKS A Comprehensive Foundation*. Prentice Hall, Hamilton, 1994.
27. Cigizoglu, H., and Ö. Kişi. Flow Prediction by Three Back Propagation Techniques Using K-Fold Partitioning of Neural Network Training Data. *Hydrology Research*, Vol. 36, No. 1 2005, pp. 49–64.
28. Marquardt, D. W. An Algorithm for Least-Squares Estimation of Nonlinear Parameters. *Journal of the Society for Industrial and Applied Mathematics*, Vol. 11, No. 2, 1963, pp. 431–441.
29. Hagan, M., and M. Menhaj. Training Feedforward Networks with the Marquardt Algorithm. *IEEE Transactions on Neural Networks*, Vol. 5, No. 6, 1994, pp. 989–993.
30. TUBITAK-107G081. *Improving Seal Coat Applications and Development of a Performance Model in Highways*. Ankara, 2011.

The Standing Committee on Pavement Preservation (AHD18) peer-reviewed this paper (18-05834).

Article

TRR
JOURNAL OF THE TRANSPORTATION RESEARCH BOARD

Transportation Research Record
2018, Vol. 2672(12) 40–50

DOI: 10.1177/0361198118758315
journals.sagepub.com/home/trr
SAGE

Universal Method to Measure Water Infiltration into Asphalt Pavements Independent of Pavement Preservation Surface Treatment

Ben C. Cox[1], Isaac L. Howard[2], and Chase A. Hopkins[3]

Abstract

In recent years, greater focus has been placed on pavement preservation treatments, where one of the primary goals is to seal an existing pavement (especially cracks) to preserve it from further oxidation (i.e. decrease its permeability). The industry needs the ability to measure the effectiveness of these surface treatments; a single universal method that could be used for this purpose would be ideal, especially if traditional asphalt pavement surfaces could also be evaluated or field and laboratory measurements successfully related. This paper's objective is to present a simple permeameter device, the Mississippi permeameter (MSP), for measuring water infiltration into asphalt pavements independent of the surface (e.g. dense-graded asphalt (DGA), chip seal). In this paper, the MSP was evaluated against traditional field and laboratory permeameters (e.g. NCAT permeameter) on a range of surfaces from open-graded friction course (most permeable) to DGA (most common) to chip and scrub seals (least permeable, roughest texture, most difficult to measure). The MSP provided measurements as reasonable as those of existing permeameters and was also successfully used to test chip and scrub sealed cracks, which is beyond existing permeameters' abilities. Additional refinements are needed for relating field and laboratory measurements, but the MSP concept appears promising and worth pursuing given the increasing pavement preservation focus.

The ability of water to infiltrate an asphalt pavement (i.e. permeability) is generally a better durability indicator than air voids (V_a) alone because oxidation and aging primarily take place through accessible, or interconnected, voids (*1*). Throughout the years, several permeameters have been developed for asphalt; however, permeability is not commonly measured in comparison to V_a for several reasons.

First, reasonable correlations between permeability and V_a do exist, and in-place V_a is already commonly measured. Second, current field and laboratory permeameters do not relate well, and field permeameters tend to be somewhat particular (e.g. use of putty to create a seal to the pavement). Lastly, permeability is typically reported by hydraulic conductivity (k) according to Darcy's law, which is violated in multiple ways for field testing.

In recent years, the focus on pavement maintenance and preservation has increased, placing a greater emphasis on treatments such as chip seals. In addition to the permeameter issues mentioned previously, another challenge with traditional permeameters is they were developed for dense-graded asphalt (DGA) rather than a diverse array of preservation treatments. One of the principle objectives of most preservation treatments is to seal a pavement (especially cracks) and prevent further oxidation and aging. The industry needs the ability to measure all surface treatments' ability to seal a pavement with one comparable method.

Another consideration is that, although V_a is generally used in place of permeability for DGA, there is no surrogate for preservation treatment permeability. For example, V_a is not applicable to chip seals and cannot be reasonably measured on thin-lift overlays. In contrast, permeability is more universal to conventional DGA and preservation treatments so long as it can be measured.

This paper's objective is to present a universal method for measuring water infiltration into an asphalt pavement independent of pavement preservation treatment. Specifically, a

[1]Airfields and Pavements Branch, U.S. Army Engineer Research and Development Center (ERDC), Vicksburg, MS
[2]Civil and Environmental Engineering, Mississippi State University, Mississippi State, MS
[3]Engineering Services, Eastman Chemical Company, Longview, TX

Corresponding Author:
Address correspondence to Ben C. Cox: benjamin.c.cox@usace.army.mil

simple and expedient permeameter device developed in Mississippi and usable in the field or laboratory is presented with data from three pavements categories (open-graded friction course (OGFC), DGA, and chip and scrub seals) specifically selected to present a broad range of applications and capabilities. For DGA, the Mississippi permeameter (MSP) was evaluated against traditional permeameters in the field and laboratory.

The MSP's appeal is two-fold. First, it can be used to measure infiltration of multiple asphalt pavement surface types. Second, the same device has the potential to be used in the laboratory and field. In essence, the MSP exhibits the potential to perform the same functions as traditional permeameters but also functions not possible with traditional permeameters.

Literature Review

Permeability Concepts

Permeability is the measure of a material's ability to allow water (or air) to pass through it and is most commonly characterized by Darcy's k (Equation 1 – for falling head test). Darcy's law is based on several assumptions, many of which are violated when testing asphalt mixtures, particularly in the field.

Darcy's law requires one-dimensional flow, but water is able to flow vertically or laterally during field testing (*2,3*). Tack coats may also disturb vertical flow. Darcy's law requires complete saturation which cannot be accurately determined in the field (*2,3*). Mallick et al. showed permeability values decreased with successive tests as voids became saturated (*4,5*). Darcy's law assumes laminar flow, which is likely not the case, especially for high-permeability mixtures; there is no means for determining flow state inside a mixture (*2,3*). Darcy's law also requires specimen dimensions to be known. Field thicknesses cannot be accurately determined without coring, and cross-sectional area is unknown because of lateral flow (*2,3*). Some of these assumptions can be satisfied in the laboratory, but none can be satisfied in the field.

$$k = \frac{aL}{At} \ln\left(\frac{h_1}{h_2}\right) \quad (1)$$

where,

k = hydraulic conductivity (cm/s)

a = inside cross-sectional area of permeameter standpipe (cm^2)

A = cross-sectional contact area (cm^2)

L = test specimen thickness (cm)

t = elapsed time between h_1 and h_2 (s)

h_1 and h_2 = initial and final head, respectively, across the test specimen (cm)

Many reasonable relationships between permeability and V_a have been established showing permeability generally increases with V_a (*3,5,6–8*). These relationships are usually specific to nominal maximum aggregate size (NMAS) and gradation type (fine or coarse) as these factors (and others such as voids in mineral aggregate (VMA) and lift-thickness-to-NMAS (t/NMAS) ratio) also affect permeability.

Hudson and Davis showed an aggregate material finer than 0.075 mm, despite exhibiting over twice the VMA, was less permeable than a clean well-graded material with a 38 mm maximum size, illustrating the significance of interconnected voids (*9*). Finer gradations, all other factors being equal, produce fewer interconnected voids and lower permeabilities (*10–12*). Generally, increasing t/NMAS decreases permeability (*4–6,13,14*) but can also increase permeability because of the greater influence of lateral permeability, which is typically 1–5 times greater (up to 30) than vertical permeability (*15*).

The critical k (k_{crit}) most frequently recommended is 100×10^{-5} cm/s (*4,5,11,16,17*), which was established by Mallick and colleagues based on the permeability of Marshall-designed 9.5 mm fine-graded mixtures that generally performed well at 8% V_a (*4,5*). The Florida Department of Transportation (DOT) initially used that k_{crit} but later increased it to 125×10^{-5} cm/s (*16*), which was also used by other researchers (*18,19*). Cooley and Maghsoodloo concluded either value would be suitable (*20*).

Permeability Measurement Devices

The two most common falling head asphalt permeameters are the FL laboratory permeameter (Figure 1*a*) and the National Center for Asphalt Technology (NCAT) field permeameter (Figure 1*b*). Other permeameters, such as the TX field permeameter (*21*) (Figure 1*c*) or variations of the NCAT permeameter, have been researched but are not as established.

Laboratory testing with the FL permeameter most fully adheres to Darcy's law because dimensions are known, a flexible membrane and confining pressure produce one-dimensional flow, and specimens are pre-saturated. Variability is relatively high with coefficients of variation (COVs) of 44% for field cores and 0–133% for laboratory-compacted specimens; therefore, a sample k_{crit} of 50×10^{-5} cm/s must be used to reasonably ensure the population k is under 100×10^{-5} cm/s (*16*).

Field testing presents more challenges. At low permeabilities (e.g. under 100×10^{-5} cm/s), field (NCAT) and laboratory (FL) k values are similar, but field results increasingly exceed laboratory results with increasing permeability because of the influence of lateral flow (*22*). Increasing the permeameter contact diameter (and cross-sectional area, A) decreases lateral flow effects, leading Harris et al. to

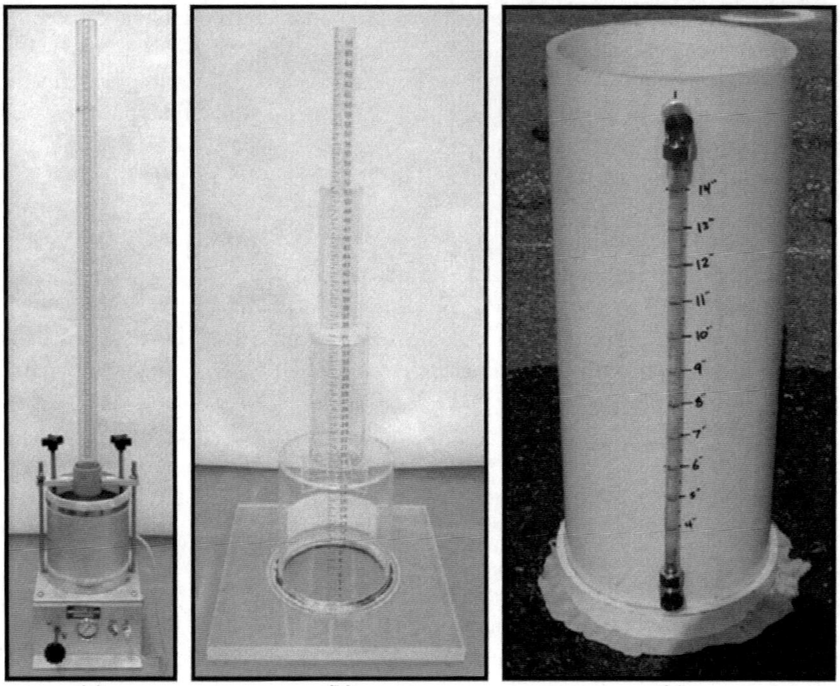

(a) (b) (c)

Figure 1. Existing permeameter devices: (*a*) FL; (*b*) NCAT; (*c*) TX.

recommend a 25 cm permeameter diameter (almost twice that of the NCAT) (*15*).

Field saturation is problematic and was neglected entirely by Mallick et al., who elected to average three successive tests per test location, arguing this was realistic (e.g. pavements are generally unsaturated prior to rain events) (*4*). Harris et al. (*15*) semi-accounted for saturation by conducting four tests and averaging the final three (permeability decreased considerably between the first two but not substantially between the last three).

The NCAT permeameter uses plumber's putty (or caulk) and several weights (9 to 47 kg) to create a seal to the pavement, which inhibits retesting at previous locations. Neoprene foam rubber gaskets have been used in place of plumber's putty (*4,6,15*), and vehicle self-weight via a hitch-mounted jack has been used as a surcharge to seal an NCAT-style permeameter to the pavement (*15*).

Williams (*23*) created 12.7 m^2 permeability maps for seven pavements, finding permeability results can rely heavily on permeameter placement. Consequently, a minimum of ten test locations per pavement was recommended to establish a representative average permeability. A round-robin with the NCAT permeameter, seven operators, eight pavements, and ten test locations each resulted in standard deviations of 10×10^{-5} cm/s for permeameter/operator reproducibility and 24×10^{-5} cm/s overall (*20*).

Permeability Values from Literature

Table 1 presents a summary of literature permeability data grouped by NMAS at typical design, performance testing, and low-end construction acceptance V_a levels. Generally, 9.5 and 12.5 mm mixtures exhibit similar permeability characteristics (*13,14*).

Experimental Program

Mississippi Permeameter (MSP) Device

The MSP concept was originally developed around 1975 by White for OGFC and was adaptable for field or laboratory use (*27,28*). Around 2009, White and Ivy upgraded the device (*29*). Further modifications were made during Mississippi DOT (MDOT) State Study 250 to produce an equipment package that consisted of a portable field system (MSP-F) and a smaller (gyratory-compacted specimens and cores) and larger (slabs) laboratory system (MSP-L_S and MSP-L_L, respectively) (*30*).

Figure 2 presents all MSP components and configurations (cost data and detailed drawings are available in Cox et al. (*30*)). The acrylic MSP standpipe (Figure 2*a*) is the key component and is sealed to the pavement via a surcharge load and a neoprene gasket. Figure 2, *b* and *c* make up the field configuration, and Figure 2, *d* and *e* are the laboratory configurations. The MSP standpipe is the same for all configurations. The

Table 1. Summary of Literature Permeability Values

NMAS (mm)	Gradation	Reference No.	LC/FC	Permeameter	R^2	k (10^{-5} cm/s) at V_a (%) 4	7	10	$V_{a,crit}$ (%)
4.75	Fine	West et al. (24)	FC	FL	—	1	10	45	11.7
9.5	Both	Mogawer et al. (11)	LC	FL	0.93	10	57	319	8.0
	Fine	Mogawer et al. (11)	FC	NCAT	0.86	4	26	173	9.1
	Fine	Cross and Bhusal (25)	FC	NCAT	0.81	24	134	399	6.4
	Coarse	Cooley (26)	FC	NCAT	0.80	1	47	693	7.7
	Coarse	Mogawer et al. (11)	FC	NCAT	0.75	1	15	201	9.2
	Coarse	Cooley et al. (13)	FC	NCAT	0.69	5	76	439	7.4
	Coarse	Cooley (26)	FC	NCAT	0.53	9	90	397	7.2
	Coarse	Cooley (26)	FC	NCAT	0.87	11	200	1289	6.1
	Avg	—	—	—	—	8	81	489	7.6
12.5	Both	Mogawer et al. (11)	LC	FL	0.52	31	180	1043	6.0
	Fine	Cooley (26)	FC	NCAT	0.67	0	12	124	9.7
	Fine	Cooley (26)	FC	NCAT	0.70	1	19	175	9.1
	Fine	Cooley (26)	FC	NCAT	0.52	1	193	6383	6.5
	Fine	Cooley (26)	FC	NCAT	0.45	12	98	367	7.0
	Coarse	Cooley (26)	FC	NCAT	0.79	3	54	376	7.8
	Coarse	Mogawer et al. (11)	FC	NCAT	0.79	6	95	1602	7.1
	—	Cooley et al. (13)	FC	NCAT	0.64	4	61	346	7.7
	Avg	—	—	—	—	7	89	1302	7.6
19	Both	Mogawer et al. (11)	LC	FL	0.16	15	33	74	11.1
	Coarse	Cooley (26)	FC	NCAT	0.78	4	310	4806	6.0
	Coarse	Cooley (26)	FC	NCAT	0.76	6	292	3333	6.0
	Coarse	Mogawer et al. (11)	FC	NCAT	0.86	13	448	14876	5.7
	Coarse	Cooley (26)	FC	NCAT	0.35	92	572	1832	4.1
	Coarse	Cooley (26)	FC	NCAT	0.48	122	2855	21279	3.9
	Coarse	Cooley (26)	FC	NCAT	0.75	620	2709	6934	2.0
	—	Cooley et al. (13)	FC	NCAT	0.42	32	341	1534	5.2
	Avg	—	—	—	—	113	945	6834	5.5
25	Coarse	Mogawer et al. (11)	FC	NCAT	0.63	421	2126	10739	1.3
	—	Cooley et al. (13)	FC	NCAT	0.50	126	1038	3987	3.8
	Avg	—	—	—	—	274	1582	7363	2.6

Note: Dashes indicate either not applicable or data not available. LC/FC = laboratory-compacted/field-compacted. R^2 = coefficient of determination. *k* at 4, 7, or 10% V_a was calculated using exponential regression fits either provided in the reference or determined from data presented in the reference. V_a was determined by AASHTO T166; $V_{a,crit}$ corresponds to V_a level at a permeability of 100×10^{-5} cm/s (i.e. k_{crit}).

MSP standpipe (alongside some means of surcharge loading) is the only truly critical component of the MSP device, meaning more economical versions could also be built.

Testing Procedure

To conduct a test, the standpipe is placed in position and loaded with 445 ± 22 N. The standpipe is filled with water to the 25.4 cm mark, and the time required for the water to fall to the 12.7 cm mark, or the distance fallen in 5 minutes, whichever occurs first, is recorded. Following the approach of Mallick et al., three replicate tests are averaged unless the test location is deemed impermeable before all three are conducted (*4*).

Given the issues presented in literature with Darcy's law, an alternative parameter was used. Of primary interest is whether water is able to penetrate the pavement and to what degree; therefore, infiltration rate (*Inf*) (Equation 2) was chosen as a reasonable index of permeability (considering the assumptions violated in the field, Darcy's *k* is also effectively an index). In addition, with the goal of using the MSP for preservation surface treatments, *Inf* was believed to be more suitable since the testing of such treatments would likely be even less representative of Darcy's law.

$$Inf = \frac{a}{At}(h_1 - h_2) \tag{2}$$

where,

Inf = infiltration rate (cm/min)

a = inside cross-sectional area of permeameter standpipe (cm²)

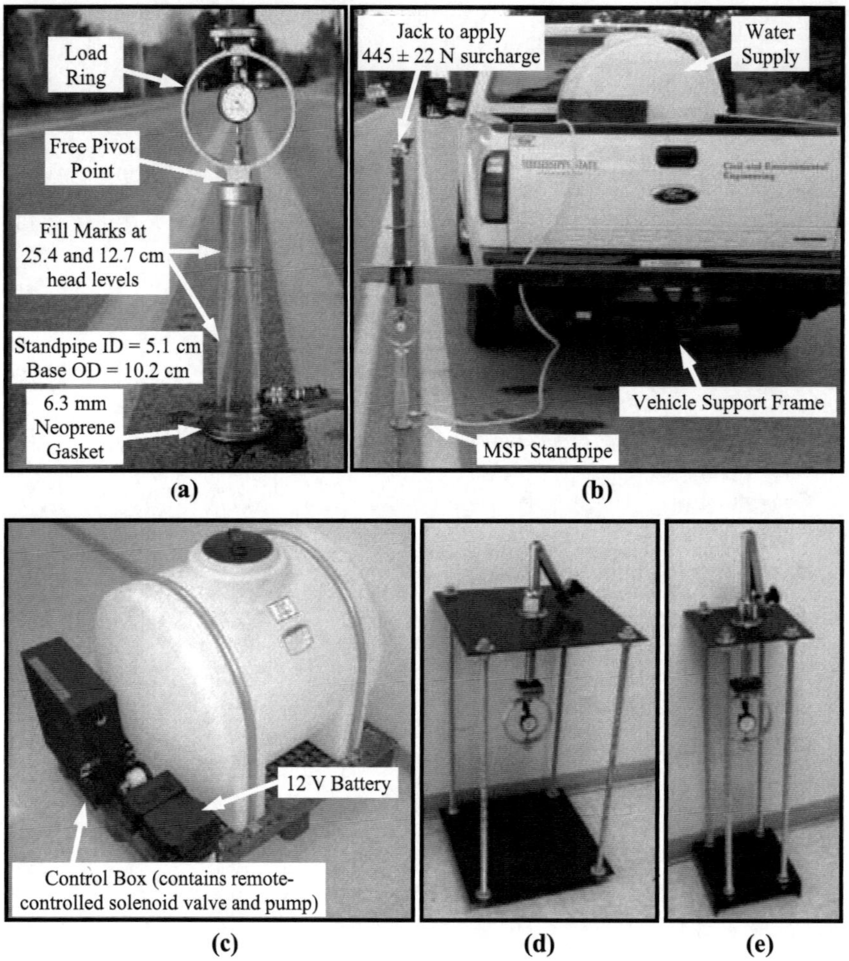

Figure 2. MSP equipment overview: (*a*) MSP standpipe detail; (*b*) MSP-F setup overview; (*c*) MSP-F water supply; (*d*) MSP-L_L test frame; (*e*) MSP-L_S test frame.

A = cross-sectional contact area (cm^2)

t = elapsed time between h_1 and h_2 (min)

h_1 and h_2 = initial and final head, respectively, across the test specimen (cm)

Additional preparations are required for test locations over chip and scrub seals because of the coarse surface texture. Figure 3 summarizes the steps taken to prepare the surface for testing. Figure 3*a* illustrates the coarse surface texture and shows a typical sealed crack in a chip or scrub seal pavement (testing cracks was of interest) with four test locations spaced approximately 0.6 m apart.

First, a 10.2 cm (inner diameter) core bit is used to cut a shallow groove through the seal treatment and just to the underlying pavement (Figure 3*b*). Next, small tins of asphalt binder are heated and poured into the groove (Figure 3, *c–e*), and a 3.8 cm tall, 10.2 cm diameter plastic ring (cut from a 10.2 by 20.4 cm cylinder mold) is pressed into the asphalt reservoir (Figure 3, *f* and *g*). The MSP standpipe is set inside the ring (Figure 3*h*), providing a watertight seal between the standpipe and pavement. Note that two neoprene gaskets were layered to further promote adequate sealing. Otherwise, testing is not altered.

Results

This section presents MSP permeability results for three surfaces: (1) OGFC—the most open and most permeable, (2) DGA—the most common, and (3) chip and scrub seals—the most closed and least permeable and also the roughest surface texture (i.e. most difficult to measure).

Open-Graded Friction Course

The MSP, originally developed for OGFC, was used to evaluate a south Mississippi test section constructed in spring 2007 on I-55 at five approximately 6-month intervals (*29*).

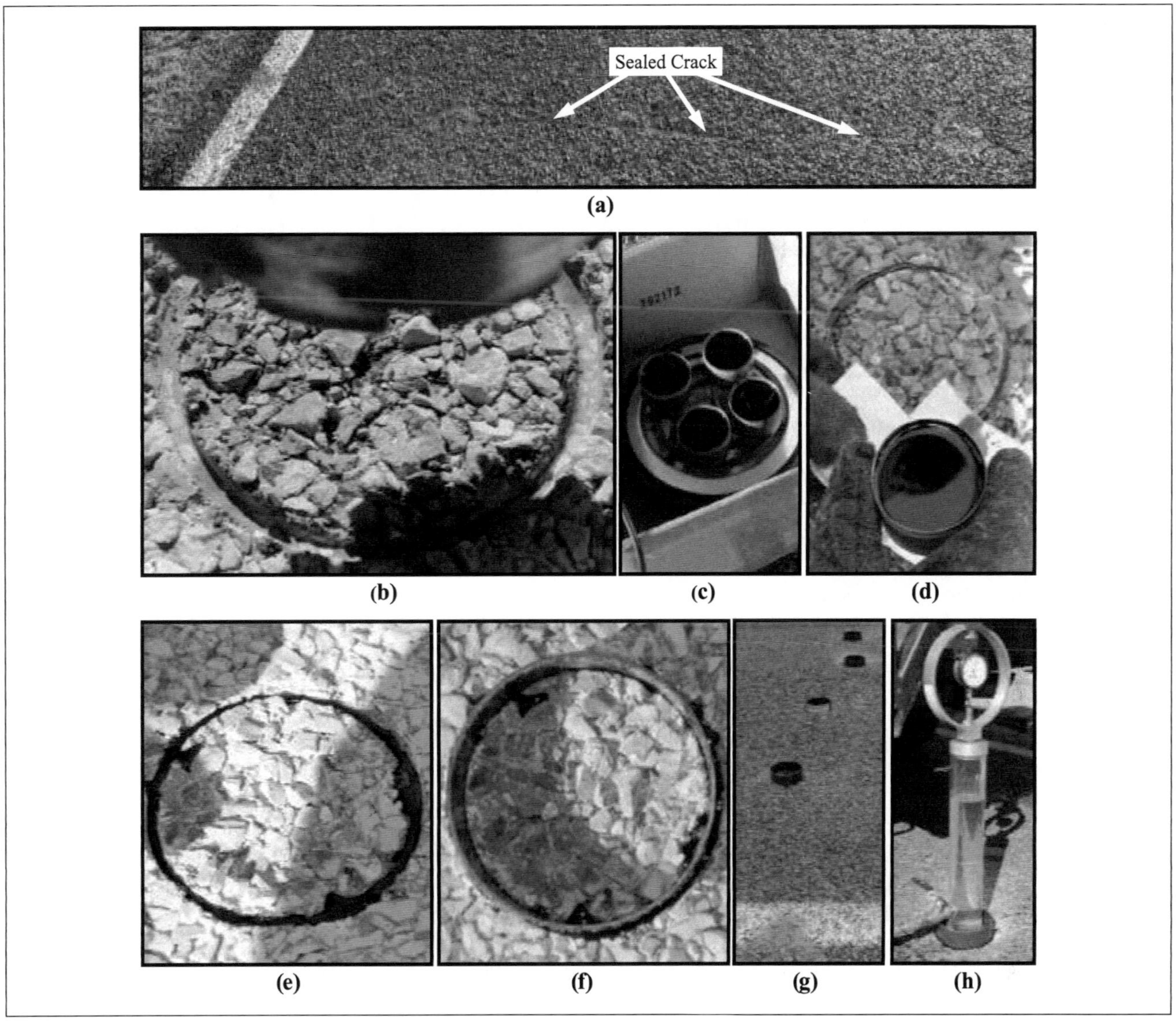

Figure 3. Chip and scrub seal testing procedures: (*a*) sealed crack in existing pavement with locations for permeability testing shown; (*b*) coring groove to underlying pavement surface; (*c*) heating binder; (*d*) filling groove; (*e*) groove filled with binder; (*f*) plastic ring set in groove; (*g*) prepared crack; (*h*) testing.

Nine to ten locations spaced 0.16 km apart were tested between and in both wheel paths for all four lanes (111 test locations, 555 total tests).

Average *Inf* decreased from 98 to 80–85 cm/min over approximately 2 years. COV at each testing interval ranged from 34 to 44%. The average *Inf* for all test intervals was 90 and 85 cm/min between and in the wheel paths, respectively. Overall, the MSP appeared to work well for this evaluation similar to other devices, such as the TX permeameter (Tex-246-F) that have also measured OGFC permeability successfully (*31*).

Dense-Graded Asphalt Overlays

DGA results with the MSP-F are presented in three sections: measurements of pavement permeability, comparisons with existing field permeameters, and comparisons with existing lab permeameters. A fourth section compares MSP-F and MSP-L results.

MSP-F Measurements of Pavement Mat and Joint Infiltration. Two DGA evaluations have been conducted with the MSP-F. The first (*30,32*) was on two thin-lift overlays (nominally 1.9 and 2.5 cm thick) which were monitored for 4 years after construction. Longitudinal joints were of primary interest, and nine test locations were selected representing a range of constructed joint qualities. Infiltration of the longitudinal joints and surrounding pavement mat were measured.

Mat infiltration was relatively consistent and low for all thin-lift locations; average *Inf* was 0.2 cm/min at 0 months and essentially impermeable at 48 months. This is reasonable

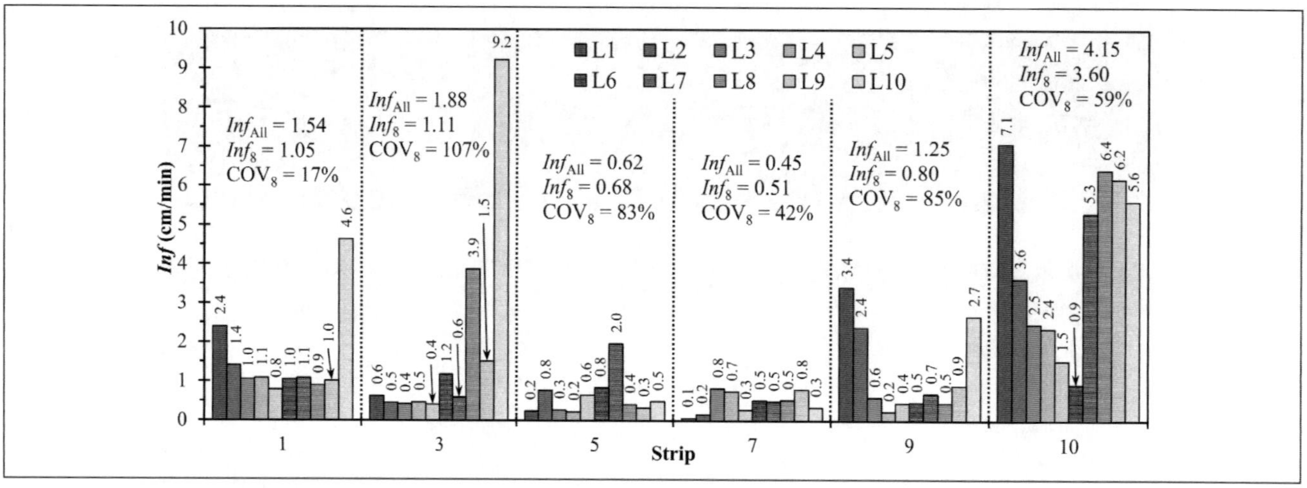

Figure 4. MSP-F results of Columbus parking lot.

for a small-NMAS mixture with high asphalt content since heat and traffic would tend to close surface voids over time. Longitudinal joint infiltration ranged from less than 1.0 cm/min to approximately 100–125 cm/min for joints which cracked during the 4-year period.

The second evaluation was on six strips of 12.5 mm NMAS asphalt placed on a parking lot in Columbus, MS (termed the Columbus parking lot) described in Howard et al. (*33*). Ten locations (L1 to L10) spaced 25 to 30 cm apart in a single transverse line were tested per strip (strips 1, 3, 5, 7, 9, and 10 in Howard et al. (*33*)). Figure 4 presents all 60 test location *Inf* results including the average for all tests in a strip (Inf_{All}) and the average and COV for the inner eight locations, L2-L9 (Inf_8 and COV_8).

Inf near the sides of a strip (e.g. L1, L10) were often considerably higher, which is characteristic of lower densities typical along the edges, especially the unsupported or free edge, and thus, a rational result. To somewhat address edge effects, Inf_8 was determined, ranging from 0.51 (strip 7) to 3.60 (strip 10) cm/min. Even with edge locations removed, COV_8 values were generally very high. However, because these strips were constructed for an emergency paving project, haul times were especially long, a material transfer vehicle was not used, one steel wheel roller was used (no rubber tire roller), and a fixed roller pattern was used that was not adjusted during compaction to prevent mat tears; therefore, this level of variability was not surprising. Strip 10 stands out as it had a large longitudinal tear down the strip approximately 60% of the way across the strip from the L1 side. This is reflected in Figure 4 *Inf* values with the abrupt transition from L6 to L7. Strip 10 *Inf* values are better interpreted alongside V_a values, which from left (L1 side) to right (L10 side) across the strip were 12.0, 9.4, 8.1, 10.1, and 11.5% (measurements obtained from cores cut a short distance from the test locations).

The key point from DGA field testing is that the MSP-F seems feasible and provided rational results. Measuring field DGA permeability with the MSP-F was considered successful.

MSP-F Comparisons with Existing Field Permeameter Devices. NCAT and TX permeameter tests were performed at the Columbus parking lot at the exact locations as the MSP-F. Based on Figure 4, nine locations (L7 to L9 of strips 1, 7 and 10) were chosen to provide three replicates of low, intermediate, and high *Inf* values. NCAT permeameter testing followed standard protocols (used approximately 0.6 kg of plumber's putty as a seal, loaded with 9.2 kg of weight, tested for 5 minutes, calculated *k* by Equation 1). TX permeameter testing followed Tex-246-F (also sealed with plumber's putty) with some calculation changes for DGA outlined in Wilson and Sebesta (*21*) and described in detail in Cox et al. (*30*); ultimately, *Inf* was reported. Cores later cut from these locations ranged from 7.0 to 7.7 cm thick.

Figure 5 presents relationships between the MSP-F, NCAT, and TX devices. The amount of data is limited, but even still, the relationships, with R^2 of 0.94 and 0.66, were fairly reasonable, especially given the physical differences (e.g. size, shape, sealing method) between each device. The primary observation was that the MSP-F's ability to measure DGA permeability/infiltration was comparable with other existing field permeameters. Comparing the MSP-F with existing permeameters was considered successful.

MSP-F Comparisons with Existing Laboratory Permeameter Devices. Slabs (0.3 m square) were sawn from the NCAT and TX permeameter test locations in the previous section and cored (150 mm diameter) for laboratory testing with the FL permeameter following ASTM PS129 (now withdrawn). Figure 6*a* shows an unimpressive relationship between the MSP-F and FL permeameter. The same is true, however, for the NCAT permeameter (Figure 6*b*). Realistically, this behavior is not surprising because, as Figure 6 highlights, one of the

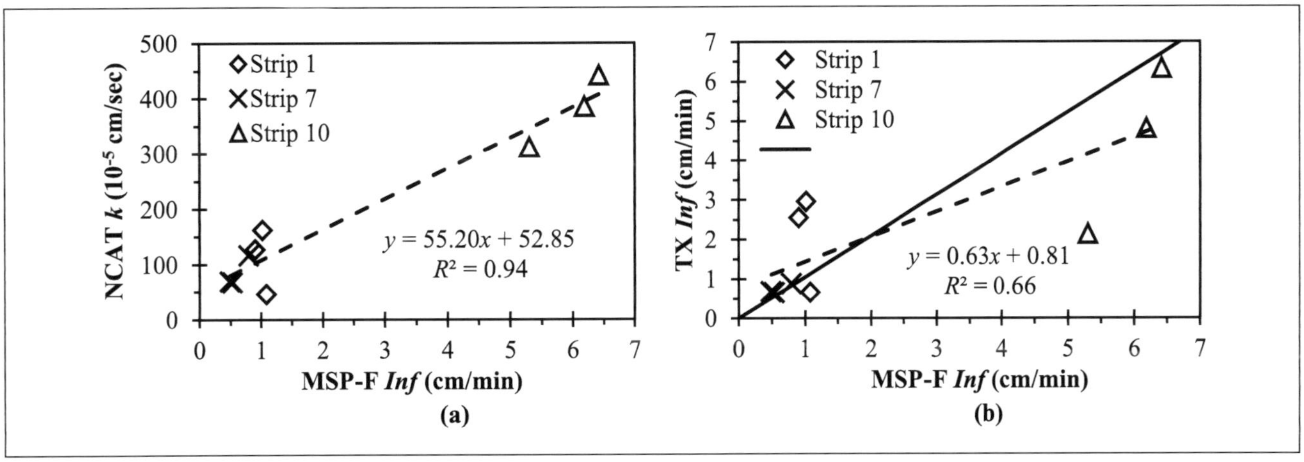

Figure 5. Relationships between the MS-P and existing field permeameters: (*a*) NCAT permeameter; (*b*) TX permeameter.

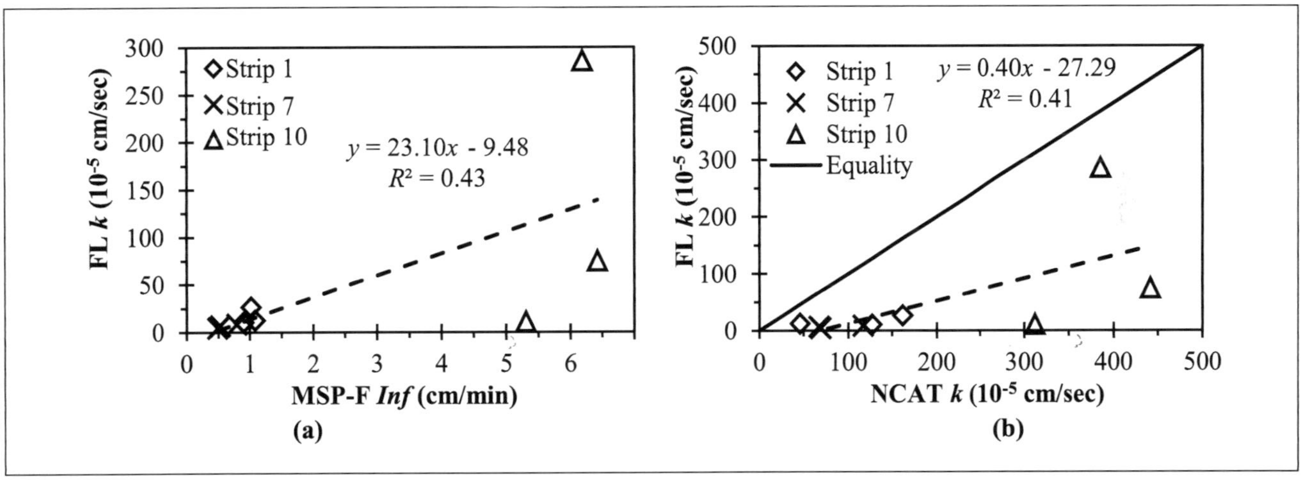

Figure 6. Relationships between field and existing laboratory permeameters: (*a*) FL versus MSP-F; (*b*) FL versus NCAT.

major differences between the FL permeameter and others is that water is forced to pass completely through the specimen rather than allowing horizontal flow just below the surface.

Relating laboratory to field permeabilities has been a longstanding challenge because of the fundamental differences in the test methods and devices. At present, the MSP has not been successful at resolving this challenge, but neither have the other methods. The MSP's design and methodology have room to be improved in the future; improving the other field methods for laboratory work appears more daunting.

Relationships between MSP-F and MSP-L Measurements. The MSP-F and MSP-L were compared for the same test locations, and MSP-L_L testing occurred on the sawn slabs prior to coring for FL permeameter (previous section) and MSP-L_S testing. MSP-L_L and MSP-L_S testing protocols were the same as for the MSP-F except the Figure 2, *d* and *e* test frames were used.

MSP-L_S results (Figure 7*a*) were not entirely unreasonable (R^2 of 0.8) but were also not remarkable, particularly because higher *Inf* values (above the equality line rather than below) would be expected because horizontal water flow paths to free edges are shorter for cores. Consequently, testing a larger specimen (i.e. slab) should further reduce *Inf*, which it did for the MSP-L_L (Figure 7*b*); however, this shifted results even further from the equality line. MSP-L_L *Inf* for strip 10 was equivalent to strips 1 and 7 despite being over six times greater with the MSP-F. This was not intuitive. No errors were observed on reviewing the data, and at present, the authors do not fully understand Figure 7, *a* and *b*.

Since the testing of cores does present a free edge difference relative to the field, laboratory conditioning was considered to better represent field conditions. Specimens were either vacuum saturated (VS) following PS129 protocols or VS and sealed with petroleum jelly along the sides. Figure 7, *c* and *d* indicate that, on average, this had no meaningful impact on *Inf*.

The primary differences between the MSP-F and MSP-L configurations are (1) the means used to provide a reaction force for providing surcharge (i.e. vehicle weight or test

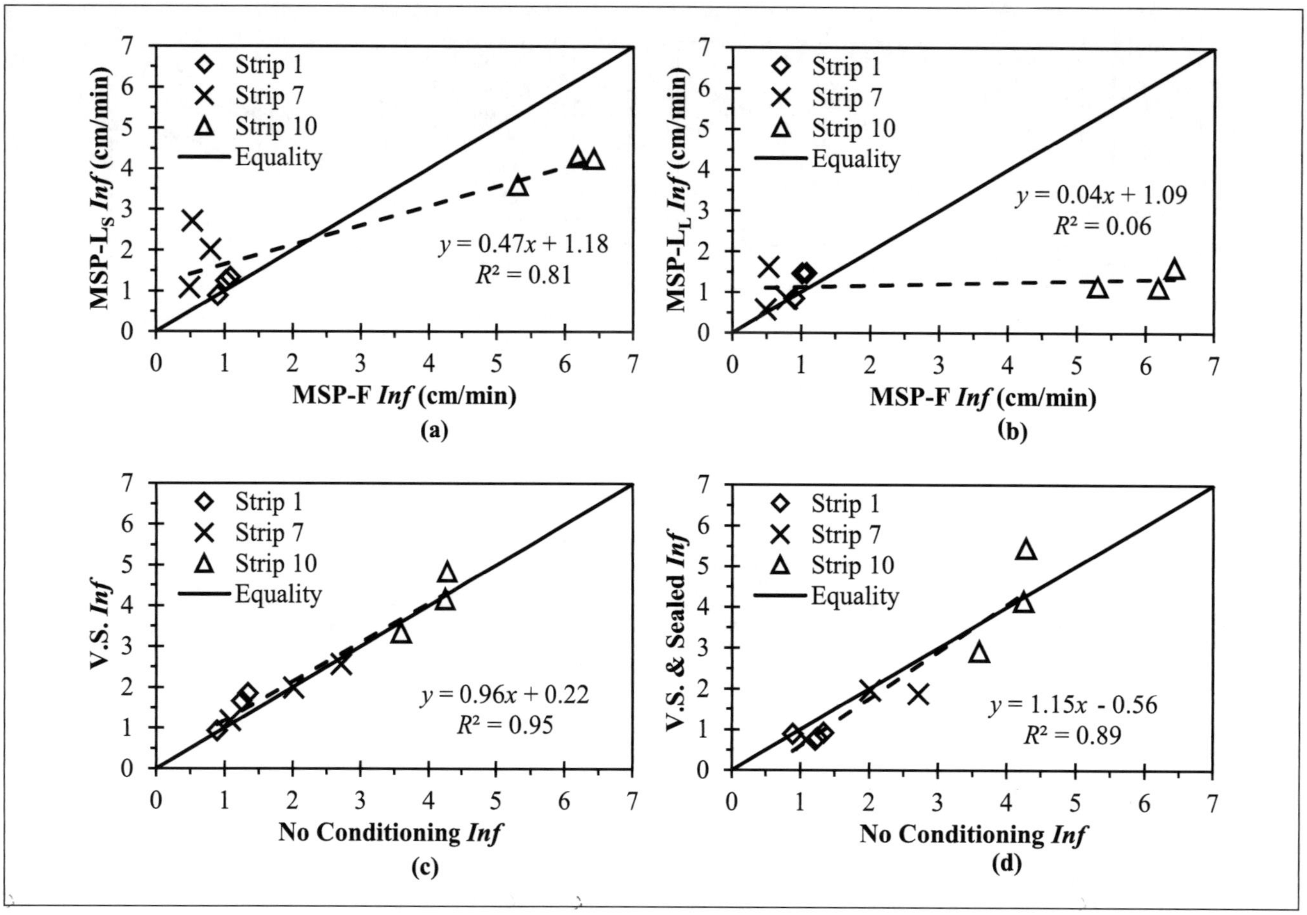

Figure 7. Relationships between field and laboratory MSP permeameters: (*a*) relationship between MSP-L_L and MSP-F; (*b*) relationship between MSP-L_S and MSP-F; (*c*) vacuum saturation effects on *Inf*; (*d*) vacuum saturation and sealing effects on *Inf*.

frame), which should not impact the test, and (2) the state of the test location (i.e. in-place or sawed/cored). Using an identical device in the laboratory or field aside from those two differences has considerable appeal and value because directly relating the laboratory and field is often a challenge. For reasons not yet understood, the MSP was unsuccessful at relating laboratory and field measurements; however, the concept appears more promising than current alternatives and is worth investigating further.

Chip and Scrub Seals

Field permeability of chip and scrub seals has not been previously measured to the knowledge of the authors. The MSP-F was used to measure and compare chip and scrub seal infiltration on six MDOT state routes (Table 2) around mid-October to mid-November of 2015 (1 to 1.5 years after the seals were placed). The most demanding role of a seal treatment is that of sealing cracks. Field permeability/infiltration of chip and scrub seals has not been previously measured successfully over cracks to the authors' knowledge. Scrub seals generally use a less viscous and slower setting emulsion that is "scrubbed" with brooms in attempts to better seal cracks. The purpose of this evaluation was to actually measure sealed cracks and test this hypothesis and simultaneously assess the MSP-F. Cracks that ran full lane widths were selected visually, targeting a low-to-medium severity level (e.g. 4–6 mm wide). Cracks on SR404 appeared slightly wider than cracks for all other pavements.

Figure 8*a* shows the *Inf* distribution for all chip and scrub seals. About 20% of chip and scrub seal test locations were impermeable (i.e. 0 cm/min). Among data up to 10 cm/min, scrub seals exhibited lower infiltration than chip seals. Above 10 cm/min, data was fairly balanced, with chip seals exhibiting the most permeable (i.e. largest) cracks overall. On average, scrub seals were slightly less permeable than chip seals (8.5 versus 12.6 cm/min). A one-tailed two-sample *t*-test yielded a *p*-value of 0.11, indicating scrub and chip seals were not significantly different at a 5% significance level.

Because SR404 cracks appeared slightly wider than for the other pavements, a repeat analysis was conducted with SR404 excluded (Figure 8*b*). Lower *Inf* values (≤ 2 cm/min) were more frequently scrubs seals, and vice versa. The scrub seal average *Inf* dropped to 3.3 cm/min, and the *t*-test *p*-value was <0.001, meaning scrub seals were significantly less permeable than chip seals at a 5% significance level.

Table 2. Summary of MDOT Chip or Scrub Sealed State Routes Evaluated

Seal type	Route	No. cracks tested	MDOT District	County	Sealed	Emulsion	Aggregate
Scrub	SR354	6	1	Tippah	May 2014	1.36 L/m^2 CMS-1PC	Size 89 Limestone
	SR403	6	1	Webster	June 2014	1.27 L/m^2 CMS-1PC	Size 89 Limestone
	SR404	6	1	Webster	June 2014	1.31 L/m^2 CMS-1PC	Size 89 Limestone
Chip	SR395	6	1	Winston	June 2014	1.63 L/m^2 CRS-2P	Size 7 Limestone
	SR436	6	3	Washington	Sept 2014	1.36 L/m^2 CRS-2P	Size 7 Limestone
	SR442	6	3	Sunflower	Aug 2014	1.45 L/m^2 CRS-2P	Size 7 Limestone

Note: Four locations tested per crack for a total of 148 locations tested for permeability with the MSP-F. SR395 chip seal was placed on an existing chip seal surface, meaning it was a double chip seal.

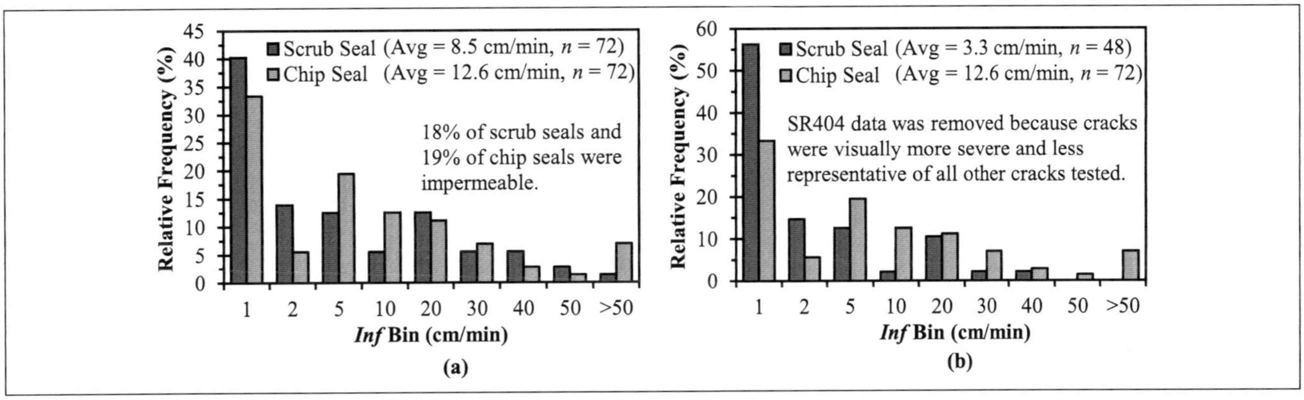

Figure 8. *Inf* distribution for chip and scrub seals: (*a*) all chip and scrub seal data; (*b*) all data except SR404.

This evaluation supported the hypothesis that scrub seals better seal cracks than chip seals. More importantly, the MSP-F was successfully used to measure the infiltration of a cracked pavement that had been chip or scrub sealed.

Summary and Conclusions

This paper's objective was to present a universal method for measuring water infiltration into an asphalt pavement independent of the pavement preservation surface treatment. The MSP device has been vetted in the field on a wide range of surface treatments from OGFC (most permeable) to DGA (most common) to chip and scrub seals (least permeable and also most difficult to measure because of coarse surface texture), and it provided reasonable measurements in all cases. With the increasing focus on pavement preservation, there is a major industry need for a device such as this, yet the authors know of no other device available. Additional refinements are needed for connecting field and laboratory MSP measurements, but overall, the concept appears promising to the degree that it is worth pursuing further.

Acknowledgments

MDOT funded State Study 250, and Ergon Asphalt & Emulsions and Paragon Technical Services, Inc. provided financial support. Thanks are due to Alex Middleton, Mark Holley, and Nan Mitchell of MDOT; Mark Ishee, Amy Walker, and Larry Tomkins of Ergon; and Dr. Thomas White and Will Crawley for their work with the Mississippi permeameter. Permission to publish was granted by the Director, Geotechnical and Structures Laboratory, US Army Engineer Research and Development Center.

References

1. Kumar, A., and W. H. Goetz. Asphalt Hardening as Affected by Film Thickness, Voids, and Permeability in Asphaltic Mixtures. *Journal of the Association of Asphalt Paving Technologists*, Vol. 46, 1963, pp. 571–605.
2. Cooley, L.A., Jr. *Permeability of Superpave Mixtures: Evaluation of Field Permeameters*. NCAT Report No. 99-1, National Center for Asphalt Technology, Auburn University, AL, 1999.
3. Cooley, L.A., Jr., and E. A. Brown. Selection and Evaluation of Field Permeability Device for Asphalt Pavements. *Transportation Research Record: Journal of the Transportation Research Board*, 2000. 1723: 73–82.
4. Mallick, R., M. Teto, and L. A. Cooley *Evaluation of Permeability of Superpave Mixes in Maine*. Report No. ME 00-1, Maine Department of Transportation, Worcester, Mass., 1999.
5. Mallick, R. B., L. A. Cooley, M. Teto, and R. Bradbury. Development of a Simple Test for Evaluation of In-Place Permeability of Asphalt Mixes. *International Journal of Pavement Engineering*, Vol. 2, No. 2, 2001, pp. 67–83.
6. Mallick, R. B., L. A. Cooley, Jr., , M. R. Teto, R. L. Bradbury, and D. Peabody. *An Evaluation of Factors Affecting Permeability*

of Superpave Designed Pavements. NCAT Report No. 03-02, National Center for Asphalt Technology, Auburn University, AL, 2003.

7. Zube, E. Compaction Studies of Asphalt Concrete Pavements as Related to the Water Permeability Test. *Highway Research Board Bulletin 358*, National Research Council, Washington, D.C., 1962, pp. 12–31.
8. Choubane, B., G. C. Page, and J. A. Musselman. Investigation of Water Permeability of Coarse Graded Superpave Pavements. *Journal of the Association of Asphalt Paving Technologists*, Vol. 67, 1998, pp. 254–276.
9. Hudson, S. B., and R. L. Davis. Relationship of Aggregate Voidage to Gradation. *Journal of the Association of Asphalt Paving Technologists*, Vol. 34, 1965, pp. 574–593.
10. Kanitpong, K., C. H. Benson, and H. U. Bahia. Hydraulic Conductivity (Permeability) of Laboratory-Compacted Asphalt Mixtures. *Transportation Research Record: Journal of the Transportation Research Board*, 2001. 1767: 25–32.
11. Mogawer, W. S., R. B. Mallick, M. R. Teto, and W. C. Crockford. *Evaluation of Permeability of Superpave Mixes*. Report No. NETCR 34. New England Transportation Consortium, Storrs, CT, 2002.
12. Bhuttacharjee, S., and R. B. Mallick. An Alternative Approach for the Determination of Bulk Specific Gravity and Permeability of Hot Mix Asphalt (HMA). *International Journal of Pavement Engineering*, Vol. 3, No. 3, 2002, pp. 143–152.
13. Cooley, L.A., Jr., , B. D. Prowell, and E. R. Brown. Issues Pertaining to the Permeability Characteristics of Coarse-graded Superpave Mixes. *Journal of the Association of Asphalt Paving Technologists*, Vol. 71, 2002, pp. 1–29.
14. Cooley, L.A., Jr., , B. D. Prowell, and E. R. Brown. *Issues Pertaining to the Permeability Characteristics of Coarse-graded Superpave Mixes*. NCAT Report 02-06. National Center for Asphalt Technology, Auburn University, AL, 2002.
15. Harris, C., L. Wang, C. Druta, T. Tan, G. Zhou, T. Zho, and L. A. Cooley, Jr. Effect of Permeameter Size and Anisotropy on Measurements of Field Pavement Permeability. *Transportation Research Record: Journal of the Transportation Research Board*, 2011. 2209: 41–51.
16. Maupin, G. W., Jr. Asphalt Permeability Testing: Specimen Preparation and Testing Variability. *Transportation Research Record: Journal of the Transportation Research Board*, 2001. 1767: 33–39.
17. Williams, S. G. *A Comprehensive Study of Field Permeability Using the Vacuum Permeameter*. Final Report MBTC-2054, 2006. Fayetteville, AR: Mack-Blackwell Transportation Center, University of Arkansas.
18. Mohammad, L. N., A. Herath, and B. Huang. Evaluation of Permeability of Superpave Asphalt Mixtures. *Transportation Research Record: Journal of the Transportation Research Board*, 2003. 1832: 50–58.
19. Brown, E. R., M. R. Hainin, A. Cooley, and G. Hurley. *Relationship of Air Voids, Lift Thickness, and Permeability in Hot Mix Asphalt Pavements*. NCHRP Report 531. Transportation Research Board, Washington, D.C., 2004.
20. Cooley, L.A, Jr., , and S. Maghsoodloo. Round-Robin Study for Field Permeability Test. *Transportation Research Record: Journal of the Transportation Research Board*, 2002. 1789: 25–35.
21. Wilson, B. T., and S. Sebasta. Comparison of Density Tests for Thin Hot-Mix Asphalt Overlays. *Transportation Research Record: Journal of the Transportation Research Board*, 2015. 2504: 148–156.
22. Prowell, B. D., and M. C. Dudley. Evaluation of Measurement Techniques for Asphalt Pavement Density and Permeability. *Transportation Research Record: Journal of the Transportation Research Board*, 2002. 1789: 36–46.
23. Williams, S. G. Sample Size Requierments for Field Permeability Measurements of Hot-Mix Asphalt Pavements. *Transportation Research Record: Journal of the Transportation Research Board*, 2007. 2001: 56–62.
24. West, R. C., M. A. Heitzman, D. M. Rausch, and G. Julian. *Laboratory Refinement and Field Validation of 4.75 mm Superpave Designed Asphalt Mixtures*. NCAT Report 11-01. National Center for Asphalt Technology, Auburn University, AL, 2011.
25. Cross, S. A., and S. Bhusal. *Longitudinal Joint Density and Permeability in Asphalt Concrete*. Final Report – FHWA-OK-08-07. Oklahoma Department of Transportation, Oklahoma City, 2009.
26. Cooley, L. A., Jr. *Evaluation of Pavement Permeability in Mississippi*. Final Report. Mississippi Department of Transportation, Jackson, 2003. http://mdot.ms.gov/documents/research/Reports/Interim%20and%20Final%20Reports/Pavement%20Permeability%20in%20Mississippi.pdf
27. White, T. D. *Porous Friction Surface Course*. Report No. FAA-RD-73-197. Federal Aviation Administration, Washington, D.C., 1975.
28. White, T. D. *Field Performance of Porous Friction Course*. Report No. FAA-RD-74-38. Federal Aviation Administration, Washington, D.C., 1976.
29. White, T. D., and J. Ivy. *I-55 OGFC Field Permeability Testing*. Report No. FHWA/MS-RD-09-201. Mississippi Department of Transportation, Jackson, 2009.
30. Cox, B. C., I. L. Howard, and J. Ivy. *Evaluation of Approaches to Improve Longitudinal Joints in Mississippi Overlay Projects*. Report FHWA/MS-DOT-RD-15-250-Volume 3, Mississippi Department of Transportation, Jackson, 2015.
31. Small Aggregate, Polymer Modifiers Key to New Texas Thin Lift Mixes. *Pavement Preservation Journal*, Vol. 7, No. 2, 2014, pp. 19–23. http://www.nxtbook.com/naylor/FPPQ/FPPQ0214/index.php#/1
32. Cox, B. C., and I. L. Howard. Time Dependent Water Infiltration Resistance of Thin-Lift Asphalt Overlay Longitudinal Joints. *Journal of Transportation Engineering, Part B: Pavements*, Technical Note, 2017, submitted June 5, 2017, in review.
33. Howard, I. L., B. A. Payne, M. Bogue, S. Glusenkamp, G. L Baumgardner, and J. M. Hemsley. *Full Scale Testing of Hot-Mixed Warm-Compacted Asphalt for Emergency Paving*. SERRI Report 70015-011. U.S. Department of Homeland Security, 2012.

The Standing Committee on Pavement Maintenance (AHD20) peer-reviewed this paper (18-01230).

Article

TRR
JOURNAL OF THE TRANSPORTATION RESEARCH BOARD

Transportation Research Record
2018, Vol. 2672(12) 51–61

DOI: 10.1177/0361198118767417
journals.sagepub.com/home/trr
SAGE

Field Evaluation and Cost Effectiveness of Crack Sealing in Flexible and Composite Pavements

Momen Mousa[1], Mostafa A. Elseifi[1], Mohammad Bashar[1], Zhongjie Zhang[2], and Kevin Gaspard[2]

Abstract

One of the most common methods used to treat longitudinal and transverse cracks is crack sealing (CS), which is categorized as a preventive maintenance method. Field performance and cost-effectiveness of this treatment widely vary depending on pavement conditions and installation of the material. The objective of this study was to evaluate the field performance and cost-effectiveness of CS in flexible and composite pavements in hot and wet climates such as Louisiana, and to develop a model that would quantify the expected benefits of CS given project conditions. To achieve this objective, 28 control sections that were crack-sealed between 2003 and 2010 were monitored for at least four years. These sections included flexible and composite pavements, sealed and unsealed segments, and varying traffic levels. The performance of these sections was evaluated for the random cracking index (RCI) and roughness index (RI). Based on the results of this analysis, it was concluded that CS only has a significant impact on random cracking. When compared with untreated segments, CS extended pavement service life (PSL) by two years. When compared with the original pavement, CS extended PSL by 5.6 and 3.2 years for flexible and composite pavements, respectively, if applied at the correct time. The cost-benefit analysis indicated that CS is cost-effective whether asphalt emulsion or rubberized asphalt sealant is used. A non-linear regression model was developed to predict the extension in PSL because of CS without the need for performance data based on the average daily traffic (ADT), pavement type, and prior pavement conditions.

Surface cracking in asphalt concrete (AC) is a major failure mechanism that develops over the pavement service life (PSL). Cracks are typically categorized into alligator, block, longitudinal, and transverse cracks. These cracks are primarily caused by traffic loading, thermal stress, and aging (*1*). It is well recognized that water infiltration through cracks may accelerate pavement deterioration, resulting in major problems such as stripping, base failure, and potholes (*2,3*). One of the most common treatments used to control surface cracks is crack sealing (CS), which is categorized as a preventive maintenance method (*4*).

Crack sealing is a surface treatment technique where hot-poured bituminous-based materials are applied into transverse and longitudinal cracks using various configurations (*5,6*). Field performance of CS is commonly evaluated for failure exhibited by the sealant material. Sealant failures result from excessive stresses in the sealant as the crack opens and closes with temperature variation and traffic loading. These failure mechanisms can be attributed to improper material selection or neglecting one or more of the construction considerations (*7*). Common sealant failures include adhesion loss, cohesion loss, spalls, pullout, extrusive, and intrusive failures (*7,8*).

Crack sealing has received considerable interest in Louisiana in recent years. The selection and construction processes for CS has been well researched and documented in previous studies (*9–15*). Yet, fewer studies were conducted to assess field performance of CS, especially in hot and wet climates such as Louisiana. Therefore, factors that may affect the field performance of CS should be evaluated and quantified to ensure that the treatment is only used when positive benefits are expected. This would also allow quantifying cost-effectiveness of this treatment method in controlling cracks and enhancing PSL.

[1]Department of Civil and Environmental Engineering, Louisiana State University, Baton Rouge, LA
[2]Louisiana Transportation Research Center, Louisiana State University, Baton Rouge, LA

Corresponding Author:
Address correspondence to Mostafa A. Elseifi: elseifi@lsu.edu

Scope and Objectives

The objective of this study is twofold. First, field performance and cost-effectiveness of CS were evaluated in flexible and composite pavements in hot and wet climates, such as Louisiana. Second, a regression model was developed to predict the increase in PSL because of CS if the pavement type, surface conditions before treatment, and Average Daily Traffic (ADT) is known. To achieve these objectives, sealed pavement sections were identified by analyzing maintenance work orders and the Louisiana Department of Transportation and Development (LaDOTD) Pavement Management System (PMS) databases. Control sections with sufficient years in service and with available untreated segments were selected for detailed performance and economic evaluation by analyzing the collected data and reviewing video crack surveys. The field performance of crack-sealed pavements was assessed in terms of random cracking and roughness data. The results of this analysis quantified the performance and cost-effectiveness of CS in extending PSL under hot and wet climates.

Background

Various studies have been conducted to evaluate the field performance of CS with a special focus on overall pavement conditions. In 1986, the Ministry of Transportation, Ontario, Canada, conducted a research study to assess the benefits of CS in extending PSL. Thirty-seven sections were selected covering different climatic conditions, traffic levels, pavement age, and thicknesses. For each sealed section, an untreated segment was identified for comparative evaluation. Performance curves, related to the pavement condition index, were drawn for sealed and untreated sections based on data collected during a 7-year monitoring period. The results indicated that CS extends the PSL by at least 2 years, depending on traffic volume, environment, and the original condition of the pavement *(16)*.

A research study was carried out by the Ohio Department of Transportation to evaluate the field performance of CS. Test sections were selected over the state, including 57 counties. In each test section, a 1,000-foot long subsection was left unsealed to serve as a control segment. Results indicated that crack-sealed pavements provided better a performance than untreated pavements, considering the pavement condition rating, in a five-year life cycle. Furthermore, it was reported that CS could prolong the PSL by up to 3.6 years *(17)*.

Based on a national survey, Eaton indicated that 70% of the states that seal cracks reported an increase in PSL by at least three years *(18)*. The short-term effectiveness of CS was evaluated for the International Roughness Index (IRI). It was concluded that CS offers an average reduction in IRI of 17 in./mile *(19)*. Yet, other studies indicated that CS has no significant impact on IRI *(20,21)*. An extensive investigation was conducted under the Long-Term Pavement Performance (LTPP) program to evaluate the field performance of different sealant materials. Key findings indicated that rubberized asphalt sealant material placed in a standard or shallow-recessed band-aid configuration provided the best field performance among other material types and placement configurations. Further findings indicated that sealant in cracks with low crack movement and low traffic performed better than crack sealants with high crack movement and traffic *(22)*.

From the published literature, there is a general agreement that CS is effective in controlling cracks in asphalt pavements, and therefore increasing PSL. Yet, this study is expected to address several knowledge gaps in the literature as follows:

- Although the main distresses that may be affected by CS are cracking and roughness, most of the studies focused on overall pavement condition indices. Generally, overall pavement condition indices include several distresses that are not influenced by CS, such as bleeding, rutting, patching, and so forth. Therefore, this study analyzed longitudinal and transverse considering the Random Cracking Index (RCI), and roughness in the context of the Roughness Index (RI) to provide a more accurate assessment of the effects of CS;
- Most of the previous studies were conducted in the Northern United States and Canada with cold climates, as CS was not typically used in the Southern United States with hot and wet climates *(18)*. Yet, this trend has changed in recent years. Therefore, it is necessary to evaluate the field performance of CS in hot and wet climates such as Louisiana;
- None of the previous studies developed a quantitative model to quantify the benefits of CS. Therefore, this study developed a simple model to predict the benefits of CS, given the limited information related to project conditions.

Methodology

Figure 1 demonstrates the general outlines of the research methodology. As shown in Figure 1, sealed pavement sections were identified by analyzing LaDOTD PMS databases. Control sections with sufficient years in service and with available untreated segments were selected for detailed performance and economic evaluation by analyzing collected data and reviewing video crack surveys. A non-linear regression model was then developed to predict the increase in PSL given the project conditions.

In Louisiana, different construction practices are adopted to ensure a successful CS installation. District surveys indicated that the most common sealant materials are hot-poured asphalt sealant and asphalt emulsion (cationic rapid setting (CRS)). CS is usually performed in late fall, early winter, and

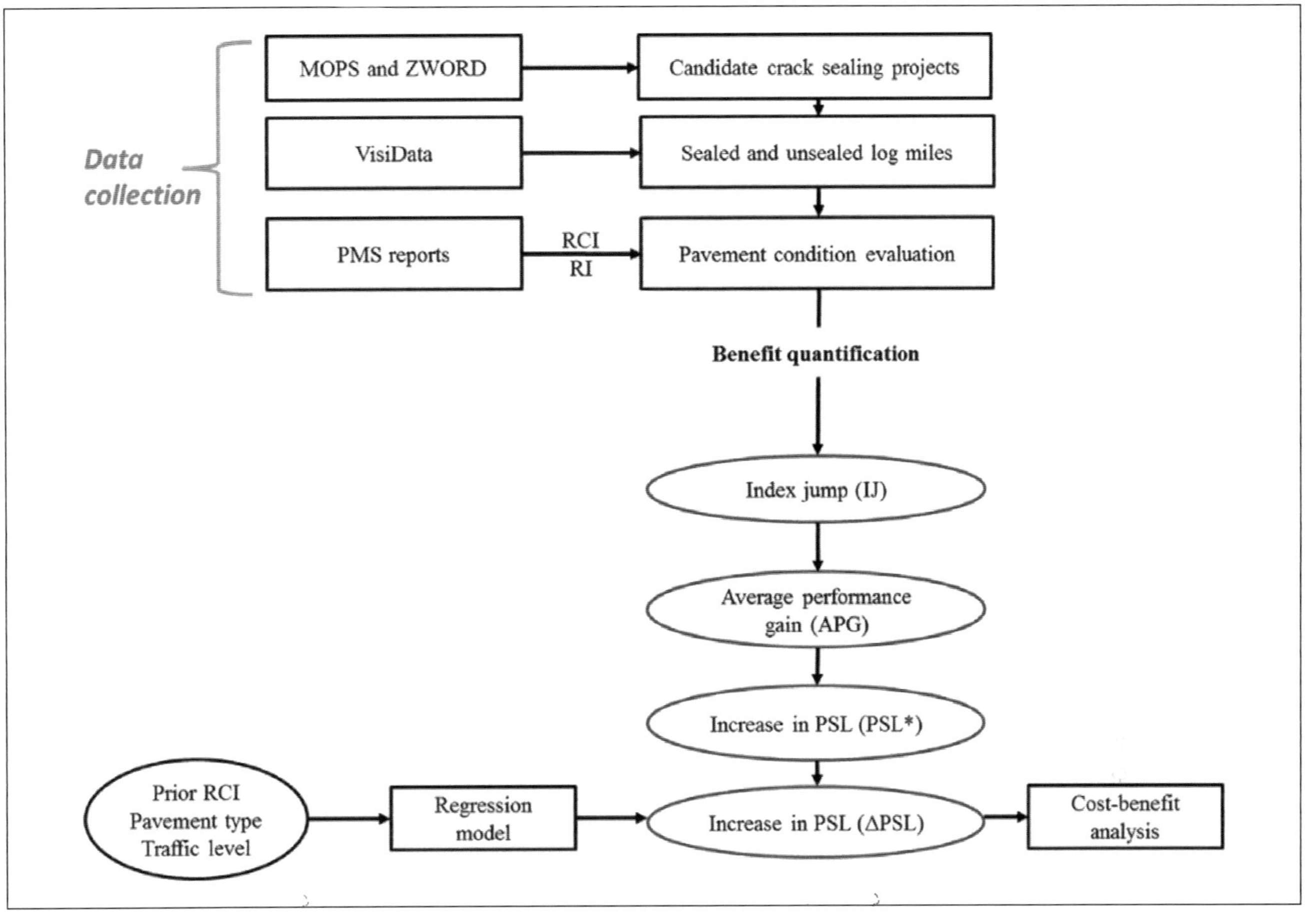

Figure 1. Overview of research methodology.

early spring when temperatures are above 50°F with dry pavements. Before CS, cracks are commonly cleaned using air an compressor rather than a hot air lance, while no routing is conducted because of limited funds. CS is placed flush fill or overband using pour pots for CRS and a melter for rubberized asphalt. A squeegee is then used for material finishing of rubberized asphalt, while CRS are blotted using a 3/16 in. lightweight expanded clay aggregate. Finally, random inspection is conducted right after construction for quality control of CS.

Data Collection

In Louisiana, CS activities are captured by the Maintenance Management System's work orders, which are included in LaDOTD databases. Pavement performance data are reported in LaDOTD PMS for the period ranging from 1996 to 2015. These data are based on the pavement condition measurements that are collected biennially using the Automatic Road Analyzer (ARAN®) system which provides a continuous assessment of the road network *(23)*. Video crack surveys are available for each state highway in Louisiana and were reviewed using VisiData™ software *(23)*.

Collected data are reported every 1/10th of a mile (log mile) and are analyzed to calculate different distress indices on a scale from zero to 100 (100 being perfect conditions). For flexible and composite pavements, RCI encompasses all random cracks, which include thermal transverse, reflective transverse, longitudinal, block, and cement-treated reflective cracks. RCI is calculated as follows *(24)*:

$$RCI = Min\{100. Max(0.100\text{-}DP_L\text{-}DP_M\text{-}DP_H)\} \quad (1)$$

where:

DP = deduct point because of random cracks;

Subscripts L, M, and H refer to the low, medium, and high severity of the cracks, respectively.

The RI is expressed on a scale from zero to 100, with 100 representing the case with a smooth pavement. It is related to IRI as follows: IRI (in/mile) = (100 − RI) * 5 + 50. In this study, work orders were reviewed for preliminary identification of sealed control sections. Unfortunately, work orders only identified the control section and not the exact location and the extent of the CS activities on the section. Furthermore, CS projects in work orders are usually defined under "other

Figure 2. Illustration of sealed (up) and unsealed (down) pavement sections before (left) and after (right) sealing date.

bituminous surface treatments," which include several surface treatments. Therefore, video crack surveys were used to filter the identified sections and to locate crack sealed control sections.

To provide an accurate presentation of the effect of CS, the analysis was conducted for every log mile, which was considered as a single data point. For the entire length of the identified crack sealed control sections, videos between 2003 and 2015 were reviewed to determine the exact date and location of CS. Furthermore, unsealed log miles were selected before and/or after the sealed log miles for comparative evaluation, see Figure 2. Several challenges were encountered that limited the number of analyzed sections. For instance, districts may use chip seal shortly after CS; therefore, these sections did not provide enough performance data to determine PSL after CS. Furthermore, many CS projects were conducted in the secondary direction, which is not rated by LaDOTD PMS.

One RCI and one RI were obtained for each of the selected sealed and unsealed log miles before the sealing date, while three RCI and RI were obtained after the sealing date. Other collected data included ADT at sealing date and pavement type. In this study, RCI^- refers to RCI collected before the sealing date, while ΔRCI refers to the difference between RCI at sealed and unsealed log miles (sealed - unsealed) at time (i). In total, 306 sealed log miles were identified and were used in the analysis; see Table 1.

Benefits Quantification

To quantify the effectiveness of CS, RCI and RI were evaluated for sealed and unsealed log miles using index jump (IJ), average performance gain, and increase in PSL.

Index Jump

The index jump for RCI and RI was calculated using data set 1 by subtracting the first collected index after sealing from RCI^- and RI^-. For all log miles, the IJ of RCI showed positive values with mean values of 7.4 ± 7.3, which indicates that CS had a significant immediate impact on RCI; see Figure 3*a*. Yet, only 22% of log miles had a positive IJ of RI with a mean value of 2.3 ± 2. This indicates that CS has minor or negligible immediate impact on roughness, as suggested by other studies (*20, 21*). Therefore, the analysis in the following sections was limited to RCI to quantify the benefits of CS.

Average Performance Gain

Data set 2 was used to calculate APG as the average of ΔRCI in different years between each pair of sealed and unsealed log miles in the same control section having similar RCI^- (*23*). Figure 3*b* presents the APG for each pair of sealed and unsealed log miles and the corresponding RCI^-. The results indicated that all pairs had positive APG, supporting

Table 1. Size and Description of Data Sets Used in the Analysis

Data set ID	Sealed segment	Unsealed segment	Type of analysis
1	306 (28)[a]	-	Index Jump (IJ) for sealed log miles
2	38 (18)	38 (18)	Average Performance Gain (APG) between sealed and unsealed log miles
3	248 (20)	125 (20)	Increase in PSL (PSL*) between sealed and unsealed segments
4	105	-	Increase in PSL (ΔPSL) of sealed log miles
5	32	-	Comparison between ΔPSL of sealed log miles under different traffic levels

Note: PSL = pavement service life.
[a]Log miles (control sections).

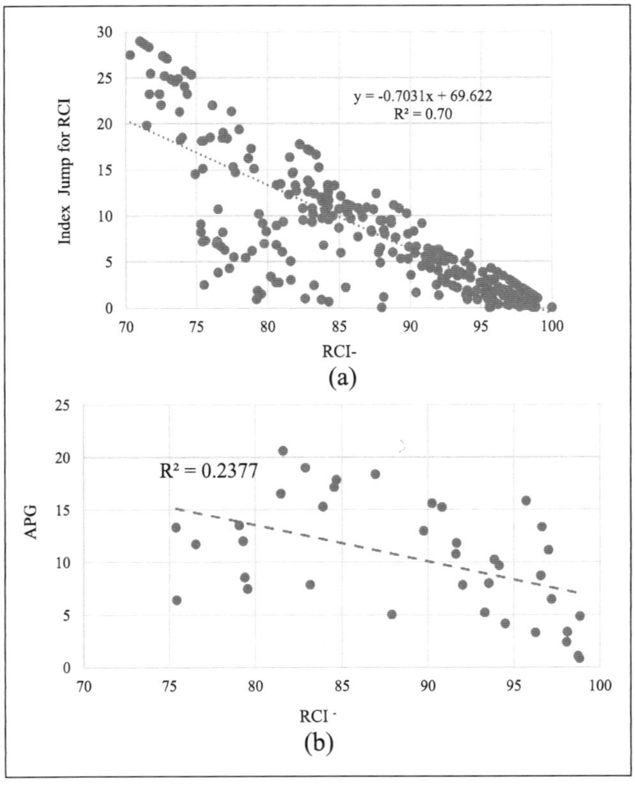

Figure 3. Correlation between (*a*) index jump and pretreatment pavement condition for RCI^- and (*b*) APG and pretreatment pavement condition for RCI^-.

the finding that CS improved pavement performance against random cracking. No strong correlation was observed between the APG and RCI^-, as supported by the coefficient of determination (R^2), which had a value of 0.25. Yet, within the evaluated range, the general trend in Figure 3*b* suggests that higher performance gains were achieved with lower values of RCI^-.

Increase in Pavement Service Life

Data Set 3 was used to evaluate the effect of CS in extending PSL when compared with the untreated segments. RCI performance curves were developed for each sealed and unsealed log mile with RCI as a function of age. RCI values at the sealing date were extrapolated and interpolated for sealed and unsealed log miles, respectively. Quadratic polynomial models provided the best fit in most cases. Using the fitted models, the number of years to reach a threshold RCI was predicted and reported as PSL for this specific log mile, see Figure 4. Unlike other surface treatments, LaDOTD PMS does not include any trigger values for CS. Based on local surveys, it was reported that Louisiana districts use visual inspection instead of PMS data to select candidate sections for CS. Therefore, a threshold RCI of 69 was assumed to match the pavement-rating scheme used by LaDOTD for other cracking distresses (*24*).

Calculated PSL for sealed and unsealed log miles were grouped by control section, and the average PSL was calculated for both segments. The average PSL of the unsealed segment was then subtracted from that of the sealed segment to obtain the increase in PSL (PSL*). Figure 5 presents PSL* and ΔRCI before sealing (ΔRCI^-) for each control section. For most of the sections, the sealed segment experienced an average increase in PSL (PSL*) of two years more than the unsealed segment, which is comparable with other studies (*16*). Statistical t-tests were conducted to compare the average PSL of the sealed and unsealed segments for the 17 sections that experienced a positive increase in PSL (PSL*). The results indicated that this increase was significant for all the control sections except for sections 8, 10, 15, 20, and 25. It is noted that sections 12 and 13 did not experience PSL* because the unsealed segment in these sections had an average RCI^- of 100. Furthermore, section 24 had severe fatigue cracks, which cannot be treated by CS (*1*); therefore, no PSL* was experienced.

Increase in Pavement Service Life

Any differences between sealed and unsealed segments, for the pavement structure, prior pavement conditions, or traffic would affect the accuracy of PSL*. Therefore, these differences were eliminated by calculating the increase in PSL for each sealed log mile when compared with the original

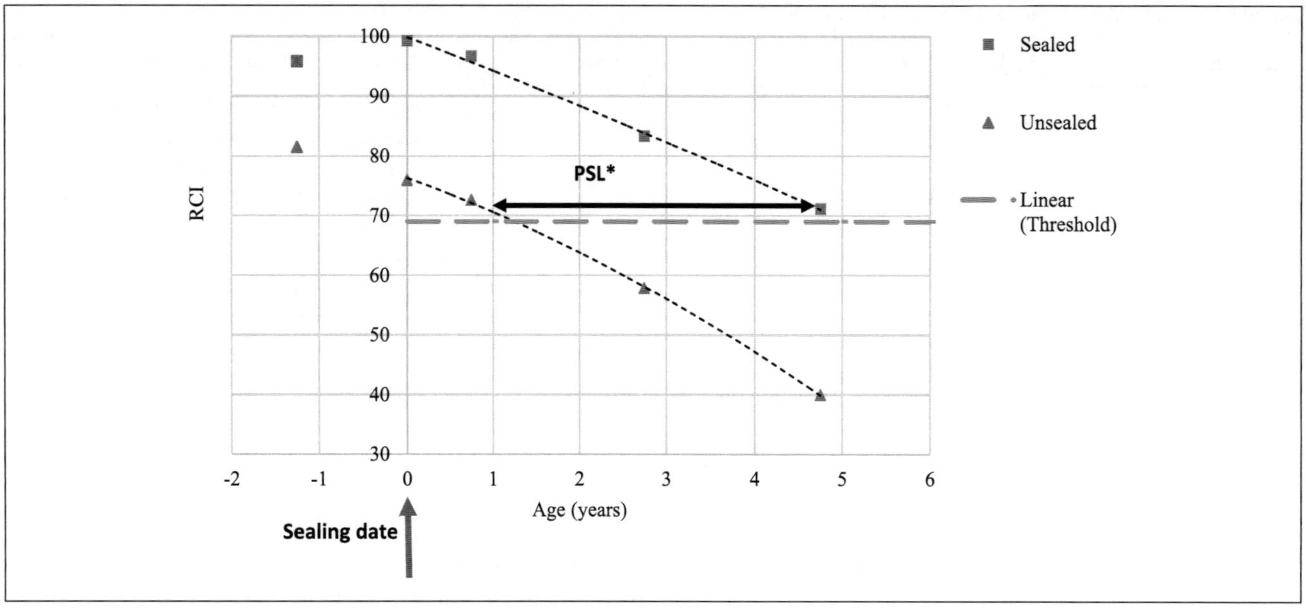

Figure 4. Variation of RCI during the monitored period for two sealed and unsealed log miles.

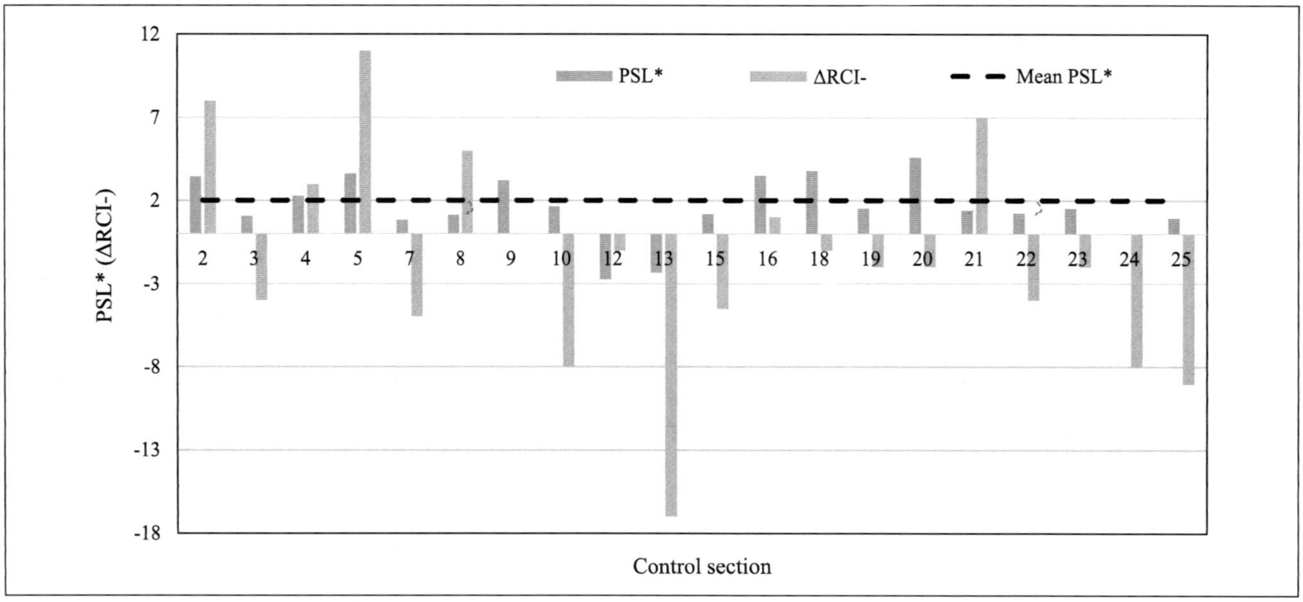

Figure 5. PSL* and ΔRCI^- for different control sections.

pavement using Data Set 4. In this analysis, RCI performance curves were plotted before and after CS, and ΔPSL was calculated as follows (Figure 6):

$$\Delta PSL = A + PSL_2 - PSL_1 \quad (2)$$

where:
A = pavement age at sealing date;
PSL_1 = PSL of original pavement before CS; and
PSL_2 = PSL after CS.

Figure 7*a* illustrates ΔPSL and RCI^- for data set 4 for both pavement types. Negative values of ΔPSL indicate that no extension in the pavement service life is achieved after crack sealing. As shown in Figure 7*a*, both pavement types followed similar trends, meaning that ΔPSL had negative values for RCI^- of more than 90. When RCI^- was less than 90, CS extended PSL by an average of 5.6 ± 1.9 and 3.15 ± 1.3 years for flexible and composite pavements, respectively. This suggests that although sealing pavements in relatively good conditions (RCI^- >90) would produce higher PSL_2, no extension in PSL would be achieved. No clear trend was observed when ΔPSL was plotted against ADT for data set 4, probably because the sections had different initial RCI^-. Therefore, the evaluation of traffic volume on ΔPSL was limited to points having exactly

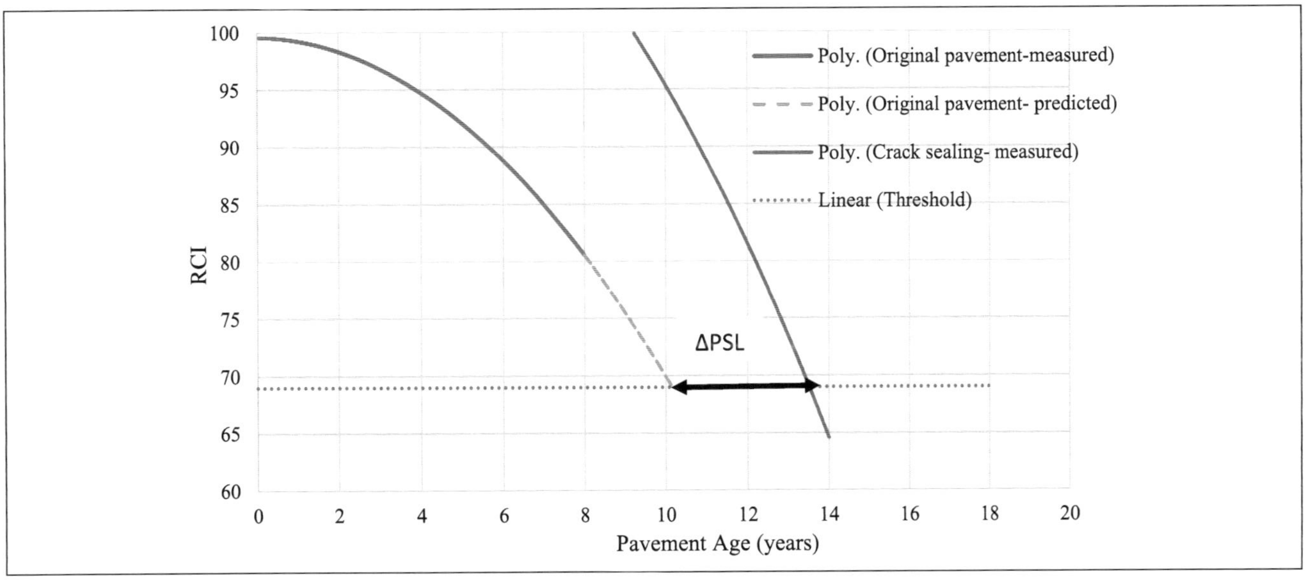

Figure 6. Calculation of ΔPSL.

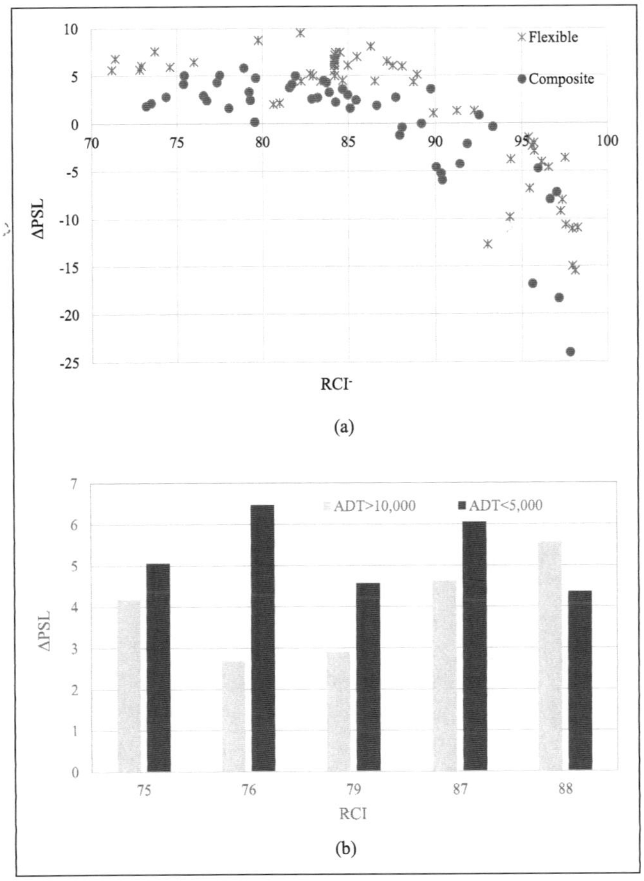

Figure 7. (*a*) ΔPSL versus RCI^- for data set 4 and (*b*) ΔPSL versus RCI^- for different traffic levels.

the same RCI^-, data set 5; see Figure 7*b*. The results indicated that the benefits of CS, in terms of ΔPSL, were greater for lower ADT, which is in agreement with previous studies (*22*).

Model Development

Based on a review of past studies, four primary variables were considered in the regression analysis, pavement type, pavement age at sealing date, RCI^-, and ADT. An Analysis of Variance (ANOVA) was conducted between ΔPSL and these four variables using SAS 9.4 software, see Table 2. RCI^- had the highest statistical correlation to ΔPSL (lowest *p*-value), followed by pavement type and ADT, while pavement age at sealing date was not statistically correlated to ΔPSL. Therefore, only RCI^-, ADT, and pavement type were considered in the regression model.

For each pavement type, 80% of the data was used to fit the model and 20% was used for validation. This resulted in 58 points for flexible pavements (46 for fitting and 12 for validation), and 47 points for composite pavements (38 for fitting and 9 for validation). The fitted models were developed after performing non-linear regression analyses on the ΔPSL as a dependent variable, and with RCI^-, and ADT as the independent variables. The models were as follows:

Flexible Pavement

$$\begin{aligned}\Delta PSL = &\left(-66.398*RCI^-\right) + \left(0.842*\{RCI^-\}^2\right) \\ &+ \left(-0.004*\{RCI^-\}^3\right) + \left(0.001*ADT\right) \\ &+\left(-3.462E-07*ADT^2\right) + \left(2.410E-11*ADT^3\right) \\ &+ 1745.0996\end{aligned} \quad (3)$$

Table 2. Results of Analysis of Variance

Variables	t-value	p-value	Interpretation
Intercept	7.36	<.001	Significant
RCI⁻	−9.56	<.001	Significant
Pavement age at sealing date	0.23	.82	Not Significant
ADT	4.39	<.001	Significant
Pavement type	−5.47	<.001	Significant

Note: RCI- = random cracking index collected before sealing date; ADT = average daily traffic.

Model is valid for ADT < = 11,100, and 70 <RCI⁻ <100

Composite Pavement

$$\Delta PSL = \left(-42.492 * RCI^-\right) + \left(0.565 * \left\{RCI^-\right\}^2\right) + \left(-0.002 * \left\{RCI^-\right\}^{33}\right) + \left(0.000316 * ADT\right) + \left(1.572E-07 * ADT^2\right) + \left(-1.174E-11 ADT^{33}\right) + 1062.824 \quad (4)$$

Model is valid for ADT < = 15,100, and 70 <RCI⁻ <100

Figure 8, *a* and *b* present the actual and predicted ΔPSL using fitting data for both pavement types. For both types, it is clear that the proposed models predicted ΔPSL with an acceptable level of accuracy as supported by the R^2 and the root mean square error (RMSE) shown in the figures. For the flexible pavement, the R^2 and RMSE were approximately 0.9 and 2.1 years, respectively, while for the composite pavement, the R^2 and RMSE were approximately 0.84 and 2.6 years, respectively. The proposed model for flexible pavements was plotted for different RCI⁻ and ADT; see Figure 9. It is noted that the developed model follows the same trends shown in Figure 7, *a* and *b*, based on the measured data.

Illustritive Applications of Predictive Model

During planning of maintenance activities, a contractor and/or state agency may be interested in determining whether CS is an appropriate treatment at the site. The proposed model is expected to help in this process by providing two main functions:

1. Deciding whether CS could be applied. Negative values of predicted ΔPSL mean that CS would provide no additional benefits and thus is not recommended.
2. Select the optimal timing for future treatments following CS through accurate predictions of positive values of ΔPSL.

Table 3 presents the application of the developed non-linear predictive models in estimating ΔPSL using validation data. It is noted that these data points were not used in the model development, and thus would reflect the model accuracy. As shown in Table 3, the model successfully satisfied the first function, since all the actual values were predicted with the correct sign (positive or negative). Furthermore, the models for flexible and composite pavements were efficient in predicting the improvement in ΔPSL, with a RMSE of 1.1 and 1.4 years, respectively.

Cost-Benefit Analysis

Life cycle cost analysis in the form of equivalent uniform annual costs (EUAC) was used in this study to evaluate the cost effectiveness of CS. The performance of the sealed pavement was compared with the performance of the original pavement without any maintenance, and the benefit-cost (B/C) ratio was calculated as follows (*25*):

$$\frac{B}{C} = \frac{EUAC_{do\,nothing} - EUAC_{treatment}}{EUAC_{CS}} \quad (5)$$

$$EUAC_{do\ nothing} = NPV1 \times \left[\frac{(1+i)^{PSL1}}{(1+i)^{PSL1} - 1}\right] \quad (6)$$

$$EUAC_{CS} = NPV2 \times \left[\frac{(1+i)^{PSL1+\Delta PSL}}{(1+i)^{PSL1+\Delta PSL} - 1}\right] \quad (7)$$

$$EUAC_{treatment} = (NPV1 + NPV2) \times \left[\frac{(1+i)^{PSL1+\Delta PSL}}{(1+i)^{PSL1+\Delta PSL} - 1}\right] \quad (8)$$

In which

NPV1 = net present value (NPV) of the initial cost of baseline pavement in \$/mile;

NPV2 = is NPV of the cost of CS in \$/mile;

i = interest rate; and

PSL_1 and ΔPSL are as previously defined.

NPV1 was selected as \$200,000/mile based on actual bid items for current projects, while i was assumed to be 6%. District surveys indicated that NPV2 for rubberized asphalt and asphalt emulsion (CRS) are \$10,296 and \$4,118 /mile, respectively. The benefit/cost ratio was then calculated for all log miles in data set 4 for points that showed positive ΔPSL (72 out of 105 points) using the corresponding ΔPSL and PSL_1, see Figure 10. Results indicated that almost all the log miles had a B/C more than 1 when emulsion was used, while 63 log miles (89%) had a B/C more than 1 when rubberized asphalt sealant was used. Therefore, it was concluded that CS is cost-effective for both sealant material types when used in hot and wet climates.

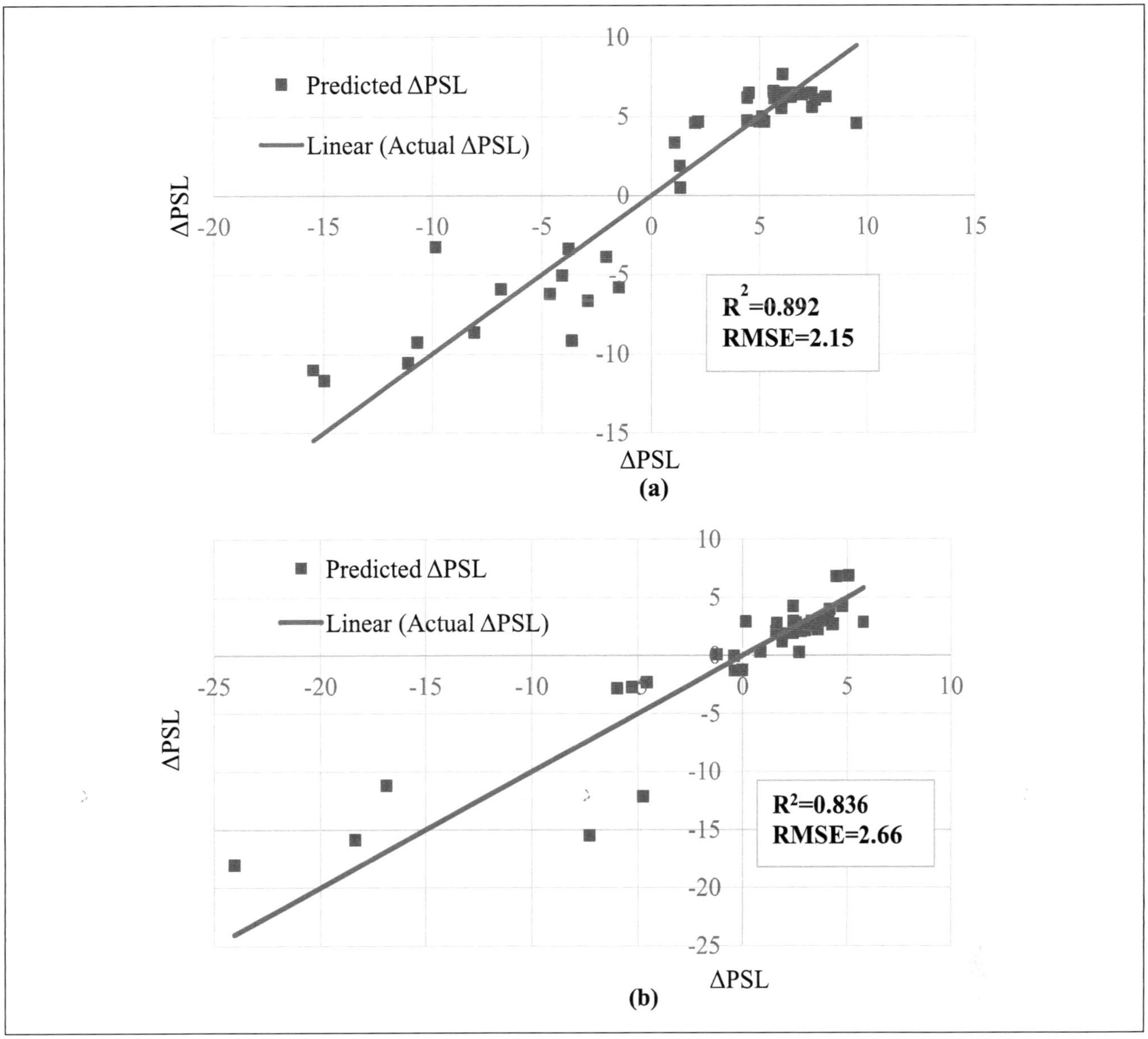

Figure 8. Predicted ΔPSL versus actual ΔPSL using fitting data for (*a*) flexible pavement and (*b*) composite pavement.

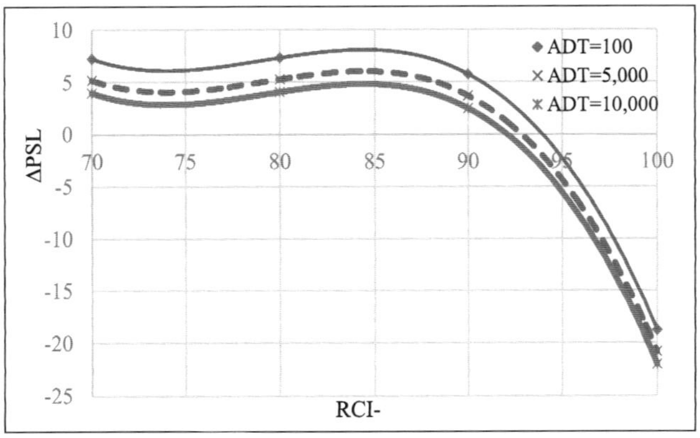

Figure 9. Model prediction of ΔPSL versus RCI⁻ for flexible pavement under different ADT.

Summary and Conclusions

The objective of this study is twofold. First, field performance and cost-effectiveness of CS were evaluated in flexible and composite pavements in hot and wet climates such as Louisiana. Second, a model was developed to predict ΔPSL caused by CS when knowing the pavement type, prior surface conditions, and ADT. Based on the results of this study, the following conclusions were drawn:

- CS resulted in a significant index jump in RCI with a mean value of 7.4 ± 7.3, whereas no significant jump in RI was observed.
- The benefits of CS, for APG, are maximized when CS is applied for relatively low values of RCI^-.

Table 3. Illustrative Applications of Proposed Models

Pavement type	RCI⁻	ADT	Actual ΔPSL	Predicted ΔPSL	RMSE (years)
Flexible	88.8	3,000	4.3	6.4	1.1
Flexible	72.9	2,600	6.1	6.1	
Flexible	79.8	2,600	8.7	7.2	
Flexible	82.3	6,500	4.4	4.6	
Flexible	84.2	11,100	4.9	6.5	
Flexible	84.2	11,100	5.5	6.5	
Flexible	84.2	11,100	7.1	6.5	
Flexible	87.3	11,100	6.5	5.9	
Composite	81.9	15,100	4.9	2.9	1.4
Composite	83.2	15,100	2.7	2.8	
Composite	84.3	15,100	2.2	2.5	
Composite	84.7	2,300	3.6	3.5	
Composite	75.4	320	5.1	2.3	
Composite	73.5	370	2.2	1.9	
Flexible	95.6	5,500	−2.3	−6.3	6.1
Flexible	98.3	3,700	−10.9	−12.8	
Flexible	97.3	1,870	−9.3	−8.3	
Flexible	93.0	6,000	−12.7	−1.2	
Composite	91.9	15,100	−2.2	−5.2	3.9
Composite	91.4	15,100	−4.3	−4.4	
Composite	96.7	3,700	−8.0	−14.2	

Note: PSL = pavement service life; RCI⁻ = random cracking index collected before sealing date; ADT = average daily traffic; RMSE = root mean square error.

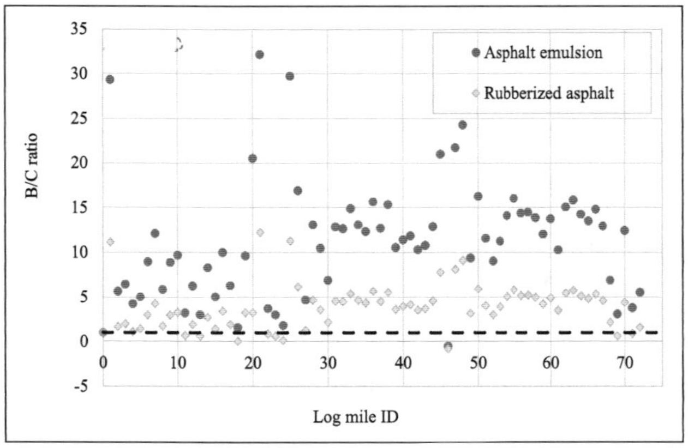

Figure 10. B/C ratio using data set 4 for different sealant materials.

- For all selected control sections, with a few exceptions, the sealed segment experienced an average increase in PSL of two years more than the unsealed segment.
- In comparison with the original pavement, CS did not extend PSL when added to pavements with an RCI⁻ above 90. When RCI⁻ was less than 90, CS extended the PSL by 5.6 and 3.2 years for flexible and composite pavements, respectively.
- CS is cost-effective, whether asphalt emulsion or rubberized asphalt is used.
- The developed regression model predicted ΔPSL with an acceptable level of accuracy based on R2 and RMSE. RCI⁻ was the most important variable in predicting ΔPSL.
- The proposed models were accurate in predicting whether CS will increase PSL, as well as the magnitude of the improvement.

Acknowledgments

The financial support of the Louisiana Transportation Research Center (LTRC) is greatly appreciated. The authors acknowledge the help of Ken Free from District 58 in Louisiana.

References

1. Hicks, G., R. , Toepfer, B., Dietz, J. *Caltrans Maintenance Technical Advisory Guide (MTAG).* California Department of Transportation, Office of Pavement Preservation, Sacramento, 2003.
2. National Cooperative Highway Research Program. *NCHRP Project 1-37A: Guide for Mechanistic–Empirical Design of New and Rehabilitated Pavement Structures.* TRB, National Research Council, Washington, D.C. 2004.
3. Apul, D. S., K. Gardner, T. Eighmy, J. Benoit, and L. Brannaka. *A Review of Water Movement in the Highway Environment: Implications for Recycled Materials Use.* Recycled Materials Resource Center, University of New Hampshire, Durham, 2002.

4. Zaniewski, J. P., and M. S. Mamlouk. *Pavement Maintenance Effectiveness-Preventive Maintenance Treatments. Participant's Handbook*. No. FHWA-SA-96-027. 1996. Washington, D.C.: U.S. Dept. of Transportation, Federal Highway Administration; Springfield, National Technical Information Service, 1996.
5. Decker, D. S. *Best Practices for Crack Treatments for Asphalt Pavements*. No. Project 20-07. Transportation Research Board, Washington, D.C. 2014.
6. Smith, K. L., and A. R. Romine. *Materials and Procedures for Sealing and Filling Cracks in Asphalt-surfaced Pavements*. Manual of Practice. No. FHWA-RD-99-147. 2001. McLean, Va.: U.S. Dept. of Transportation, Federal Highway Administration [Pavement Performance Division]: LTPP; Springfield, Va. : Available to the public through the National Technical Information Service, 2001.
7. Department of Defense. *Standard Practice for Sealing Joints and Cracks in Rigid and Flexible Pavements*. Unified facilities criteria, UFC3-250-08FA. Department of the Army, Washington, D.C., 2004.
8. Federal Highway Administration and the Foundation for Pavement Preservation. *Pavement Preservation Checklist Series: Crack Seal Application*. FHWA-IF-02-005. Federal Highway Administration, 2001.
9. Wang, C-P., and F. E. Weisgerber. Effects of Seal Geometry on Adhesive Stresses in Pavement Joint Seals. *Transportation Research Record: Journal of the Transportation Research Board*, 1993. 1392: 64–70.
10. Khuri, M. F., and E. Tons. Comparing Rectangular and Trapezoidal Seals using the Finite Element Method. *Transportation Research Record: Journal of the Transportation Research Board*, 1992. 1334: 25–37.
11. Chong, G. J. Rout and Seal Cracks in Flexible Pavement- A Cost-Effective Preventive Maintenance Procedure. *Transportation Research Record: Journal of the Transportation Research Board*, 1990. 1268: 8–16.
12. Chehovits, J., and M. Manning. Materials and Methods for Sealing Cracks in Asphalt Concrete Pavements. *Transportation Research Record: Journal of the Transportation Research Board*, 1984. 990: 21–30.
13. Yang S.-H, I. L. Al-Qadi, J. McGraw, J.-F. Masson, and K. M. McGhee. *Transportation Research Record: Journal of the Transportation Research Board*, 2010. 2150: 87–95.
14. Carter, S., K. Ksaibati, and G. Huntington. Field and Laboratory Evaluations of Hot-Poured Thermoplastic Bituminous Crack Sealing of Asphalt Pavements. *Transportation Research Record: Journal of the Transportation Research Board*, 2005. 1933: 113–120.
15. Johnson, D. R., R. B. Freeman, and J. R. Stevenson. Cost-Effectiveness of Crack Sealing Materials and Techniques for Asphalt Pavements. *Transportation Research Record: Journal of the Transportation Research Board*, 2000. 1697: 31–40.
16. Ponniah, J., and G. Kennepohl. Crack Sealing in Flexible Pavements: A Life-Cycle Cost Analysis. *Transportation Research Record: Journal of the Transportation Research Board*, 1996. 1529: 86–94.
17. Rajagopal, A. *Effectiveness of Crack Sealing on Pavement Serviceability and Life*. FHWA/OH-2011/ 14. Ohio Department of Transportation, Columbus, 2011.
18. Eaton, R. A., and J. Ashcraft. *State-of-the-Art Survey of Flexible Pavement Crack Sealing Procedures in the United States*. No. CRREL-92-18. Cold Regions Research and Engineering Lab, Hanover, 1992.
19. Lu, P., and D. Tolliver. Pavement Treatment Short-Term Effectiveness in IRI Change Using Long-Term Pavement Program Data. *Journal of Transportation Engineering*, Vol. 138, No. 11, 2012, pp. 1297–1302.
20. Chuanxin, F., G. Khaled, W. David, and H. John. Paurdue University.
21. Ong, G. P., T. E. Nantung, and K. C. Sinha. *Indiana Pavement Preservation Program*. Publication FHWA/IN/JTRP-2010/14, Joint Transportation Research Program, 2010 Purdue University 2010-01-01T08:00:00Z.
22. Smith, K. L., and A. R. Romine. *LTPP Pavement Maintenance Materials: SHRP Crack Treatment Experiment*. Final report FHWA-RD-99-143, Champaign, 1999.
23. Khattak, M. J., G. Y. Baladi, Z. Zhang, and S. Ismail. A Review of the Pavement Management System of the State of Louisiana-Phase. *Transportation Research Record: Journal of the Transportation Research Board*, 2008. 2084: 18–27.
24. Khattak, M. J., G. Y. Baladi, and X. Sun. *Development of Index Based Pavement Performance Models for Pavement Management System (PMS) of LADOTD*. FHWA/LA. 08/460, 2009. Louisiana Transportation Research Center, Baton Rouge, 2009.
25. Morian, D. A. *Cost Benefit Analysis of Including Microsurfacing in Pavement Treatment Strategies & Cycle Maintenance*. FHWA-PA-2011-001-080503, Harrisburg, 2011.

The Standing Committee on Sealants and Fillers for Joints and Cracks (AHD25) peer-reviewed this paper (18-00328).

Article

TRR
JOURNAL OF THE TRANSPORTATION RESEARCH BOARD

Transportation Research Record
2018, Vol. 2672(12) 62–71

DOI: 10.1177/0361198118784172
journals.sagepub.com/home/trr
SAGE

Laboratory Investigation of Pavement Backfilling Materials for Micro-Trenching in Cold Regions

Leila Hashemian[1], Vinicius Afonso Velasco Rios[2], and Alireza Bayat[1]

Abstract

This study investigated the performance of different materials in a micro-trench composite backfilling design. Laboratory tests were conducted to evaluate the effect of cold temperatures and freeze/thaw cycles on a cement grout and seven preparatory cold asphalt mixes. To compare the performance of cold mix asphalt and epoxy grout with hot mix asphalt as the host material, rutting tests and dynamic modulus tests at different loading frequencies and temperatures were conducted. Finally, laboratory scale micro-trench samples were prepared using different backfilling materials and were loaded using a wheel tracker after freeze/thaw conditioning. The results showed that cement grout could effectively be used to secure the conduit inside the trench. It was also concluded that using high-quality cold mix asphalt, a compatible material with hot mix asphalt, could improve micro-trench durability compared with epoxy grout.

One of the main challenges for fiber optic (FO) service providers is accessing homes [known as fiber-to-the-home (FTTH)] because of possibly damaging existing utilities, disrupting traffic, and disturbing the surrounding environment (*1*). Using traditional trenching methods, such as open cut, especially for the last 200 meters of FO deployment, can be uneconomical, time-consuming, and disruptive to the local area. In addition, it can cause major traffic delays and disturb citizens. In contrast, the use of a micro-trench (MT) for FO deployment could be a fast and cost-effective alternative to traditional methods (*2*).

The MT process includes three main steps. The first step is cutting a narrow trench in the pavement. The dimensions of the trench are usually selected according to the size of conduit, pavement thickness, and regional considerations (usually deeper in freeze/thaw climates). In general, MTs are <40 mm wide and up to 300 mm deep. The second step is cleaning the trench and laying the cable or conduit inside, and the third and final step is backfilling the trench (*2*).

Replacing an open cut trench (minimum 300 mm wide) with a narrower MT (20–40 mm wide) makes it possible to reduce the excavation cost by 60% (*3*). FO service providers have found MT construction is up to 75% less expensive, and construction time is 50% lower, when utilizing MT techniques (*2*). However, this method has not been widely accepted by internet service providers because municipalities and road authorities are not willing to accept the risks, including potential pavement damage. There is also a concern for service providers that, because of the shallow depth of the trench, installations may not be stable. In cold regions subjected to several freeze/thaw cycles, MTs are prone to damage, which may lead to damage to the FO cable.

Similar to other utility installations, the quality of the backfilling material has an important role in MT performance. It has been proven in utility installations that major pavement failures may happen using unsuitable or improperly installed backfilling materials, which can significantly decrease pavement life (*4*). Backfilling material secures the cable in the MT and protects the cable or conduit against traffic loads. Considering the small size of the MT, there are some special requirements for backfilling material, such as being flowable enough to penetrate the tiny trench, being easy to apply, and easy to compact by multi-skilled installer teams. The material should also be compatible with the existing pavement type and have appropriate bonding with the host pavement. It should also cure fast to allow traffic to pass as quickly as possible (*5*).

In different MT projects, several materials have been used for backfilling purposes. For example, in a project in the Caribbean that involved backfilling a 24 mm × 305 mm trench, beach sand was used to fill the bottom part of the

[1]Donadeo Innovation Centre for Engineering, University of Alberta, Edmonton, Alberta, Canada
[2]Markin/CNRL Natural Resources Engineering Facility, University of Alberta, Edmonton, Alberta, Canada

Corresponding Author:
Address correspondence to Alireza Bayat: abayat@ualberta.ca

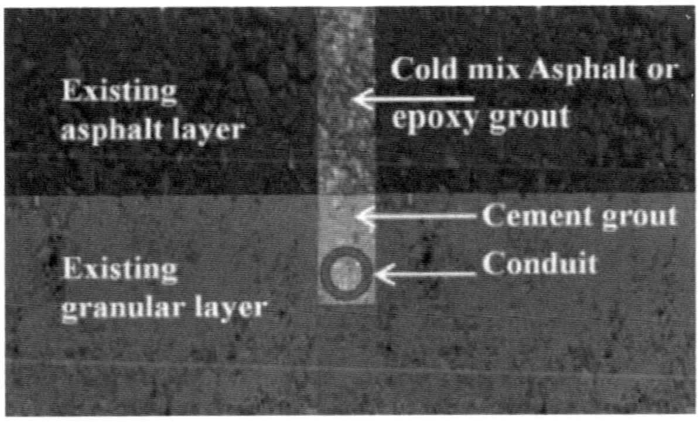

Figure 1. Micro-trench backfilling methods.

trench and cold mix asphalt (CMA) was applied and compacted on top of the sand as the final layer (*6*). In New Zealand, a pilot MT project was conducted to deploy FO cables in an 80 mm × 430 mm trench using a specially designed mortar as backfilling material. After filling the trench with the mortar, the top part of the trench was milled (400 mm wide and 25–30 mm deep) and was filled with hot mix asphalt (HMA) (*7*). The milling operation resulted in pavement disturbance, dust and sound generation, and prolonged traffic detours. Some researchers proposed filling the whole trench with grout in heavy traffic areas (*8*). However, differential stiffness of grout and asphalt in the pavement may generate cracks and failures under dynamic traffic loads and severe weather conditions. Additionally, future road rehabilitation operations may damage the integrity of bonding between grout and conduit, and between grout and original pavement.

In another pilot study for FO installation in Edmonton, Alberta, Canada, two different MT techniques were utilized. With the first method, MTs with dimensions of 15 mm × 230 mm were cut in the pavement and, after placing of the conduit, were backfilled with sand at the bottom and CMA on top. With the second method, the MT dimensions were 9 mm × 76 mm and trenches were backfilled using sand with sealant on top. FO cable locations were monitored using ground penetrating radar (GPR) immediately after deployment and again each subsequent winter and spring. GPR results revealed that significant movement was observed in the installed conduits, especially in heavier traffic locations and bends (*9*). Visual observation showed some premature failures in the backfilling material, such as settlements and cracks in the CMA and sealant. Laboratory investigations concluded that the backfilling materials were not suitable for cold region applications (*10*).

There are a variety of other materials available for MT backfilling, such as CMA, fly ash based grouts, and epoxy grouts (*11*, *12*). However, the performance of these materials for MT backfilling has not been evaluated and monitored. The aim of this research is to investigate and compare the performance of different materials for backfilling of MTs in cold regions. For this purpose, a composite backfilling section was designed using different materials. Laboratory investigations were conducted to evaluate backfilling material performance in flexible pavements.

Backfilling Method and Material Properties

To investigate and compare the performance of different MT backfilling materials in cold regions, a backfilling section was designed. Previous studies proved that sand is not an appropriate material to secure the conduit inside of the trench in cold regions (*10*). Therefore, as Figure 1 shows, to prevent the conduit moving inside the trench, it was stabilized in the soil using a cement grout. Because the cement grout is a rigid material and is not compatible with the existing HMA, the upper part of the trench in the asphalt layer was backfilled using CMA or epoxy grout.

Cement Grout

To shorten the curing time and prevent the formation of shrinkage cracks in the upper backfilling layer, a nonshrink, fast set cement grout was selected to stabilize the conduit in the soil. The grout's water content was selected because, as it becomes fluid, it easily penetrates the narrow trench. According to the manufacturer's data sheet, the addition of 25% of the grout's weight in water is enough to make the grout flowable. At normal temperature (21°C) with 25% water, the grout working time and setting time were approximately 20 and 45 min, respectively. Compressive strength of the grout was 13.8, 27.6, 34.5, and 44.8 MPa after 3 hours, 1 day, 7 days, and 28 days of curing (*13*).

Because MTs are not always constructed at normal temperatures, several laboratory tests were conducted to investigate the effect of lower ambient temperatures on the grout setting time, and the results are summarized in Table 1*a*.

For this purpose, at different ambient temperatures ranging from 0°C to 20°C, the effect of two different water temperatures of 21°C and 45°C on the grout setting time were examined. As the results show, using warmer water could significantly (about 50%) decrease the setting time of the grout in colder temperatures.

For cold season applications, the grout may freeze when the temperature is below zero. To study the effect of freezing on grout compressive strength after thawing and resetting, a compressive test was conducted on a frozen and reset grout according to ASTM C109/C109M–07 (*14*). After mixing, grout was poured into the molds and the molds were transferred to an environmental chamber with a temperature of −18°C for 24 hours. After the grout became completely frozen, the molds were placed in a moist room. After 24 hours of thawing and resetting in the moist room, the samples were

Table 1. Results of Grout Testing

The effect of temperature on grout setting time		
Water temperature (°C)	Grout temperature (°C)	Setting time (min)
21	0	155
	8	111
	21	50
45	0	91
	8	60
	21	22
Freezing effect on grout compressive strength		
Grout condition	Curing age	Compressive strength (MPa)
Normal	3 hr	14
	1 day	24
	7 days	31
	28 days	40
Frozen	3 hr	12
	1 day	17
	7 days	34
	28 days	40.5

unmolded and stored for another 24 hours and then tested. The reflected results in Table 1*b* show that the compressive strength 1 day after resetting is slightly lower than nonfrozen grout, but after 7 days, the strength was comparable with the nonfrozen samples.

Cold Mix Asphalt (CMA)

Seven CMAs with different mineral aggregate grain size distribution types (open and dense) were selected from available proprietary materials. Table 2 summarizes mix properties according to the manufacturers' data sheets. To investigate the performance of the CMAs and select the best mix for trench backfilling, different laboratory tests were conducted.

Marshall Stability. To evaluate and compare the strength and workability of CMAs, Marshall Stability tests were conducted on overnight cured mixes. It was observed that uncured samples were not strong enough to be tested in the laboratory; therefore, the test was conducted on the cured CMA samples. The samples were cured at 135°C overnight to simulate the field conditions after several weeks of patching (*22*). After curing, CMA materials were poured in a 100 mm diameter mold with a height of 65 mm and were compacted using 75 blows of a standard Marshall Hammer on both faces. Table 3 shows the result of the Marshall Stability test. It can be observed that all the mixes have lower densities and higher air void contents compared with typical HMA. As can be concluded from the table, the lowest stability value obtained was for CMA-1 and the highest stability value obtained was for CMA-7. CMA-5 and CMA-7 had comparable values with a typical HMA mix. Marshall Quotient (MQ), the ratio of Marshall Stability over flow number, is an indicator of the mixes' workability (*23*). It was noted that the higher the MQ value, the lower the workability of the mix. From Table 3 it can be concluded that CMA-7 had the lowest workability and CMA-1 was the most workable mixture. Compared with typical MQ values for HMA, CMA-2, -3, -4, and -6 all had acceptable workability.

Indirect Tensile Strength Test. Indirect tensile strength (ITS) measures the sensitivity of asphalt mixes to moisture damage (*24*). Moisture resistance of a material affects the durability of patching, especially in cold regions where the materials are subject to precipitation and freeze/thaw cycles.

For this test, three sets of samples were prepared from each material with the same curing and compaction procedure that was explained for the Marshall Stability test. Each set consisted of three samples. The first set was tested in dry conditions, the second set was 70–80% saturated in 25°C water before testing for ITS. The third set was tested after 70–80% saturation, followed by wrapping the samples in plastic bags and freezing them at −18°C for 16 hours and thawing in a water bath at 60°C for 24 hours. Finally, the samples were put in a 25°C water bath for 2 hours before testing.

According to the AASHTO T 283 standard, moisture sensitivity of the samples was evaluated by calculating tensile strength ratio (TSR) according to Equation 1. The minimum acceptable TSR value for HMA samples is 0.8:

$$\text{TSR} = \frac{S_2}{S_1} \tag{1}$$

where

S_1 = tensile strength of dry sample in kPa and
S_2 = tensile strength of conditioned sample in kPa.

Table 3 shows the results of ITS tests on CMA mixes. Similar to Marshall Stability, minimum and maximum dry ITS values were obtained for CMA-1 and CMA-7. Comparing dry ITS values of the cold mixes with typical values for HMA shows that CMA-6 and CMA-7 had acceptable values. The calculated TSR values for saturated samples show that, except for CMA-3, all the samples could be considered insensitive to moisture. However, TSR values after freeze/thaw cycles show that only CMA-1 was not sensitive to freeze/thaw cycles. CMA-5, -6, and -7 had TSR values of more than 70%, which was close to the required value.

Adhesiveness Test. Adhesiveness is an indicator of the bonding strength between the original pavement and new patching materials, which is mainly dependent on mortar composition (asphalt binder and fines). It is a very influential parameter for MT performance because poor adhesion

Table 2. CMA Material Properties

Mix	Description	Nominal maximum aggregate size	Bitumen content	Bitumen type
CMA-1 (*15*)	An open graded mix consisting of 85% mainland sand and gravel plus 15% washed sand	9.5 mm	3.8%	Polymer modified cut-back
CMA-2 (*16*)	An open graded mix composed of 100% crushed limestone aggregates	9.5 mm	5.5%	Modified cut-back
CMA-3 (*17*)	An open graded mix consisting of a minimum 80% crushed, washed or screened limestone, or an approved equal, that was free from excessive fine dust and foreign material	9.5 mm	4.5–6.6%	Modified cut-back with special additives
CMA-4 (*18*)	An open graded mix composed of 100% crushed stone or a laboratory approved equivalent	9.5 mm	4–6%	Modified cut-back; no volatile-organic-content liquid blend
CMA-5 (*19*)	An open graded mix designed for winter patching with a temperature range of −26 to 38°C	9.5 mm	5–6.5%	Modified cut-back
CMA-6 (*20*)	A dense graded mix consisting of sand and stone with a high percentage (>90%) of graded recycled asphalt pavement	9.5 mm	3–7%	Cut-back bitumen and proprietary additives
CMA-7 (*21*)	A dense graded mix consisting of limestone aggregates	12.5 mm	2–8%	Polymer modified cut-back with special adhesives

of a CMA material to the existing asphalt causes the formation of narrow gaps at the trench wall and, consequently, water ingress into the trench, which is followed by trench failure.

In this study, the adhesiveness test was conducted according to the Virginia DOT procedure (*22*). For preparing samples, 500 g of loose cold mix was placed on top of an HMA sample (75 mm height) in a 100 mm diameter Marshall Mold and compacted with 10 blows of a standard Marshall Hammer. The sample was extruded from the mold and held inverted until the compacted cold mix separated from the HMA surface. The time required for the separation of cold mix from the HMA surface was measured as the adhesion time. For cold mixes, optimum adhesion time has been recommended as 5–30 seconds (*22*).

Table 3 reflects the adhesiveness test results on different CMAs. As can be seen, the highest adhesion time obtained was for CMA-7. Except for CMA-6, all the CMAs showed values lower than the minimum limit. However, because the test result for CMA-6 is still close to the minimum value, the addition of a tack coat layer at the trench walls can improve the binding of the new material to existing asphalt.

Comparing all the CMA test results shows that the most appropriate cold mix could be CMA-7 because it had the best test results. However, as the MQ values showed in Table 3, CMA-7 had low workability at room temperature; it was hard to remove the mix from the bucket and compact it. Because workability is one of the major parameters for MT backfilling material, CMA-7 should not be considered as an appropriate option for MT backfilling in normal temperatures. Among the other CMAs, CMA-6 showed the best adhesive properties, acceptable workability, high tensile strength, and moisture resistance. Consequently, CMA-6 was selected as the backfilling material for further testing. It should be noted that the Marshall Stability and freeze/thaw susceptibility of this material were still lower than the required values.

Epoxy Grout

Epoxy grout is another alternative material that has been used in some micro-trenching projects. In this study, the epoxy grout used was a thermosetting product resulting from reaction of epoxy resin (component A) with a curing agent (component B) together with recycled glass beads with a nominal size of 0.63 mm. The grout was a flowable, easy to apply, and self-compacting material designed to be installed in clean, dry, and sound areas. According to the manufacturer's data sheet, it can be applied in temperatures ranging from 4°C to 43°C, with working time between 3 and 90 min (adjustable using a kicker in cold weather). It has appropriate adhesion properties with a primarily substrate failure >200 kPa according to ASTM D7234 (*12*, *25*).

Because epoxy grout was used as an alternative to CMA to fill the trench inside the existing asphalt layer, to investigate compatibility of this material with HMA as the host material, regular asphalt tests were performed on grout samples. To prepare the samples, epoxy grout was prepared and poured into cylindrical 100 mm diameter standard molds. The temperature of the mix after adding component B was around 50°C, and working time at ambient temperature of 21°C was almost 10 min. The samples were extruded from the mold after cooling down and were tested.

Table 3. Results of CMA Testing

Test result	CMA-1	CMA-2	CMA-3	CMA-4	CMA-5	CMA-6	CMA-7	Typical HMA
Density (g/cm^3)	2.12	2.25	2.26	2.26	2.21	2.25	2.26	2.36
Void content	13%	12%	11%	10%	13%	8%	8%	4%
Marshall Stability (N)	3,200	5,400	3,400	5,450	10,100	6,800	18,950	11,500
Flow (mm)	1.5	1.25	1.25	1.25	1.25	1.75	1.5	2.0–3.5
MQ (N/mm)	2,133	4,320	2,720	4,360	8,080	3,885	12,633	3,285–5,750
Dry ITS (kPa)	276	522	370	546	592	790	1,568	800
TSR (after saturation)	1.13	0.95	0.64	0.85	0.89	1.05	1.06	≥ 0.8
TSR (after freeze/thaw)	0.92	0.56	0.36	0.62	0.76	0.7	0.75	≥ 0.8
Adhesiveness time (sec)	1.21	1.08	1.3	2.62	1.69	5.46	49.9	Recommended value 5 ≤ time ≤ 30

Note: ITS = indirect tensile strength; TSR = tensile strength ratio.

Marshall Stability of the material was calculated to be 1,850 N, which was quite high and comparable to HMA. However, the flow number obtained around 14 mm was almost 10 times that of HMA. It was observed that the epoxy grout material showed a large deformation under the loading before failure; however, after removing the load, most of the deformation disappeared, and only a small part remained. This can be explained as the behavior of the epoxy grout as a thermoplastic material compared with asphalt, which is a visco-elastic material.

The ITS of the epoxy grout was calculated to be 2,940 kPa, almost triple that of HMA samples. To condition the samples, they were placed for 48 hours under water at 25°C and were then wrapped in plastic bags and put in a freezer, following the same thawing and testing procedure as the asphalt samples. The samples showed no sensitivity to moisture and freeze/thaw cycles.

Evaluation of Micro-Trench Samples

Similar to other utilities' backfilling, the best results of MT backfilling performance could be achieved by using material compatible with the existing pavement material. To compare the behavior of backfilling material to HMA as the MT host material under different loading and thermal conditions, dynamic modulus tests were performed at different loading frequencies and temperatures. For this purpose, a regular dense graded HMA sample, composed of nominal aggregate size of 12.5 mm and PG 58-28 bitumen, was prepared and compared with CMA-6 and epoxy grout. To investigate the permanent deformation of the backfilling materials and compare it to HMA, wheel tracking tests were conducted on the HMA, CMA-6, and the epoxy grout sample. To investigate the performance of the composite backfilling material in the trench, small trench slabs were prepared in the lab and loaded using the wheel tracker in the same condition and the results were compared.

Dynamic Modulus Test

Because HMA and CMA are visco-elastic materials, their dynamic modulus or $|E^*|$ is dependent on loading frequency and temperature. Epoxy grout, as a thermoset material, is expected to be less sensitive to temperature and loading. To determine the dynamic modulus of these materials, a cylindrical gyrator compacted the HMA, and cured CMA-6 samples were prepared with dimensions of 150 mm in diameter and 170 mm in height according to AASHTO TP79 (*26*). Epoxy grout samples of the same size as the CMA-6 samples were also prepared in the lab. All the samples were cored and cut to 100 mm diameter × 150 mm height specimens for dynamic modulus tests at temperatures of −10°C, 4°C, 20°C, and 35°C and frequencies of 10, 5, 2, 1, 0.5, 0.2, 0.1, and 0.01 Hz as per AASHTO PP61 (*27*). CMA-6 was the only exception as, because of large deformations at high temperatures, the test could not be completed at 35°C.

Figure 2*a* compares the results of a dynamic modulus test on HMA and CMA-6 samples. As the figure shows, both HMA and CMA-6 samples are sensitive to temperature and loading time, but dynamic modulus of CMA-6 is significantly lower than HMA, which could be a result of using softer cut-back binder to increase the mix workability. Figure 2*b* shows the epoxy grout material was also lower than HMA. Although the epoxy grout was a thermoset material, it was quite sensitive to temperature changes, but not too sensitive to the loading time.

Permanent Deformation

To investigate permanent deformation of different backfilling material under loading and compare it to HMA, a rutting test was conducted using a Hamburg wheel tracking (HWT) device. HWT measures the rutting of asphalt mixtures by rolling a steel wheel over the sample inside a temperature-controlled chamber according to AASHTO T324-11 (*28*). For this test, two cylindrical samples of different backfilling

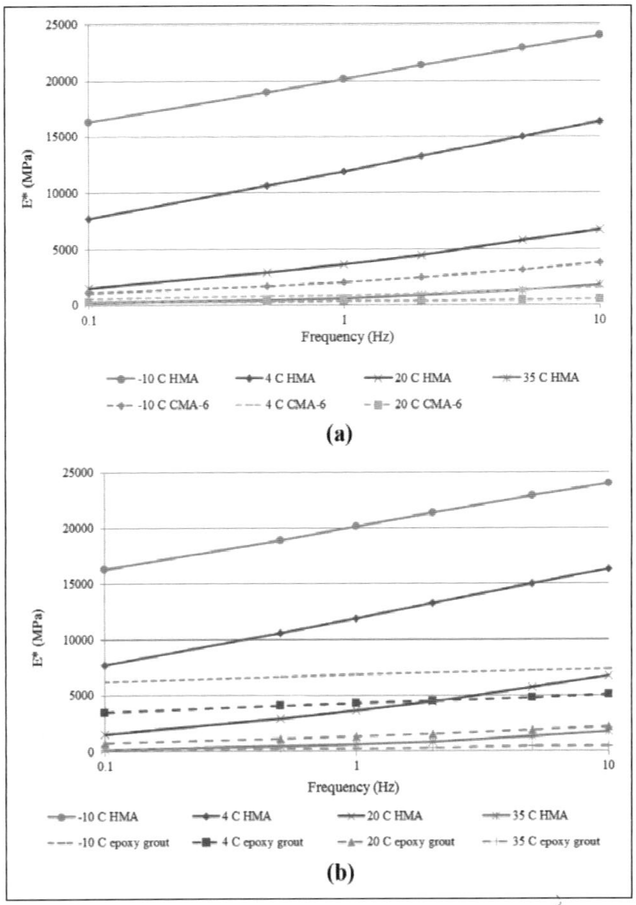

Figure 2. Comparison of E^* of HMA with (*a*) CMA-6 and (*b*) epoxy grout.

material (CMA-6, epoxy grout, and HMA) were prepared and fitted in a 60 mm deep high-density polyethylene mold and inserted beneath the wheel. The wheel moved at a rate of 52 passes per minute, which simulates loading frequency almost equal to 1 Hz at a standard loading of 705 ± 4.5 N. The vertical deformation induced at the middle of the sample versus the number of passes was recorded as the indicator of rutting resistance of the samples. The test was terminated after 20,000 passes, or when vertical deformation of the samples reached 12 mm.

Figure 3 compares permanent deformation of different backfilling material against HMA. It should be mentioned that the samples were conditioned under 50°C water, except for the CMA-6, which was tested in two different temperatures. As can be seen in Figure 3, permanent deformation of this mix tested in 50°C (CMA-6 overnight cured [OC], tested in hot water [HW]) was reached at 12 mm, after 3,750 loading cycles. Comparing this result with the dynamic modulus test shows that this mixture was very weak at high temperatures and low frequencies. Therefore, for this mix, another test was conducted and the sample was conditioned under cold water at a temperature of 21°C (CMA-6 OC, tested in cold water [W]). The results showed significant decrease in the rutting depth of this material.

Compared with the asphalt samples, the epoxy grout shows completely different behavior under loading. As Figure 3 shows, 1 mm deformation occurred after applying the wheel load on top of the sample, which was almost constant during the testing procedure. It was also observed that the deformation was not permanent for the epoxy sample as it returned to its original condition after removing the load and cooling down the sample.

HWT Tests on Simulated Trench

To study the evaluate MT composite backfilling, different small slabs with dimensions of 400 mm × 300 mm were prepared in the laboratory. Because the maximum height of the slab molds was limited to 120 mm, 80 mm of the height was filled with regular base material and the 40 mm top part was filled with HMA material. A laboratory roller compactor was used to compact the layers. MTs with dimensions of 20 mm wide and 80 mm deep were created in the middle of the slabs.

To compare the stability of the conduit in the cement grout with sand that had been used in many studies under freeze/thaw conditions, two similar trenches were prepared using both materials. After placing a 15 mm diameter conduit at the bottom of the trenches, the soil part (lower 40 mm) of one trench was backfilled with the cement grout and the other one with sand. The top 40 mm part inside the asphalt slab was backfilled with CMA-6 for both trenches. Both trenches were saturated in water for 24 hours and then placed in an environmental chamber at −18°C for 24 hours, followed by thawing at the room temperature for 24 hours. Finally, the slabs were put under 20,000 wheel load passes, perpendicular to the MT.

The maximum rutting depths for the sand and cement grout backfilled trenches were obtained at 9 and 6 mm, respectively. Significant conduit movement was observed in the trench backfilled with sand because of the gradual movement of the sand around the conduit under the wheel pressure, and also because of penetration and washing out of tiny sand particles into the surrounding coarse soil material. In contrast, no movement was observed in the stabilized conduit in the grout.

To compare the performance of CMA and epoxy grout in the trench, one more MT and one traditional trench sample were prepared. The lower part of both trenches was backfilled with the cement grout, and the top parts filled with the epoxy grout. After curing the epoxy grout, the MT sample was conditioned in one freeze/thaw cycle using the same procedure that was used for the backfilled trench with cement grout and CMA, and HWT was performed on the sample. Permanent deformation of the backfilled trench was 4 mm, which only occurred in the HMA layer, whereas no permanent deformation was observed in the epoxy grout. Figures

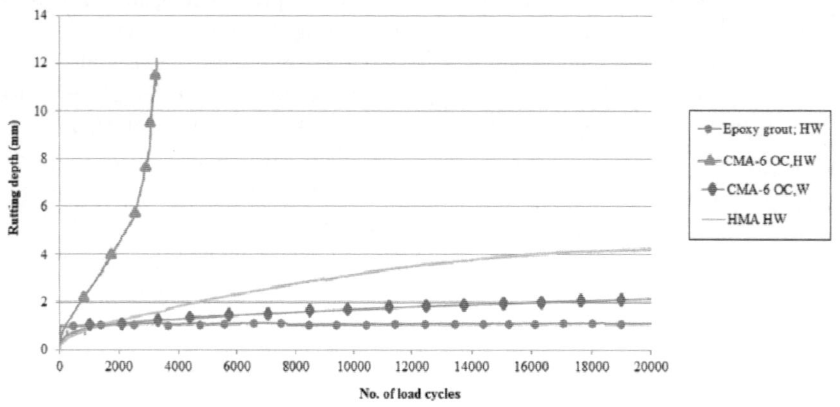

Figure 3. Permanent deformation of backfilling material.

4*a* and *b* show the MT samples before and after 20,000 loading passes, respectively. Figure 4*c* shows the cross section of backfilled MT with cement grout at the bottom and epoxy grout on top and Figure 4*d* shows the same sample after 20,000 loading passes. As Figure 4*b* shows, the deformation of MT samples backfilled with cement grout and CMA is uniform all over the trench. In contrast, according to Figure 4*d*, the different behavior of HMA and epoxy grout under loading caused cracks around the trench edges, which may cause water ingress into the trench and, consequently, trench failures.

Comparing the results shows that minimal rutting was observed in the backfilled trench using cement and epoxy grout. However, this method of backfilling may cause cracking in the existing asphalt layer. In conclusion, for MT backfilling in flexible pavements, using a high-quality CMA mix on top of a cement grout could be a better solution compared with epoxy grout. Cement grout or epoxy grout could be appropriate backfilling materials for MTs in rigid pavements.

Cost Comparison of Backfilling Material

Table 4 compares unit prices of the used backfilling materials. Because all the materials were from commercial off-the-shelf products, the recent prices were used for the cost calculation. The total backfilling cost would depend on the MT dimensions and thickness of each backfilling layer. As Table 4 shows, the cheapest backfilling option could be a combination of CMA and sand. Sand can be easily poured into the trench and CMA application and compaction may start immediately. The trench can be open to traffic right after CMA compaction. However, the HWT simulated trench test showed this was not an appropriate method for cold regions. The second most cost-effective backfilling option is the combination of cement grout and CMA that could potentially have an acceptable performance in cold climates. According to Table 1*a*, grout curing could last between 22 and 155 min, depending on the ambient, grout water, and trench temperature. CMA compaction should start after the curing time, after which the trench might be open to traffic. Epoxy grout is almost six times more expensive than CMA and it may have an adverse effect on the trench performance in flexible pavements. However, the application of self-compacting epoxy grout is much easier than CMA. Compaction of CMA in a tiny trench could be very challenging, whereas epoxy grout can be poured directly into the trench right after cement grout curing (22–155 min), and its curing time is short (3–90 min).

It should be noted that for this analysis, only material cost has been considered. To have a better estimation about the total backfilling cost, other important parameters, such as required backfilling time, labor, and machinery expenses should also be considered.

Conclusions

This study was conducted to evaluate different materials for MT backfilling in cold regions subjected to freeze/thaw cycles. The results of this study can be summarized as follows:

1. Among seven preparatory mixes, CMA-6 was selected as the most appropriate mix for MT backfilling. However, it did not show satisfactory Marshall Stability and freeze/thaw resistance.
2. Epoxy grout showed three times higher ITS, and higher or comparable Marshall Stability for HMA samples. However, the flow number of this material was 10 times higher than HMA under loading. It was not sensitive to moisture damage and freeze/thaw cycles.
3. Dynamic modulus test results showed that epoxy grout stiffness was almost three times lower than

Figure 4. Laboratory MT samples: (*a*) backfilled with cement grout and CMA before loading; (*b*) backfilled with cement grout and CMA after loading; (*c*) backfilled with cement grout and epoxy grout cross section; (*d*) backfilled with cement grout and epoxy grout after loading.

Table 4. Cost Comparisons of Different Backfilling Materials

	Material cost (USD/kg)	Unit weight (kg/m^3)	Material cost (USD/m^3)	Opening to traffic
Play sand	0.23	1,500	345	After pouring
CMA 6	0.45	2,250	1,012	After compaction
Cement grout	0.83	1,700	1,411	22–155 min
Epoxy grout	6.08	1,120	6,810	3–90 min

Note: USD = U.S. dollars.

HMA and was very sensitive to temperature variation, yet not very sensitive to loading frequency variation.

4. Rutting test results showed that CMA-6 was very sensitive to deformation at high temperatures. Despite its low resilient modulus, epoxy grout did not show any permanent deformation under wheel loading.
5. HWT tests were simulated on trenches backfilled with cement grout with sand at the bottom and CMA on top. Comparisons were made after subjecting the samples to a freeze/thaw cycle. The results showed that deformation of the trench backfilled with sand increased from 6 to 9 mm after a freeze/thaw cycle. Using a cement grout could enhance the trench stability and prevent conduit movement inside the trench.
6. HWT tests on simulated trenches backfilled with cement grout at the bottom, and CMA at the top showed acceptable permanent deformation in the HMA and CMA under loading. For the MT sample backfilled with cement grout at the bottom and epoxy grout at the top, HMA had acceptable permanent deformation. However, no permanent deformation was observed in the epoxy grout material, which could cause cracking in the existing HMA layer.
7. Cost comparison of backfilling material showed that using CMA is much more economical than epoxy grout. However, it should be considered that compaction of CMA is more challenging and could increase the time and final cost of the backfilling operation.

Further Study Recommendations

This study was limited to laboratory investigations on MT performance in cold regions, which could be completed by comparing the results with the long-term performance evaluation of real MT samples in the field. Backfilling materials were limited to commercial off-the-shelf products, and laboratory test results showed the selected material may not be ideal for this kind of application. Therefore, further study is recommended to design appropriate material for MT backfilling that considers the project location and the existing pavement type. A comprehensive cost comparison between different backfilling options, considering all influential parameters such as required material, machinery and labor cost, and project duration would be valuable to optimize the cost of MT backfilling methods.

Author Contributions

The authors confirm contribution to the paper as follows: Study conception and design: Leila Hashemian, Alireza Bayat; Data collection: Leila Hashemian, Vinicius Afonso Velasco Rios; Analysis and interpretation of results: Leila Hashemian, Alireza Bayat, Vinicius Afonso Velasco Rios; Draft manuscript preparation: Leila Hashemian; All authors reviewed the results and approved the final version of the manuscript: Leila Hashemian, Alireza Bayat, Vinicius Afonso Velasco Rios.

References

1. Atalah, A., C. Chang-Jin, and K. Osburn. Comparison Study of Installing Fiber Optic Cable in University Campuses Using Trenchless Techniques Relative to Open Cut. *Pipelines 2002*, American Society of Civil Engineers (ASCE), August 4–7, pp. 1–17. https://doi.org/10.1061/40641(2002)70
2. Duraline Technical Report, Recommended Procedures for Conduit Microtrenching and Restoration. *Technical Bulletin: DCEB-10002*, 2016, pp. 1–21.
3. Trezise, S. *Best Practice for Installing Fiber Through Micro Trenching*, 2017. http://www.ppc-online.com/blog/best-practice-for-installing-fiber-through-micro-trenching. Accessed June, 2017.
4. Jensen, K. A, V. R. Schaefer, M. T. Suleiman, and D. T. White. Characterization of Utility Cut Pavement Settlement and Repair Techniques. *Proc., 2005 Mid-Continent Transportation Research Symposium*, Ames, Iowa, August 2005.
5. Stirling Lloyd Polychem Ltd. *Technical Advice Note 11-09, Micro-Trenching Reinstatement (Technical Advice Note)*, 2011. http://www.highwaymaintenanceproducts.co.uk/downloads/TAN11_09_Micro_Trench_Reinstatement_(Issue_3)_1352371156.pdf. Accessed June 1, 2015.
6. Schultz, B. *How Microtrenching Cuts Fiber-Optic Cable Costs*, September 2011. http://www.forconstructionpros.com/article/10327940/thinner-cuts-slice-costs. Accessed October 8, 2015.
7. Chorus. *Deployment Standards Initiative Kelson Micro & Mini Trenching Pilot*. DSI-28042011 Final, Ministry of Business, Innovation & Employment, Broadband and Mobile Initiatives, New Zealand, 2011.
8. Trawick, K. W. Improving Micro-Trenching: The Vision Becomes Reality. *Broadband Magazine*, Vol. 30, No. 6, 2009, pp. 92–93.
9. Vaseli, H. *Application of Micro-Trenching for Fiber to the Home*. MSc thesis. University of Alberta, Edmonton, Alberta, 2015.
10Hashemian, L., M. Rezaee, and A. Bayat. Field and Laboratory Investigations on Pavement Backfilling Material for Micro-Trenching in Cold Regions. *International Journal of Pavement Research and Technology*, Vol. 10, No. 4, 2017, pp. 333–342.
11. Konczak, J. J. *Grout for Filling a Micro-Trench*. Patent No. WO2012027534 A1, 2012.
12. *FastPatch DPR Rev 01 Technical Data Sheet*. Willamette Valley Company, Eugene, OR, 2016.
13. *Commercial Grade FastSet™ Non-Shrink Grout*. QUIKRETE® Companies, LLC, Atlanta, GA, 2009.
14. *ASTM C109/C109M-16a Standard Test Method for Compressive Strength of Hydraulic Cement Mortars (Using 2-in. or [50-mm] Cube Specimens)*. ASTM, West Conshohocken, PA, 2016, pp. 1–10.
15. *EZ Street Asphalt Product Information*. EZ Street Asphalt, Canada, 2014, pp. 1–10.
16. *MCA Perma-Patch: Cold Mix, Product Data Sheet*. Mac Asphalt, Toronto, ON, 2015, pp. 1–2.
17. *Product Information and Specification Sheet ProPatch™ (Permanent High Performance Cold Patch*. Innovative Surface Solutions, Ajax, ON, 2011, pp. 1–2.
18. *QPR High Performance Permanent Pavement Repair Material, Material Specification*. Quality Pavement Repair, Chattanooga, TN, 2014, pp. 1–3.
19. *Data Sheet UNIQUE® High Performance Fast Set Mortar*. Unique High Performance Fast Set Mortar, Cleveland, OH, 2015, pp. 3–4.
20. *Quikrete Asphalt Cold Patch*. QUIKRETE® Companies, LLC, Atlanta, GA, 2009, pp. 1–2.
21. *Instant Road Repair® Safety Data Sheet*. International Roadway Research, Humble, TX, 2015, pp. 1–6.
22. Wilson, T. *Long-Term Monitoring of Pavement Maintenance Materials Test Sites*. Report no. FHWA-RD-98-063. Office of Engineering Research and Development, Federal Highway Administration, McLean, VA, 1998.
23. Hattatoglu, F., S. Hinislioglu, and H. F. Bayata. Modeling of Marshall Quotient of Hot Mix Asphalts by Artificial Neural Networks. Presented at the 1st International Conference on Engineering and Natural Sciences (ICENS), Skopje, Republic of Macedonia, 2015.
24.AASHTO (American Association of State Highway and Transportation Officials). *Standard Method of Test for Resistance of Compacted Hot Mix Asphalt (HMA) to Moisture-Induced Damage*. T 283-14, Washington, D.C., 2014.
25. ASTM (American Society for Testing and Materials). *Standard Test Method for Pull-Off Adhesion Strength of Coatings on Concrete Using Portable Pull-Off Adhesion Testers*. D7234, West Conshohocken, PA, 2012.
26AASHTO (American Association of State Highway and Transportation Officials). *Determining the Dynamic Modulus and Flow Number for Hot Mix Asphalt (HMA) Using the*

Asphalt Mixture Performance Tester (AMPT). AASHTO TP79-09, Washington, D.C., 2009.

27. AASHTO (American Association of State Highway and Transportation Officials). *Developing Dynamic Modulus Master Curves for Hot Mix Asphalt (HMA) Using the Asphalt Mixture Performance Tester (AMPT)*. AASHTO PP61-09, Washington, D.C., 2009.

28. Federal Highway Administration Research and Technology. *Bituminous Mixtures Laboratory*. U.S. Department of Transportation. https://www.fhwa.dot.gov/research/tfhrc/labs/materialscomplex/mixtures/. Accessed June, 2017.

The Standing Committee on Sealants and Fillers for Joints and Cracks (AHD25) peer-reviewed this paper (18-05340).

Article

TRR
JOURNAL OF THE TRANSPORTATION RESEARCH BOARD

Transportation Research Record
2018, Vol. 2672(12) 72–81

DOI: 10.1177/0361198118780825
journals.sagepub.com/home/trr
SAGE

Field Application of UAS-Based Bridge Inspection

Junwon Seo[1], Luis Duque[1], and James P. Wacker[2]

Abstract

The use of Unmanned Aerial Systems (UASs), commonly known as drones, has significantly increased over recent years in the field of civil engineering. In detail, the need for a more efficient alternative for bridge inspection has risen because of the increased interest from bridge owners. The primary goal of this paper is to evaluate the efficiency of a drone as a supplemental bridge inspection tool. To complete this study, a glued laminated (glulam) girder with a composite concrete deck bridge was chosen in South Dakota, and a Dà-Jiāng Innovations (DJI) Phantom 4 drone, was employed to perform the bridge inspection. Based on the literature review, an inspection procedure with a drone was developed to efficiently identify damage on the bridge. A drone-enabled inspection was performed following the procedure, and resulting images were checked with those available in the past inspection report from South Dakota Department of Transportation (DOT). This study includes UAS-based bridge inspection considerations to capture appropriate image data necessary for bridge damage determination. A key finding demonstrated throughout this project is that different types of structural damage on the bridge were identified using the UAS.

Unmanned Aircraft Systems (UASs), referred to as drones, have assisted in visual inspection of different types of structures such as bridges in recent years (*1*). The use of UASs has become more appealing to bridge owners, researchers, and stakeholders because of their efficiency and effectiveness to gather relevant data in shorter time and at a lower cost compared with traditional inspection methods (*2*). Recent research efforts have been made to not only investigate but also to monitor structures. For instance, Chen et al. (*3*) developed a small-format aerial photography (SFAP) methodology to monitor cracks on the deck surface. To accomplish this, geo-referenced pictures were taken at an altitude of 305 m (1,000 Ft) to visually observe the damage. Chen et al. concluded that SFAP is a promising tool for bridge construction monitoring. Another investigation of the UASs' capabilities by Chen and Hutchinson (*4*) was conducted to monitor bridges. Their main goal was to monitor cracks' propagation through an image-based approach. The methodology used included images taken at different periods of time to observe the crack location, propagation, and geometric quantification. Due to its effectiveness, this emerging technology presents great potential for bridge inspection, as these structures often present inaccessible areas for inspectors (*5*).

Currently, approximately 9.1% of the United States' bridges are classified as structurally deficient (*6*). Despite the quantity of deficient bridges declining in recent years, visual inspection of such bridges needs to be conducted to effectively recognize damage and determine the appropriate retrofit methods. Several state departments of transportation (DOTs) have investigated UASs as a bridge inspection tool. For example, Caltrans developed a twin-motor, single-duct, electric-powered UAS designed to carry cameras for visual inspection of bridges (*7*). Similarly, Florida DOT and Otero used a multi-rotor helicopter-based UAS with high-definition cameras to transmit video data of structural components of bridges, including timber bridges (*8*). Some stress cracks on bearing areas and guardrail supports were detected by investigating the high-resolution images captured from the UAS. Recently, Minnesota DOT, in partnership with Lovelance and Zink, performed a research project with regard to visual inspection using UAS technology on four different types of bridges in its state (*9*). A rotor aircraft UAS with fixed wings was used for the bridge inspections. The research demonstrated the capability and advantages of the UAS, enabling damage identification on critical areas in the selected bridges in a more cost-effective and safe manner. The United States Department of Agriculture—Forest Service (USDA-FS) —Region 10 (Alaska) also developed a drone system, and its associated inspection protocols were applied to the Placer River Trail Bridge located in the Alaskan Kenai Peninsula (*5*). It was found that the drone was able to

[1]Department of Civil and Environmental Engineering, South Dakota State University, Brookings, SD
[2]Forest Service, US Forest Products Lab, Madison, WI

Corresponding Author:
Address correspondence to Junwon Seo: junwon.seo@sdstate.edu

efficiently inspect the bridge using its high-resolution scans and 3-D virtual model.

The main objective of this study was to investigate the abilities of UASs for bridge inspection. To effectively identify structural damage, a bridge inspection approach was identified and applied to an in-service bridge located in South Dakota following state and government regulations (e.g., South Dakota DOT and FAA). This paper is structured in five sections exclusive of this introduction section. The second section details the UAS and includes a bridge description to perform this application. The third section presents the bridge inspection approach based on the findings from the literature review and various guidelines from South Dakota DOT and FAA. The fourth section presents the obtained results from the damage identification and its comparison with historical inspection reports from South Dakota DOT. The fifth section presents conclusions based on findings from this work, whereas the final section details challenges raised by this UAS application to the bridge.

UAS and Bridge Choice

For this study, an appropriate UAS necessary for bridge inspections was initially selected considering different factors, including flight time, upward viewing camera, camera resolution, video resolution, and others. Then, a bridge structure was chosen to evaluate the efficiency of the UAS-based inspection. The bridge selected is a glulam, three-span timber girder bridge located near Keystone, South Dakota. The details of the UAS and structure selection are presented below.

Considered UASs

A suitable UAS needed to be chosen for the visual inspection of the bridge. While performing the literature review on UAS technology, several researchers investigated various drones to compare their capabilities with reference to data acquisition. Based on the knowledge from the literature review, the following considerations were checked to select the most suitable inspection UAS for this study:

1. **Flying time over 20 min**: A relatively long flying time is beneficial for a more efficient structure inspection, limiting the need of additional batteries and allowing for longer inspection times;
2. **Additional camera on top of UAS**: A second camera facing straight up to inspect underneath the bridge will allow for a comprehensive inspection;
3. **Camera resolution with low illumination**: Low illumination reduces the image quality and slight damage would be difficult to detect. Additional flashlights can be attached to a UAS to enhance illumination;
4. **Video resolution**: High-resolution video is required to observe details of damage;
5. **Payload capacity**: It would be beneficial for attachments that might be required to be carried by a UAS;
6. **UAS lights**: Light-emitting diode lights attached to a UAS will provide some extra illumination, which is required for efficient damage observation underneath a bridge;
7. **Remote range**: As some bridge structures might not be relatively close to the pilot location, long-range modules for remote control will allow inspection of the structure at long distances;
8. **Ability to fly without global positioning system (GPS) signal**: The ability to fly without a GPS signal allows the UAS to inspect the underside of the deck more efficiently without accidents caused by GPS signal failure.

Based on the above considerations, four UASs were deemed suitable. The appropriate UASs for structure inspection include a DJI Matrice 100, DJI S900, DJI Phantom 3 pro, and DJI Phantom 4. Other UASs studied include the Voyager 3, Yuneec Typhoon H, Yuneec Typhoon 4K, Blade Chroma, and Autel Robotics X-Star Premium. The DJI Phantom 4 was selected over the others because its performance and versatility met the aforementioned specifications at a reasonable cost. Additional equipment, such as obstacle avoidance (OA) technology, will be beneficial while approaching a bridge to prevent damage to both bridge and UAS components (*10*). Another key consideration is the ability to fly without a GPS signal. This enables the UAS to inspect underneath the bridge without the problem of losing the satellite connection. A list of specifications for the DJI Phantom 4 is presented in Table 1. Detailed information related to the UAS selection for the bridge inspection can be found elsewhere (*11*).

Studied Bridge

For the bridge inspection using the UAS, a glulam girder with a composite concrete deck bridge was selected. The bridge is presented in Figure 1. It is located on US16 to US16A highway, near the city of Keystone in Pennington County, South Dakota. The timber girder bridge has three simply supported spans with four girders spaced at 2.3 m (7.5 Ft) on center and a clear width of 7.9 m (26 Ft). The bridge is curved horizontally at an estimated radius of 116.4 m (881.97 Ft) and is 51.8 m (170 Ft) long with steel guardrails along the edges of the superstructure.

Considerations for UAS-Based Bridge Inspection

For the field bridge inspection using the UAS, there are several factors to consider before and during an inspection to ensure the collected data are gathered and analyzed correctly. For instance, both the FAA and South Dakota DOT

Table 1. DJI Phantom 4 UAS Specifications (12)

DJI Phantom 4 specifications	
Aircraft	
Weight (including battery)	1,380 G
Max. flight time	Approximately 28 min
Satellite systems	GPS/GLONASS
Camera	
Sensor	1/2.3" Effective pixels: 12 M
ISO range	100–3,200 (video) and 100–1,600 (photo)
Electronic shutter speed	8 s to 1/8,000 s
Max. image size	4,000 × 3,000
Max. video bitrate	60 Mbps
Photo	JPEG, DNG (RAW)
Video	MP4/MOV (MPEG—4 AVC/H.264)
Supported secure digital cards	Micro SD, max capacity: 64 GB. Class 10 or UHS-1 rating required
Gimbal	
Controllable range	Pitch: –90° to +30°
Remote controller	
Operating frequency	2.400 GHz to 2.483 GHz
Max. transmission distance	FCC compliant: 3.1 mi (5 km)
Battery	6,000 mAh LiPo 2S
Intelligent flight battery	
Capacity	5,350 mAh
Voltage	15.2 V
Battery type	LiPo 4S
Net weight	462 G
Max. charging power	100 W

Note: Max. = maximum.

Figure 1. Overview of the selected timber girder bridge captured with the UAS (captured by Luis Duque).

have limitations for the use of UASs, including no flying over 122 m (400 Ft) and no flying over traffic, respectively. A detailed description of the inspection procedure is presented as follows.

Before Inspection

To complete a bridge inspection using the UAS, it is important to consider different factors before flying the UAS. Before arriving at the bridge site, documents including construction plans and inspection reports should be reviewed to ensure efficient data collection. The construction plans review allows a pilot to recognize general information of the bridge, including structural elements locations and dimension, enhancing inspection capabilities of the UAS. For example, identifying structural elements (e.g., girders) with difficult accessibility allows an inspector to develop an approaching strategy to safely inspect the bridge using the UAS. Moreover, the review of historical inspection reports permits the location of critical damage for either monitoring or updating the damage on the bridge using the UAS.

After the bridge documentation is reviewed, the surrounding areas of the bridge must be observed to identify potential risks to flight safety (e.g., adjacent trees and traffic volume). Conducting a visual observation is necessary to identify any critical landing/takeoff zones for approaching the bridge safely. Additionally, the bridge should be visually inspected from the ground to identify critical damage areas to be inspected using the UAS. It is also required that any restrictions that may apply to the bridge location are considered. These restrictions include, but are not limited to, FAA certifications, FAA airspace class restriction [e.g., flying within 8.05 km (5 miles) of an airport], DOT flying restrictions, and DOT safety measures. Traffic controls must also be prepared for the inspection on the right-of-way of any highway per DOT regulations by displaying warning signs and cones to guide approaching traffic, to avoid work zones.

The final step before conducting the bridge inspection using a UAS is to perform a preflight check of the UAS platform to ensure the equipment is working correctly. Both the FAA and UAS manufacturers recommend a preflight check for the first flight of the day to ensure satisfactory flying performance. This check list includes, but is not limited to, propellers and rotors inspection, full charging of all instruments (e.g., a remote controller, storage batteries, and a monitor), remote controller adjustments, gimbal inspection, and firmware updates. During the bridge inspection, the UAS may lose GPS signal, especially while flying under the deck; thus, it is necessary to calibrate the compass to avoid an unanticipated signal failure and possible flyaway.

During Inspection

After the bridge documentation, visual observation, and UAS preflight check are conducted, the inspection of the bridge using the UAS can be completed. The flight strategy must consider weather limitations, FAA and DOT regulations, and others as applicable (e.g., company or organization specific limitations). For instance, it is recommended to fly the UAS during calm wind conditions [e.g., wind speeds less than 24.1 km/h (15 mph)] to avoid turbulence and complications when approaching the bridge. The data collection should be completed by gathering an overall view first before proceeding to specific structural components to ensure the capture of all damage. Limitations from DOT, including not flying over traffic, may affect data collection of the deck and other sections over any adjacent roadway due to not being able to approach them closely. It can be noted that the pilot-in-command (PIC), who must pass an initial aeronautical knowledge test at an FAA-approved knowledge testing center, can be continuously assisted by an observer as recommended by the FAA. The observer can either assist the PIC when flying close to structural components or by observing the UAS camera; therefore, the PIC avoids distractions while inspecting the bridge using the UAS.

UAS-Based Bridge Inspection

This section presents the inspection conducted on the selected bridge in accordance with the proposed procedure.

Before Inspection

The detailed review of the as-built bridge plan and corresponding past inspection report provided by South Dakota DOT was conducted. During the review, it was found that the joints of the bridge were the most critical structural component as water leakage coming from the deck had caused concrete to spall off and corrosion on the exposed rebar. For ease of damage identification, each bridge component was marked, as seen in Figure 2, for the bridge schematics. Details for the component marks on the plan and elevation views can be seen in Figure 2, *a* and *b*, respectively. A brief description of the bridge based on the construction plan is shown below.

Timber Girder Bridge

- 51.8 m (170 Ft) long;
- Glulam timber;
- Three spans, simple supported girders;
- Two bents with four columns each; and
- Curved concrete deck.

A visual observation of the bridge surroundings was conducted to identify potential risks to the operation of the UAS. The bridge was not located in a high-risk zone as there were not many large trees or structures that may have caused difficulty in approaching the bridge or may have affected the flight safety. As such, two regions adjacent to the bridge were identified as potential landing zones for approaching specific components. The UAS-based inspection of the bridge over the roadway was conducted from the edge of the roadway by the landing zones; thus, the traffic was not halted during the inspection.

Further regulations, including South Dakota DOT and FAA flying limitations, needed to be considered before conducting the inspection. For instance, South Dakota DOT specified that the UAS operation was not permitted over traffic, and liability insurance had to be obtained to protect the UAS and bridge in case of an unexpected accident. To fulfill these requirements, warning signs were placed at 228.6 m (750 Ft) from the bridges and cones near inspection zones to warn approaching traffic as requested by the South Dakota DOT. Apart from South Dakota DOT regulations, there were no other limitations for this specific bridge location. Note that the FAA regulations (i.e., Part 107 FAA Rules), to operate the UAS regardless of the use, still apply. Details of the restrictions are presented as follows:

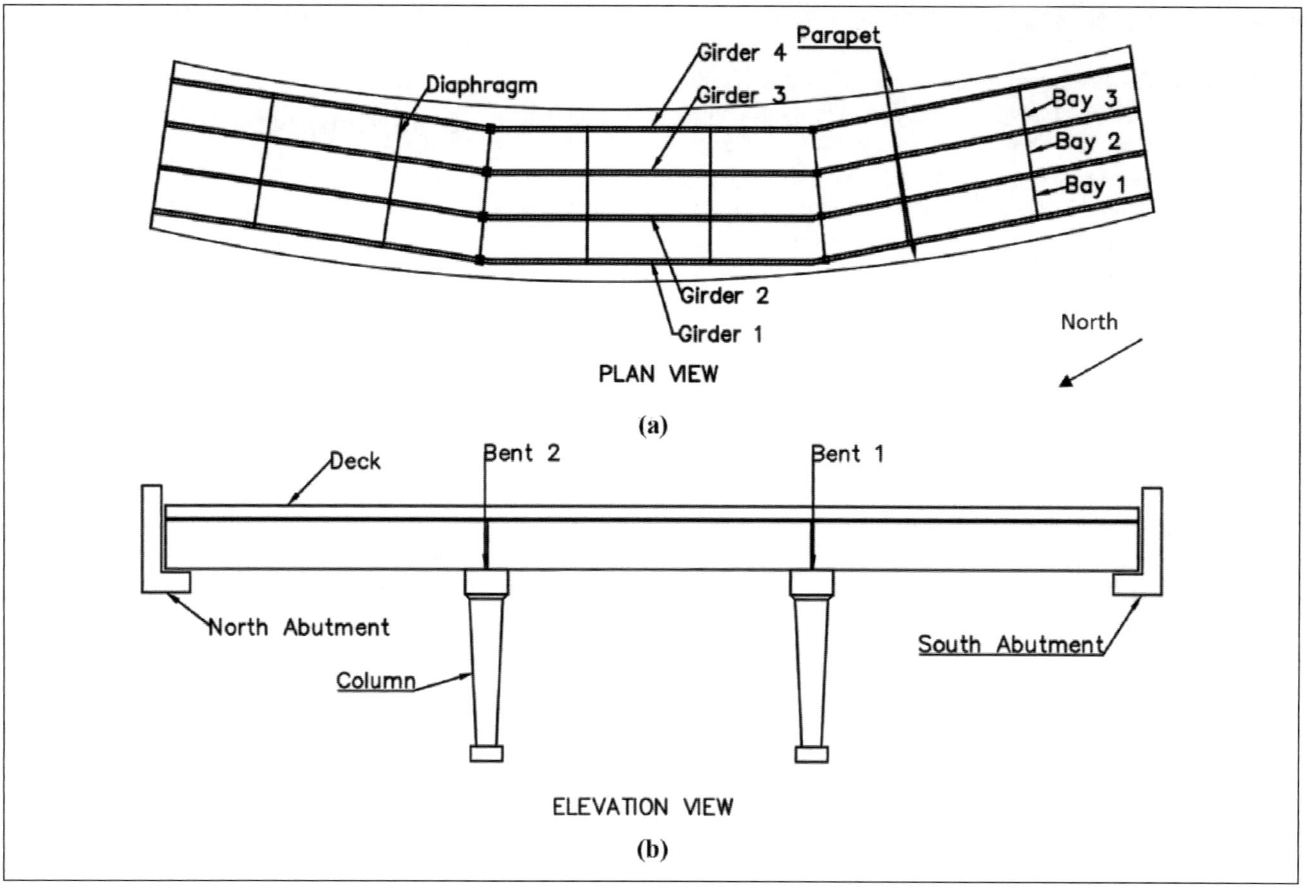

Figure 2. Bridge overview and component marks: (*a*) component marking on plan view and (*b*) component marking on elevation view.

- No restriction for Class G airspace; need air traffic control tower permission otherwise;
- Must keep the aircraft in sight (visual line-of-sight);
- Must fly under 122 m (400 Ft);
- Must fly during the day;
- Must fly at or below 100 mph;
- Must yield right-of-way to manned aircraft;
- Must not fly over people; and
- Must not fly from a moving vehicle.

The final step before the inspection of the bridge using a UAS was to conduct a preflight check. The inspection of the DJI Phantom 4 was conducted according to the manufacturer and FAA requirements. For example, rotors, propellers, batteries, iPad, remote controller, gimbal, and software updates were checked to ensure high flight operation performance. The equipment was found to be in excellent condition, and no defects that may have caused unsafe flying conditions were observed. Additionally, the compass of the UAS was successfully calibrated to avoid any signal loss or potential flyaway. To complete the calibration, the UAS was rotated counterclockwise, while being held horizontally. Then, this same process was vertically performed once more with the camera facing down. The UAS setup was completed without any issues that could have a negative influence on the flight and inspection performance.

During Inspection

After the reviews of information and precautionary actions are taken, the UAS-enabled bridge inspection can take place. The inspection using DJI Phantom 4 was completed over the course of 2 days, February 16 and 17, 2017. During the first inspection day, February 16, the weather conditions were favorable. A sample image captured using the UAS can be seen in Figure 3. To complete the inspection, the overall view of the structural elements was gathered and then specific damage was observed. During this flight, the PIC was continuously assisted by an observer to ensure the flight was safely conducted. Figure 3 shows the UAS aviation near structural components for the bridge.

The weather conditions during the second day of inspection, February 17, were not favorable because of high wind speeds of 24.1 km/h. (15 mph) and wind gusts of 43.5 km/h. (27 mph). A limited number of pictures was obtained during this day using the UAS. For this inspection, a different inspection approach was explored by using video-based

Figure 3. Sample image obtained from the timber girder bridge inspection on February 16, 2017: UAS flying near column of timber girder bridge (captured by Junwon Seo).

data acquisition to eliminate any unnecessary distractions from the picture-taking process while flying under such harsh weather conditions. Using the recorded videos, sample images were obtained to identify damage as seen in Figure 4*a*. Figure 4*a* shows concrete cracks and high moisture resulting from water leakage at the North Abutment of the bridge. It has been proved that the video-based inspection facilitated the acquisition of image information necessary for the damage identification. Figure 4*b* shows the UAS approaching components of the bridge.

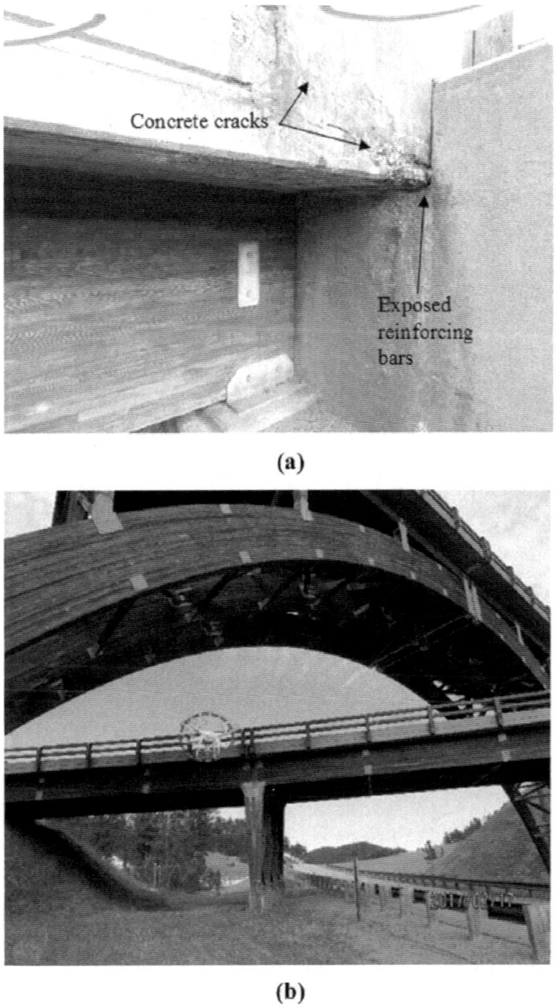

Figure 4. Sample images obtained from the timber girder bridge inspection on February 17, 2017: (*a*) concrete cracking and water damage at abutment location and (*b*) UAS approaching the bridge (captured by Junwon Seo).

Results and Discussion

This section includes the damage identified using the UAS and its comparison with images provided by South Dakota DOT. The first subsection includes a detailed damage identification for two structural elements, underside of deck and girder, whereas the second subsection provides a comparison to South Dakota DOT damage identification. Comprehensive results of the bridge inspection for the columns and abutments as well as other components can be found in Duque et al. (*11*).

UAS-Image-Based Damage Identification

The damage identification presented below aims to demonstrate the capabilities of the UAS to observe different types of damage including cracks, corrosion, and spalling. Sample results for the underside of deck and girder damage are presented as follows.

Underside of Deck. Overall, the deck had several water leakage areas causing moisture-related damage such as corrosion. Sample images of the damage are presented in Figure 5. Figure 5*a* corresponds to corrosion, spalling and exposed rebar at Girder 4 and Joint 1, and Figure 5*b* shows efflorescence in Bay 2 between South Abutment and Joint 1.

Girders. The girders were generally in good condition, with some moisture caused by water leakage resulting from the deck. Some stains and discoloration were also apparent on some girders, possibly caused by calcium deposits and corrosion coming from the chemical reaction of the water with concrete and steel. Sample images of the damage identified on the

Figure 5. Underside of deck damage detected using UAS: (*a*) concrete spalling and corrosion with exposed rebar at Girder 4 and 1 and (*b*) efflorescence in Bay 2 between South Abutment and Joint 1.

girders are shown in Figure 6, *a* and *b*. Figure 6*a* shows visible moisture corresponding to Girder 4 between South Abutment and Joint 1, and Figure 6*b* shows stains under Girder 3 at Joint 1 caused by corrosion of the steel connection bracket.

Other elements. The inspection of other structural components, including abutments and pier columns, was also conducted. In detail, these elements were found to be in good condition aside from minor damage such as cracking and discoloration. For example, damage such as minor discoloration, spalling, and moisture caused by water coming from the deck at North Abutment near Girder 4 was observed, whereas some efflorescence on some column pedestals at Bent 2 was observed.

Comparison to Images from Traditional Inspection

After completing the UAS-enabled inspection, a comparison of results for the underside of the deck and girder was conducted to demonstrate the effectiveness of data obtained with the UAS. A sample image provided by South Dakota DOT is included for each structural component to visually compare the results obtained.

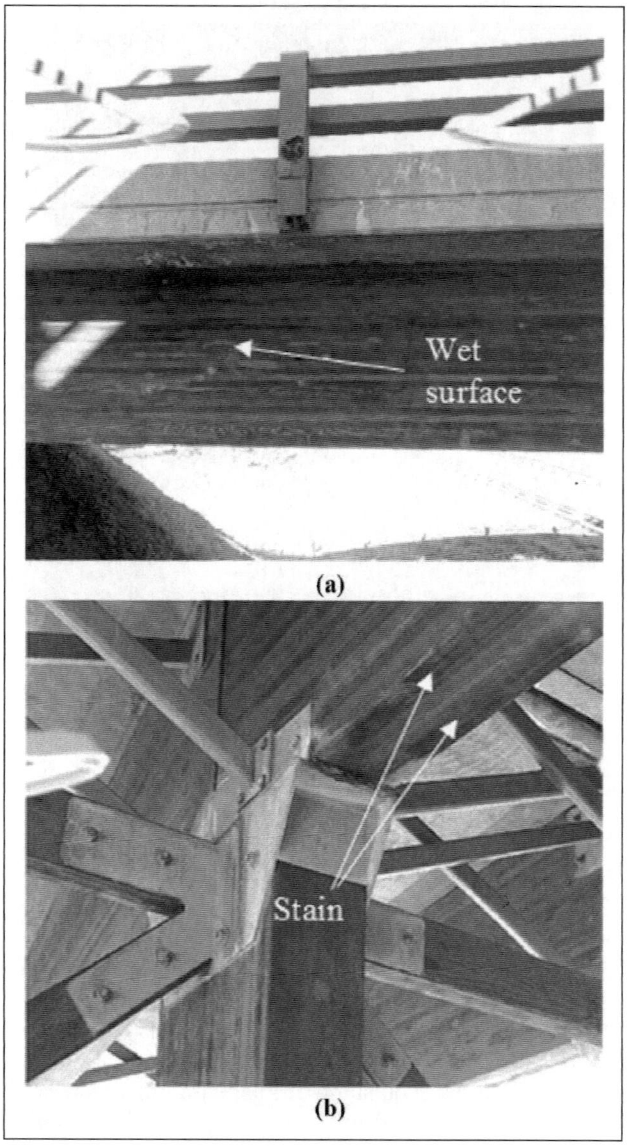

Figure 6. Girder damage detected using the UAS: (*a*) moisture damage on the side of Girder 4 between South Abutment and Joint 1 and (*b*) stains under Girder 3 at Joint 1.

Deck. The UAS was able to identify the damage underneath the deck as identified from the South Dakota DOT inspection report. As reported before, some concrete spalling, delamination, and cracking were observed near joints and abutments. Figure 7 shows sample damage on the deck captured using the UAS. A representative picture provided by the South Dakota DOT (see Figure 7*a*) is included to visually compare the quality of images between the South Dakota DOT and the UAS-enabled bridge inspection. The partial

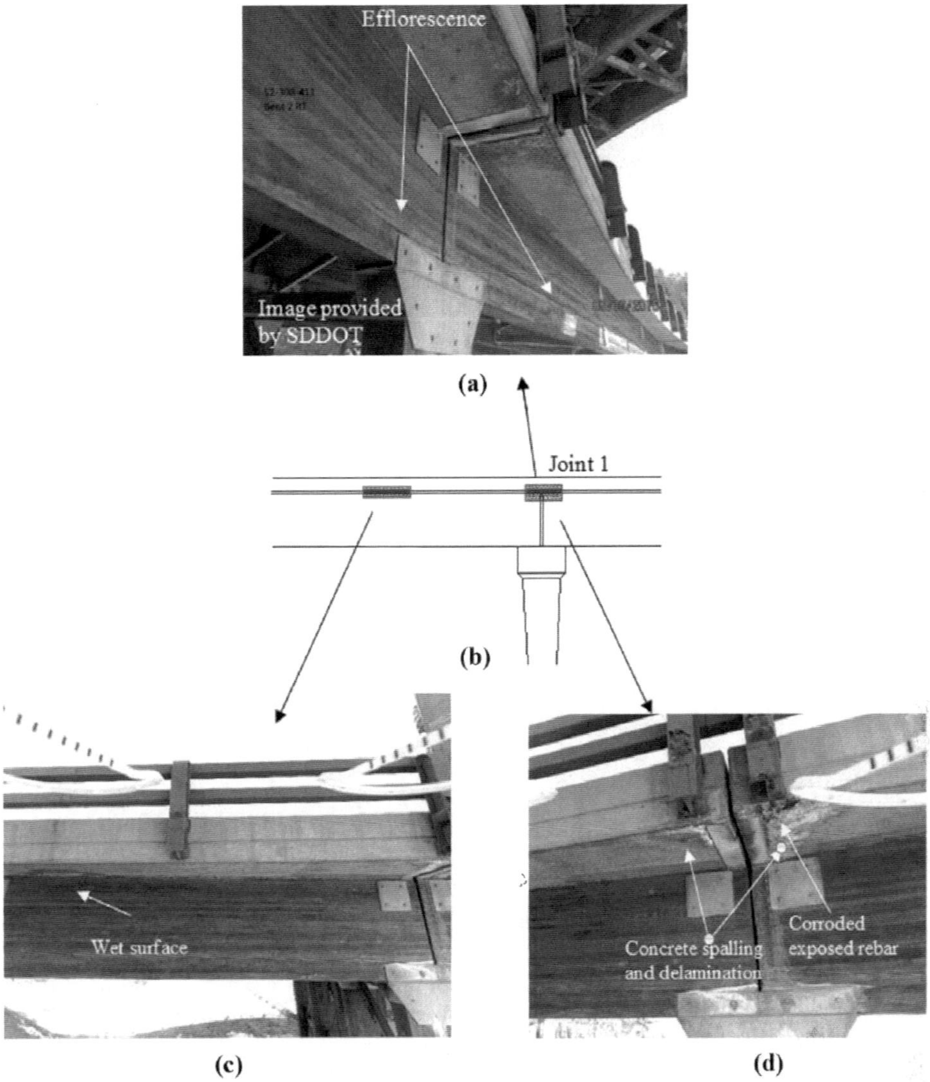

Figure 7. Sample damage on deck near Girder 4: (*a*) image provided by South Dakota DOT (*13*), (*b*) damage locations, (*c*) water leakage under deck between South Abutment and Joint 1, and (*d*) concrete spalling and corrosion with exposed rebar near Girder 4 at Joint 1.

bridge plan containing the specific damage locations represented by red hatched boxes is included in Figure 7*b*. It is apparent that the images are comparable, and the damage is identified. The water leakage under the deck between South Abutment and Joint 1 (see Figure 7*c*) and cracking, spalling, and exposed rebar (see Figure 7*d*) were identified using the UAS. Only one quantified area of damage was provided by the South Dakota DOT. The damage is identified as a 0.3 m (1 Ft) by 0.6 m (2 Ft) spalled area with exposed rebar at Joint 1. The same area was calculated using an image-based methodology detailed in Duque et al. (*11*). The results for the measurement was 0.18 m^2 (1.96 Ft^2). It can be observed that the results are identical, confirming the ability of the UAS to identify and quantify damage on the structure.

Girders. The damage detected on the girders using the UAS coincided with the report provided by South Dakota DOT. A picture provided by South Dakota DOT (see Figure 8*a*) is included to compare with those taken by the UAS. Several damaged areas caused by high moisture content were localized on the girders (see Figure 8*b*). For example, some moisture on Girder 4 between South Abutment and Joint 1 was identified, as seen in Figure 8*c*. Girders damaged because of high moisture were also recognized in other areas, especially near joints, as seen in Figure 8*d*. Note that the South Dakota DOT does not specify moisture-related damage of Girder 4 between South Abutment and Joint 1 and Girder 4 between North Abutment and Joint 2. Finally, the South Dakota DOT did not report quantifiable damage apart from a 5.08-cm (2-in.) by 5.08-cm

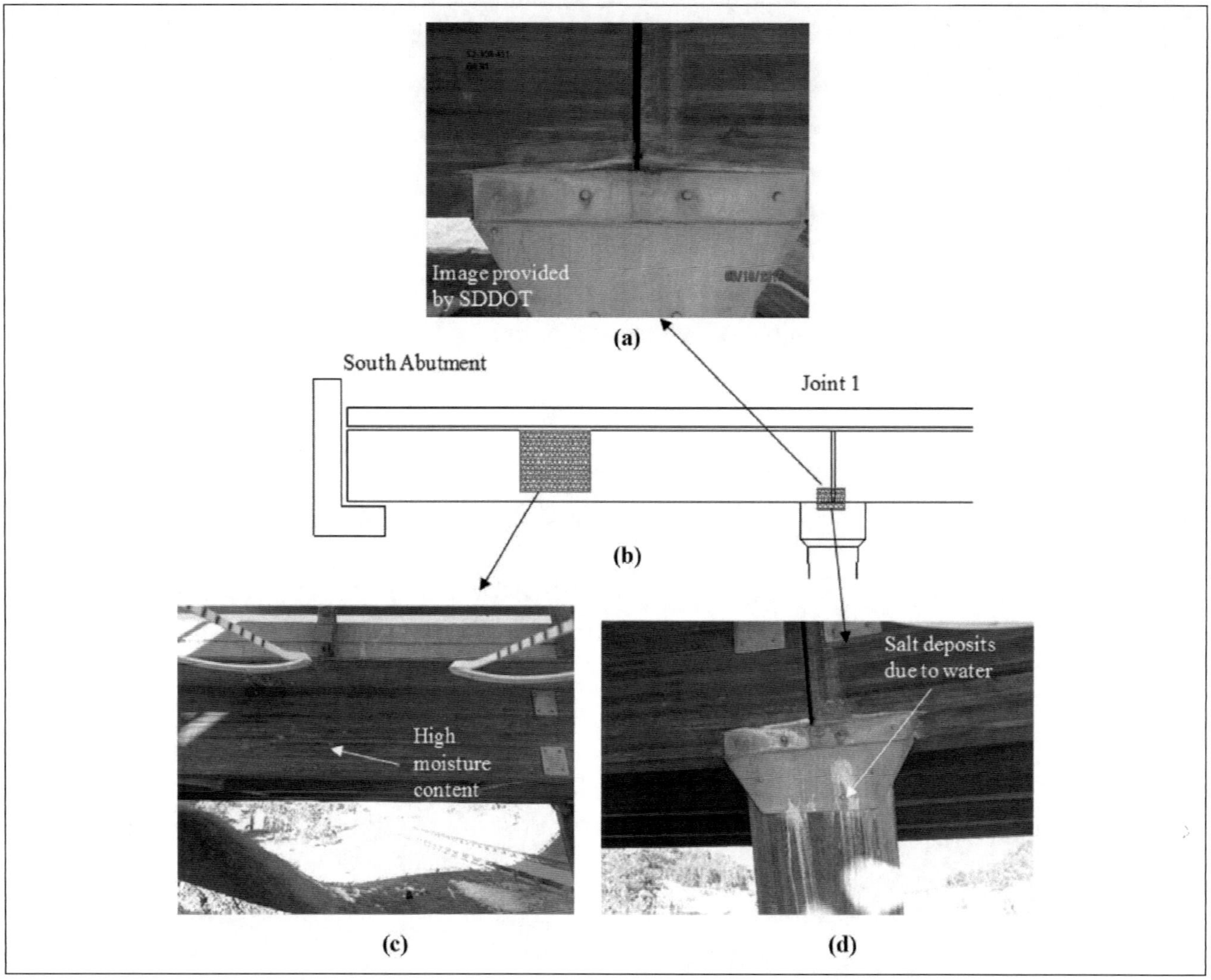

Figure 8. Sample damage on girder: (*a*) image provided by South Dakota DOT (*13*), (*b*) damage location, (*c*) some moisture on the side of Girder 4 between South Abutment and Joint 1, and (*d*) salt deposits caused by water coming from the deck at the support of Girder 4 at Joint 1.

(2-in.) surface scratch on the bottom of Girder 1 between Joints 1 and 2 which was also measured using the data from the UAS.

Conclusions

The following conclusions can be drawn from the literature review, UAS selection, and UAS-based inspection of the selected bridge located in South Dakota.

1. UASs present an efficient alternative for bridge inspection because of their high-resolution cameras and increasing technology development (e.g., OA).
2. Appropriate weather conditions must be considered to safely operate the UAS. For example, calm wind conditions with wind speeds below 15 mph.
3. The UAS used for the selected bridge was able to perform an underside of deck inspection without GPS signal failure. Also, the high-resolution camera mounted on the UAS allowed for a detailed damage identification under the deck while tilted up at an angle of 35 degrees.
4. The UAS was effective in identifying different types of bridge damage, especially for concrete cracks, spalling, and moisture on concrete decking, and salt deposit and moisture on timber glulam girders.
5. The comparison of images captured from the UAS-enabled inspection and traditional inspection methods demonstrated the accuracy of the damage identification using UAS data.
6. The use of UASs can minimize the risk of injury to inspectors, helping to reduce safety concerns during field inspection work. In addition, the UASs allow inspectors to view areas of the bridge not visible during routine bridge inspections.

Challenges

During this study, several limitations, including light conditions on the underside of the deck, high-contrast on sunny days, and some obstacles under the bridge deck, were identified. Such challenges could be overcome by stringent flight and inspection planning to avoid undesirable image overexposure, and attaching additional flashlights to the UAS to enhance the illumination under the deck. It is expected that the DOT will incorporate the use of UASs to supplement routine inspections in the near future.

Acknowledgments

Financial support for this research was provided by the Federal Highway Administration through the Forest Products Laboratory (USDA-Forest Service). The assistance and cooperation of the South Dakota DOT is gratefully acknowledged.

References

1. Eschmann, C., C.-M. Kuo, and C. Boller. Unmanned Aircraft Systems for Remote Building Inspection and Monitoring. *Proc., 6th European Workshop on Structural Health Monitoring*, Dresden, Germany, Vol. 2, 2012, pp. 1–8.
2. Koonce, J., T. Demski, M. Rowe, and N. Morriss. *Bridge Inspection Access to Minimize Operational Impacts*. Research Project, Final Report 2015-40. Collins Engineers Inc., Chicago, IL, 2011.
3. Chen, S., C. Rice, C. Boyle, and E. Hauser. Small-Format Aerial Photography for Highway-Bridge Monitoring. *Journal of Performance of Constructed Facilities*, Vol. 25, No. 2, 2011, pp. 105–112.
4. Chen, Z., and T. C. Hutchinson. Image-Based Framework for Concrete Surface Crack Monitoring and Quantification. *Advances in Civil Engineering*, Vol. 1, 2010, pp. 1–18.
5. Khaloo, A., D. Lattanzi, K. Cunningham, R. Dell'Andrea, and R. Mark. Unmanned Aerial Vehicles (UAV) Inspection of the Placer River Trail Bridge through Image-Based 3D Modeling. *Structure and Infrastructure Engineering*, 2017. http://dx.doi.org/10.1080/15732479.2017.1330891.
6. ASCE Infrastructure Report Card. *ASCE News*, 2016.
7. Moller, P. *CALTRANS Bridge Inspection Aerial Robot*. Publication Report No. CA08-0182. Caltrans, 2011.
8. Otero, L. D. *Proof of Concept for using Unmanned Aerial Vehicles for High Mast Pole and Bridge Inspections*. Publication Final Report, DOT, Florida, 2015.
9. Lovelace, B., and J. Zink. *Unmanned Aerial Vehicle Bridge Inspection Demonstration Project*. Publication Report No. MN/RC 2015-40. DOT, Minnesota, 2015.
10. *DJI Phantom 4 Specifications*. DJI Phantom Company. www.dji.com/product/phantom-4/info#specs. Accessed May 23, 2017.
11. Duque, L., J. Seo, and J. Wacker. Synthesis of Unmanned Aerial Vehicle Applications for Infrastructures. *Journal of Performance of Constructed Facilities*, In press.
12. Seo, J., J. Wacker, and L. Duque. Evaluation of Unmanned Aircraft Systems as a Bridge Inspection Tool. U.S. Department of Agriculture, Forest Service, Forest Products Laboratory, Madison, WI. (currently under review).
13. South Dakota DOT. *Bridge Inspection Report*, South Dakota DOT, USA, 2016.

The Standing Committee on Structures Maintenance (AHD30) peer-reviewed this paper (18-00122).

Article

TRR
JOURNAL OF THE TRANSPORTATION RESEARCH BOARD

Transportation Research Record
2018, Vol. 2672(12) 82–92

DOI: 10.1177/0361198118781396
journals.sagepub.com/home/trr
SAGE

Life Cycle Assessment Framework for the U.S. Bridge Inventory

Saeed Babanajad[1], Yun Bai[1], Helmut Wenzel[2], Moritz Wenzel[3], Hooman Parvardeh[1], Ali Rezvani[4], Robert Zobel[5], Franklin Moon[1], and Ali Maher[1]

Abstract

The effective management of bridges requires a good understanding of their life expectancies. Improved prediction of bridge service life is required to be developed in order to better understand bridge deterioration and to find more effective maintenance and repair strategies. These models are integral components of the Long-Term Bridge Performance Program (LTBP), a 20-year research effort initiated by the U.S. Federal Highway Administration (FHWA) to improve the understanding of bridge performance. In this paper, the development of a life expectancy model framework, as part of the research effort in this program, is presented. The framework is established based on a semi-probabilistic approach to adherently maintain the advantages of both deterministic and probabilistic techniques. The modeling follows a step-by-step process which incorporates data collected from historical records, training the data, creating a model based on the most suitable approach, and reducing the associated uncertainties. The basic model is first trained by the network of bridge inventory and the uncertainties are reflected by determining lower and upper margins. Then the model is improved by introducing the new knowledge gained from the external attributes influencing the structure. Finally, the condition states of the bridge components are employed directly to refine the model for realistic assessment. The developed model is later automated into the Bridge Portal, the main core of the bridge-performance data warehouse. A detailed example using the Mid-Atlantic cluster bridge inventory data is presented in this paper to illustrate the application of the method described above.

There are over 614,000 aging bridges in the United States with an average age of 38 years. Due to the economic hardships facing federal and local agencies, it is inevitably critical to extend the lifespan of the nation's bridges. Accurately predicting the future conditions of bridge facilities is a key input to effective maintenance and rehabilitation decision making for bridge owners. Indeed, the precise forecasting of bridge behavior creates a twofold benefit for the infrastructure owners: to efficiently manage the available budget for rehabilitation or reconstruction, and to properly monitor the bridge's functionality in both short- and long-term horizons. One way to predict future conditions of bridges is to utilize infrastructure deterioration models. Deterioration is a stochastic and complex process that varies widely depending on various factors, many of which are generally not captured by available data (*1*). Three different approaches have been routinely taken by researchers to simulate bridge deterioration, including deterministic, probabilistic, and artificial intelligence (AI). Review of the technical literature reveals that each technique has its own advantages and disadvantages that eventually lead to failure in practical responses for deterioration modeling purposes. Deterministic models basically correlate age and a few other parameters with the component's condition using a simple mathematical formulation, such as the mean, standard deviation, and regression (*2*). Despite the ease of model development and interpretation, deterministic models neither account for the stochastic characteristics of bridge deterioration process nor consider the effects of unobserved explanatory variables (*3*, *4*). In contrast with deterministic models, probabilistic deterioration models view bridge deterioration as a stochastic process that is

[1]Center for Advanced Infrastructure and Transportation (CAIT), Rutgers University, Piscataway, NJ
[2]Wenzel Consulting Engineers GmbH, Vienna, Austria
[3]University of Trento, Trento, Italy
[4]Moffatt & Nichol, New York, NY
[5]Turner Fairbanks Highway Research Center, Federal Highway Administration (FHWA), McLean, VA

Corresponding Author:
Address correspondence to Saeed Babanajad:
saeed.babanajad@rutgers.edu

affected by various external parameters. The probabilistic models rely on transition probabilities to capture the nature of the evolution of condition states from one discrete time point to the next. Despite the obvious advantages of being able to capture uncertainties, these models suffer from underlying limitations. Several issues include difficulties in estimating transition probabilities, incomplete capture of various explanatory variables, complex computations, lack of historical condition data for training, and data censorship. Alternatively, AI models have also been proposed. AI models utilize modern computer techniques to automate intelligent data "learning" process of bridge deterioration behaviors, such as artificial neural networks (ANN) (*5*) and machine learning (ML) (*6*). These techniques are mostly inspired by natural rules and present the solutions based on experience and development of various discriminators. In spite of the novelties, there are major drawbacks which disable these models from creating sufficient knowledge and transparency. As a result, those models act as black boxes and cannot explicitly provide a transparent function correlating the output to the given inputs. The computations must be conducted in a-priori format requiring significant trial-and-error operations. In essence, the consequences of the limitations associated with the various methods discussed above will directly lead to poor predictions of future infrastructure condition, thus compromising maintenance and rehabilitation decision making.

In 2008, the U.S. Federal Highway Administration (FHWA) launched its largest bridge research effort, the Long-Term Bridge Performance (LTBP) Program. The LTBP Program is a 20-year research endeavor which collects scientific performance field data from a representative sample of bridges across the U.S.A. This will assist the bridge community to better understand bridge deterioration and performance. The output of this program includes a collection of data-driven decision-making tools, including predictive deterioration and life cycle analysis models that will assist bridge owners with their bridge management. In a collaboration with the individual campaign from Wenzel Consulting Engineers GmbH Consult under the course of the Integrated European Industrial Risk Reduction System (IRIS) project, a novel life cycle assessment framework has been developed. The objective of the research described herein is to present the backbone of the proposed framework. The methodology is established on the basis of a semi-probabilistic approach to adherently maintain the advantages of both deterministic and probabilistic techniques. The model is organized in a step-by-step structure to incorporate the associated uncertainties. The basic model is first trained by the user-specified network of bridge inventory and the uncertainties are reflected by determining lower and upper margins. In the second step, the model is improved through introduction and evaluation of new knowledge gained about the external attributes influencing the structure. Finally, the direct condition of bridge components is employed straightforwardly to refine the model for realistic assessment. The developed model is later automated into the LTBP Bridge Portal, the main core of the bridge-performance data warehouse. The Bridge Portal was primarily developed by LTBP to act as an intelligent web platform containing data sets including the National Bridge Inventory (NBI), the American Association of State Highway and Transportation Officials (AASHTO) bridge elements, traffic, environmental parameters, inspection reports, and maintenance data as well as data acquired through LTBP Program field testing. The LTBP Bridge Portal is publicly accessible at www.fhwaapps.fhwa.dot.gov/ltbpp. In the following section, the core elements of the developed model for life cycle assessment are introduced, followed by a case study for completeness.

National Bridge Inventory (NBI) Database

To unify the database system and information regarding bridges nationwide, the FHWA has routinely compiled the NBI since 1972. The NBI is an extensive database system relating to the critical characteristics of bridges, for which data is continuously collected by states and bridge owners. The NBI mainly includes structural identification information, bridge types and specifications, operational conditions, geometric data with functional descriptions, and inspection data (*7*).

Proposed Methodology

The formulations described in this paper pertain to the long-term prediction of a bridge's NBI general condition rating as a whole (including deck, superstructure, and substructure). The model is established on a multi-level approach that incorporates multiple structural and external attributes. The general model is derived on the base of the general deterioration process of the bridges in the inventory and provides a rough estimation of the expected structural service life. In the second step, an adapted model considers certain structural and external (environmental and traffic) information and adjusts the lifeline progression. Finally, in the last step, the model is refined based on performance-specific information, including but not limited to non-destructive evaluation (NDE), structural health monitoring (SHM), and/or advanced inspections.

Global Model

In general, the structural assessment has been done first and foremost by compulsory visual inspections. For critical cases, in-depth studies are initiated, mostly in terms of numerical simulations or SHM. To merge the results from all individual investigations and to obtain a more detailed picture of structural performance over the entire service life, Equations 1 and 2 have been utilized, as follows:

$$h_i^t = h_1 + a_n(S_i - S_1)^c \quad (1)$$

$$a_n = \frac{h_f - h_1}{(S_f - S_1)^c} \quad (2)$$

where h_i^t = current condition; h_1 = initial condition; h_f = final condition; S_i = current year of service life; S_1 = initial year of service life condition; S_f = final year of service life; c = deterioration power exponent which is a constant empirical values derived from sensitivity analysis and is taken to be 3 for bridge components; and a_n = slope of deterioration.

The model was initially established by Miyamoto and Ishida (*8*) and was later adapted by Wenzel et al. (*9*); it is shown schematically in Figure 1. In principle, this model covers all the major sources of deterioration and describes the lifeline progression within a stated service life expectancy. The primary goal of the global model is to address the overall performance of the bridge in a long-term horizon. In other words, for a given bridge, the model could not be employed for short-term predictions as it needs more detailed information associated with the bridge itself. The ranges for service life expectancy and the total deterioration capacity must be known in advance. To estimate the a_n and S_f, a large number of NBI general condition ratings from the selected bridge inventory have to be analyzed thoroughly.

The input terms in Equations 1 and 2 are constants that embody the performance specifications of each specific bridge. In essence, the c and a_n parameters define the general shape and rate of the deterioration curve in terms of bridge age and $m\,(=S_f - S_1)$ indicates the average life expectancy of bridges in the cluster. The accuracy of the model will depend on the accurate determination of the discussed inputs. To accurately compute the above parameters, it is necessary to calibrate them through a large-scale database with sufficient numbers of bridge records. During the calibration process, the input parameters are determined by defining an error function (ψ), which represents the difference between measured (historical NBI general condition rating records) and theoretical condition ratings. ψ is minimized with respect to m for which the minimization function is given below:

$$\psi = \{\sum_{i=1}^{n} [h_i^r - h_i^e]^2 \quad (3)$$

where ψ is the objective function, and h_i^r and h_i^e are the ith recorded and estimated condition ratings, respectively. n denotes the total number of available data points from the selected bridge inventory. Once m is computed during the calibration process, the general life cycle expectancy curve can be back-quantified by using Equation 1. The acquired model will then act as a general representative of the bridge inventory used throughout the calibration. The lower and upper boundaries are further determined on the basis of 5% and 95% percentiles. The boundaries ensure that 90% of the bridges perform within the prediction range and eliminate the outliers that could be caused by inspection errors or a special type of bridge, such as signature cable-stayed bridge. According to the initial trial-and-errors, the lower and upper boundaries are set to 0.5 m and 1.33 m, respectively. It basically means that 90% of bridges reach the end of their service life (condition rating 4) sometime between 0.5 m and 1.33 m of their age. For example, given m equals 64 years (mean value), 90% of bridges reach the end of their service life at some age between 32 and 85 years.

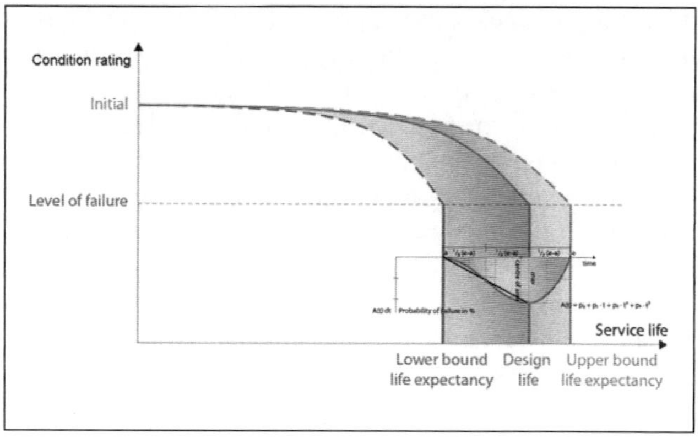

Figure 1. General concept of structural aging (*9*).

Adapted Model

The general model generates the mean result derived from the benchmark data. However, the model does not encompass different characteristics of individual bridges. Among others, external attributes such as environmental conditions or traffic volume, as well as the bridge type and design specifications, can significantly alter the performance of the bridge from the average expected life. Similar to Equation 1, Equations 4–6 provide the fundamental formulations for calculating the lower, mean, and upper boundaries by means of statistical analysis using probability density function. More details for the basics of formulation can be referred to in Wenzel et al. (*9*)

$$h_t^L = 9\left(1 - \left(\frac{t - Yr_{Built}}{a'}\right)^3\right) \quad (4)$$

$$h_t^m = 9\left(1 - \left(\frac{t - Yr_{Built}}{m'}\right)^3\right) \quad (5)$$

$$h_t^U = 9\left(1 - \left(\frac{t - Yr_{Built}}{e'}\right)^3\right) \quad (6)$$

where Yr_{Built} corresponds to the year of bridge construction or reconstruction, whichever is the most recent. h_t^u, h_t^m, and h_t^l denote condition rating corresponding to upper, mean, and lower boundaries, respectively. a', m', and e' indicate the slopes of deterioration curves and can be determined by the following expressions:

$$a' = a.K \quad (7)$$

$$m' = a' + 0.6(e' - a') \quad (8)$$

$$e' = e.K \quad (9)$$

where m', a', and e' correspond to the adapted mean, lower, and upper bound life expectancies, respectively. As noted earlier, the values of a and e are estimated by assuming $0.5m$ and $1.33m$, respectively, where m indicates the average expected life. Computation of Equations 7–9 requires the determination of the K parameter which is a constant value signifying different characteristics of the bridge and the environment where the bridge is located. The calculation of adjustment factor, $K(= k_1.k_2.k_3.k_4.k_5.k_6)$, corresponds to the determination of $k_1 \ldots k_6$ and is thoroughly discussed in a later section of this paper. k_1 denotes the variation in average life expectancy based on the year of construction or reconstruction. k_2 and k_3 demonstrate the expected service life on the basis of design and materials types, respectively. Alternatively, $k_4 - k_5$ and k_6 respectively quantify the effects of traffic load and environmental condition where the bridge is placed. Similar to the general model, the accuracy of the adapted model will depend on the accurate determination of the K factor. A statistical approach has been utilized to individually compute each of the k_i factors during the calibration process. Throughout the process, for a given k_i parameter, all other k_i parameters were kept as constants and non-linear optimization method has been employed to minimize the error function (ψ), as follows:

$$\psi = \{\sum_{i=1}^{n} [h_i^r - h_i^e]^2 \quad (10)$$

where, ψ is the objective function, and h_i^r and h_i^e are the ith recorded and estimated condition ratings, respectively. n denotes the total number of available data points from the selected bridge inventory. To calculate h_i^e, the expression established by Equation 5 has been employed.

Determination of Adjustment Factor (K). In the proposed adapted model, two major types of data have been incorporated into the modeling to customize the general model for individual bridges. The first type corresponds to the structural, material, geometrical, and construction properties of the bridge. It primarily includes:

k_1 **Year of construction/reconstruction:** embodies the periodic construction quality changes imposed throughout the last decades. Design specifications and construction practices have been extensively modified in the United States with two important milestones. The first period includes the time before 1975 in which bridges were designed and built by traditional code specifications and construction techniques ($k_1 = 0.90$). Around 1975, new codes were released to consider incorporating the concepts of earthquake and performance-based designs into actual bridge projects. The second period pertains to the years between 1975 and 1985, sometimes called the transition decade. Throughout this period, new materials were introduced to the construction industry as well as the revised design and construction codes being employed by the civil engineering community ($k_1 = 0.95$). After 1985, it may be postulated that the design codes and construction periods have reached a stable plateau where the inconsistency between construction and design is minimal ($k_1 = 1.0$).

k_2 **Type of design:** reflects the associated redundancies imposed by different design portfolios (Table 1).

k_3 **Types of Materials:** different materials are employed for bridge fabrication, including but not limited to reinforced concrete, steel, pre/post-tension, etc. The functionality and expected service life depend significantly on the types of materials being used for the construction of deck and superstructure. Table 1 demonstrates the final values of k_3 designated for each type of material mainly used by the bridge industry. It should be noted that the given k values were computed using a clustered dataset (case study), however, those could be altered based on the agency's dataset.

The second type of data used in adjusting the adapted model is the external attributes influencing the bridge's performance, including environmental conditions and traffic volume.

$k_4 - k_5$ **Weather and Environmental Characteristics:** To evaluate the environmental effects on bridge performance, two weather data sets (numbers of snowfalls and freeze-thaw cycles) were extracted from the National Oceanic and Atmospheric Administration database. The number of recorded snowfalls and the temperature histograms (freeze-thaw) represent the average of the recorded values from nearby weather stations. The number of snow falls is determined based on the number of days in a year where snowfalls of more than 0.2 inches (0.5 mm) occurred. The number of freeze-thaw cycles is the number of days in a year for which the following conditions were met:

Table 1. Designation of k_2 and k_3

Type of design (k_2)		Type of material (k_3)	
Design description	k-factor	Design description	k-factor
Slab	1.05	Concrete	1.1
Stringer/Multi-beam or girder	0.95	Concrete continuous	1.1
Girder and floorbeam system	0.95	Steel	1
Tee beam	0.95	Steel continuous	1
Box beam or girders - Multiple	1	Prestressed concrete	0.9
Box Beam or girders - Single or Spread	1	Prestressed concrete continuous	0.9
Frame	0.95		
Orthotropic	0.95		
Truss - Deck	0.95		
Truss - Thru	0.95		
Arch - Deck	0.95		
Arch - Thru	0.95		

- A minimum temperature of $\leq -2.2°C/28.04°F$
- A maximum temperature of $\geq 0°C/32°F$

The following expressions direct the magnitude of snowfall and freeze-thaw cycles to the associated $k_4 - k_5$ factors:

$$K_4 = \begin{cases} -0.04 \times Snowfall + 1.2 & if\ Snowfall \leq 5 \\ 1.0 & if\ 5 < Snowfall \leq 15 \\ -0.025 \times (Snowfall - 15) + 1.0 & if\ 15 < Snowfall \leq 25 \\ 0.75 & if\ 25 < Snowfall \end{cases} \quad (11)$$

$$K_5 = \begin{cases} -0.00571 \times Freeze\text{−}thaw + 1.2 & if\ Freeze\text{−}thaw \leq 35 \\ 1.0 & if\ 35 < Freeze\text{−}thaw \leq 55 \\ -0.0167 \times (Freeze\text{−}thaw - 55) + 1.0 & if\ 55 < Freeze\text{−}thaw \leq 70 \\ 0.75 & if\ 70 < Freeze\text{−}thaw \end{cases} \quad (12)$$

k_6 **Traffic volume:** In general, daily traffic does not have a severe impact on the quality of the bridge, however, the truck traffic considerably degrades the bridge components. Overweight trucks and improper suspension systems are among the most critical effects causing steep degradation of the structural elements. To portray such effects, average daily truck traffic (ADTT) has been taken into account within the modeling scheme, as

$$K_6 = \begin{cases} -0.002 \times ADTT + 1.2 & if\ ADTT \leq 100 \\ 1.0 & if\ 100 < ADTT \leq 600 \\ -2.66E(-5) \times (ADTT - 600) + 1.0 & if\ 600 < ADTT \leq 10,000 \\ 0.75 & if\ 10,000 < ADTT \end{cases} \quad (13)$$

Refined Model

As more information becomes available about the structural condition of a bridge, predictions of the remaining service life have the potential to be more accurate by incorporating additional structural information into the life expectancy model. The campaign of detailed visual inspection, SHM, and non-destructive testing is an example of performance-specific information which could play a significant role in modifying the prognosis of remaining service life. The current study only pertains to the formulation of non-destructive testing results into the life expectancy model. Currently, three different NDE tests have been implemented through the formulation, and the refined model remains capable of incorporating additional tests. As pinpointed by Gucunski et al. (*12*), deck is the most vulnerable bridge component due to its direct exposure to traffic, environmental loading, and application of deicing chemicals. Therefore, the NDE tests employed concentrate on assessment of bridge deck condition via three main components: evaluation of the corrosive environment and corrosion processes, evaluation of concrete degradation, and assessment with respect to deck delamination. To do so, three complementary NDE tests have been selected so far to be included in the refined modeling scheme. These tests include: impact echo (IE) to detect and characterize delamination, ground-penetrating radar (GPR) to describe the corrosive environment, measurement of concrete cover and description of its overall condition, and electrical resistivity (ER) to estimate the existing corrosion rate. In order to capture the results of the NDE tests quantitatively, Guncuski et al. (*10*) defined four different condition indexes for IE, ER, GPR, and Combined-NDE on the scale of 0 (worst) to 100 (best). In essence, the established conversion builds a unified and scale-less determination factor of deck condition assessment no matter the NDE test type. The following sub-sections yield the relationships used to determine the condition indexes.

Impact Echo (IE). IE is the primary NDE technique used for detection and characterization of deck delamination. The test is conducted using an impact source and a single nearby receiver. Depending on the signal spectrum and energy perturbation, the condition is described as good, fair, poor, or serious (*10*). The condition index with respect to delamination was calculated according to:

$$\text{IE index} = \frac{A_{Good} \times 100 + A_{Fair} \times 50 + A_{Poor} \times 50 + A_{Serious} \times 0}{A_{Total}} \quad (14)$$

where A_{Good}, A_{Fair}, A_{Poor}, and $A_{Serious}$ = areas in good, fair, poor, and serious conditions. A_{Total} corresponds to the total surveyed area.

Ground-Penetrating Radar (GPR). GPR technology has been conventionally implemented as part of the concrete quality assurance. The GPR can be described as providing a qualitative assessment of concrete condition, describing potential concrete deterioration, delamination, and corrosive environment (*11*). The GPR assessment is based on the measurement of attenuation of electromagnetic waves as they propagate through concrete layers. The condition index from the GPR surveys is calculated according to:

$$\text{GPR index} = \frac{A_{Good} \times 100 + A_{Fair} \times 70 + A_{Poor} \times 40 + A_{Serious} \times 0}{A_{Total}} \quad (15)$$

Where A_{Good}, A_{Fair}, A_{Poor}, and $A_{Serious}$ = areas with GPR signal attenuation (normalized dB) ranges of > −15, −15 to −17, −17 to −20, and < −20, respectively. A_{Total} corresponds to the total surveyed area.

Electrical Resistivity (ER). ER technology has been used as to assess corrosion inside concrete specimens. Low resistivity of concrete is often a consequence of high moisture level and chloride concentration, while high electrical resistivity allows only a low corrosion current to pass between the anodic and cathodic areas of the reinforcing steel and thus does not facilitate corrosive processes. To quantify the overall deck condition with respect to the corrosive environment and anticipated corrosion rate, the following expression has been introduced:

$$\text{ER index} = \frac{A_{Low} \times 100 + A_{Moderate} \times 50 + A_{High} \times 0}{A_{Total}} \quad (16)$$

where A_{High}, $A_{Moderate}$, and A_{Low} = areas with ER in ranges of <40 kΩ.cm, 40 to 70 kΩ.cm, and >70 kΩ.cm, respectively. A_{Total} corresponds to the total surveyed area.

Combined-NDE Index. To tandem all different NDE test results into a single indicator, the following expression has been introduced to aggregate a weighted average:

$$\text{Combined-NDE index} = \frac{IE\ index \times W_{IE} + ER\ index \times W_{ER} + GPR\ index \times W_{GPR}}{W_{IE} + W_{ER} + W_{GPR}} \quad (17)$$

Where *IE index*, *GPR index*, and *ER index* denote the calculated indexes from Equation 14 through Equation 16, respectively. W_{IE}, W_{GPR}, and W_{ER} assign the weights defined based on the significance of a particular type of deterioration in the bridge management practices of the bridge owner.

Aggregation of NDE Results and Life Cycle Model. For a given year of bridge testing, the Combined-NDE index is calculated by employing Equation 17. The following formulas are then implemented for estimation of mean, and lower and upper boundaries,

$$h_t^L = 9\left(1 - \left(\frac{t - Yr_{Built}}{a''}\right)^3\right) \quad (18)$$

$$h_t^m = 9\left(1 - \left(\frac{t - Yr_{Built}}{m''}\right)^3\right) \quad (19)$$

$$h_t^U = 9\left(1 - \left(\frac{t - Yr_{Built}}{e''}\right)^3\right) \quad (20)$$

where m'', a'', and e'' are explicitly estimated by the following expressions

$$m'' = \left(Yr_{Testing} - Yr_{Built}\right).\left(\frac{9}{SR}\right)^{1/3} \quad (21)$$

$$a'' = m'' - \varepsilon.\sigma \quad (22)$$

$$e'' = m'' + \varepsilon.\sigma \quad (23)$$

$$SR = (-0.075) \times (Combined\text{-}NDE\ index) + 1.5 \quad (24)$$

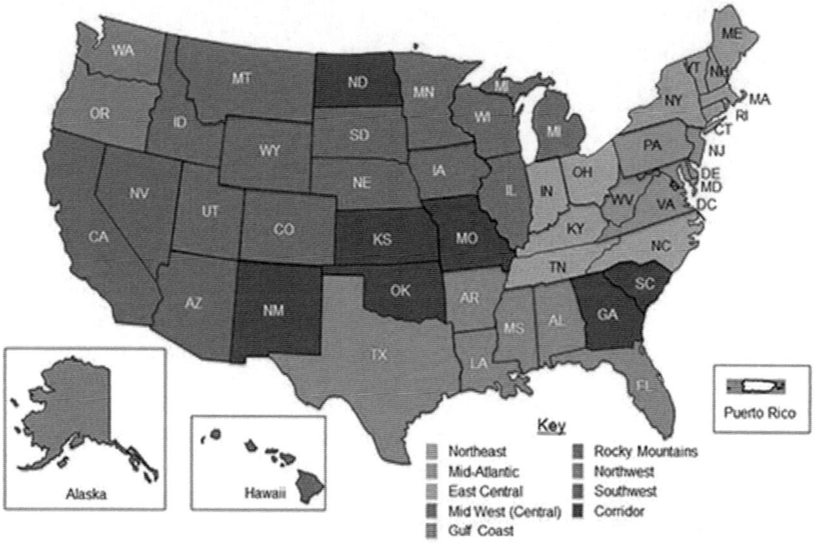

Figure 2. Classification of LTBP Program clusters.

For a given year of NDE testing ($Yr_{Testing}$), SR denotes the Combined-NDE index which is back-calculated by using Equation 24. Yr_{Built} corresponds to the year of bridge construction or reconstruction, whichever is the most recent. The ε and σ are statistical components mapping the outliers of lower and upper boundaries where $\sigma =$ is a constant determined by multiplication of $|m''|$ and V, where V equals 0.11 (refer to (9) for details). Similarly, ε is a constant equal to 1.645, signifying the 5% to 95% percentiles of the bridge inventory.

Results and Discussion

Case Study

As per the requirements of the LTBP Program, a representative sample of bridges nationwide are periodically field-tested to ensure the collection of high-quality performance data over an extended period of time. In this program, the U.S.A. is divided into multiple clusters and corridors based on their common environmental conditions (Figure 2).

Seven pilot bridges and twenty-four bridges from the Mid-Atlantic cluster were field-tested in 2013, 2015, and 2017. In the work presented here, one of the surveyed bridges, located in the Mid-Atlantic cluster, is selected as the sample bridge for this case study. The bridge was constructed in 1973 and comprises of a two-span continuous steel girder superstructure with monolithic reinforced concrete deck, each span is 13.3 m (44 ft) long, with a total structure length of 27.7 m (91 ft). According to historical data, the bridge deck received an NBI rating of 6 during the 1997–2016 period. The average daily traffic (ADT) and ADTT rates in the prior five years were an average of approximately 8,000 and 640, respectively.

Figure 3 illustrates a projected 85-year life expectancy curve for the surveyed bridge, derived from the LTBP Bridge Portal web-based platform. The developed general model starts from a basic model reflecting the expected global service life. The calibration of Equation 3 on the entire inventory of the state (e.g., State of Maryland), where the surveyed bridge is located, reveals that the average, lower, and upper life expectancies fall around 64, 32, and 85 years, respectively. The general model generates a very rough estimation of bridge life cycle trend for the entire bridge inventory.

While the general model is remarkable in concept to independently compare two different entities, it does not provide a deep insight into the individual bridges. It considers neither the influence of environmental conditions nor traffic volume, or the bridge type and design specifications. To compensate for such circumstances, Figure 4 demonstrates the adapted model derived for the surveyed bridge. By completing the calibration, the k_1 through k_6 parameters are computed to be 0.9, 0.95, 1.0, 1.0, 0.89, and 0.99, respectively. As formulated earlier, the K parameter was then obtained by multiplication of all k_i parameters, equal to 0.75. Subsequently, the estimated K parameter (smaller than 1.0) decreases the expected average life by 25%, pointing out that additional information enhances the prognosis of remaining service life. Figure 4 concurrently shows the general model (in black) and the adapted model (in green) for the selected bridge. It is clear that the estimated performance of the bridge based

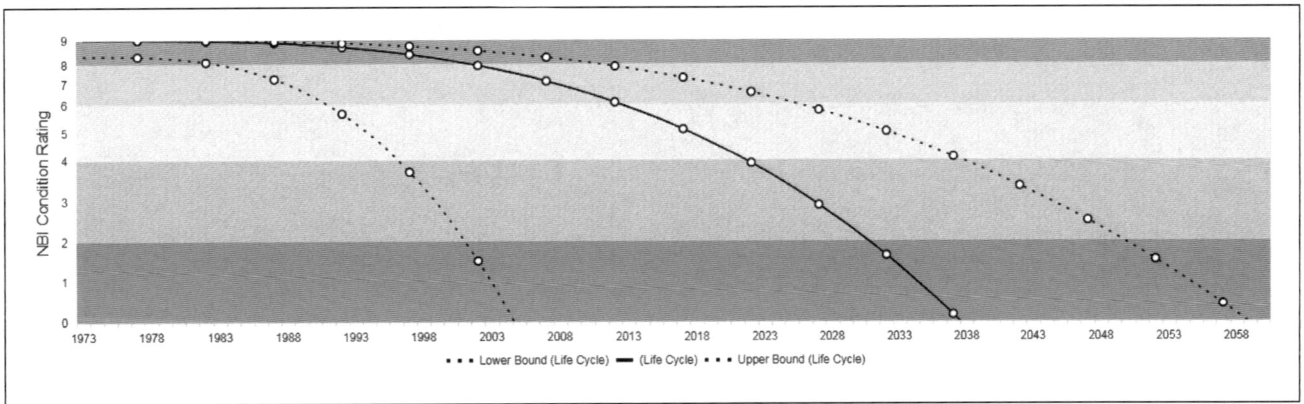

Figure 3. Theoretical global life cycle curve.

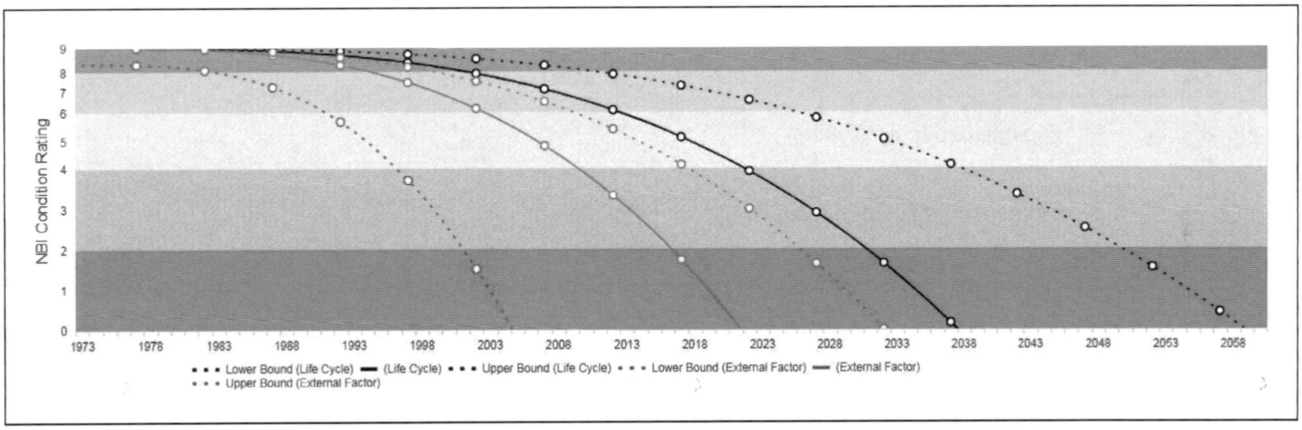

Figure 4. Theoretical global and adapted life cycle curves.

on the adapted models is dramatically worse due to a higher ADTT volume as well as higher rate of freeze-thaw cycles.

In contrast to the surveyed bridge used in this case study, it is possible to obtain a K factor larger than 1.0, meaning an extended expected life. In principle, the precise prediction of the adapted model has a twofold benefit. First, if the bridge is recognized as having an over-performance capacity, the associated maintenance and preservation costs would reduce. Second, in the case of an under-performance capacity or any extreme environmental conditions, the bridge will be red-flagged and must be meticulously evaluated.

Multiple advanced nondestructive evaluation (NDE) testing methods are conducted on the surveyed bridge to detect flaws and instances of delamination inside the structures due to fatigue, corrosion, overloads, and/or environmental conditions (*10*). As an example, Figure 5 represents the heat maps for different NDE tests conducted in 2013. Comparing the heat maps for 2013 and 2015 reveals that delamination and degradation progressions demonstrate a clear transition from dominantly sound areas in 2013 to areas receiving fair to serious grades in 2015. The transition is quantified by the corresponding condition index, where the Combined-NDE index has been shifted from 57 in 2013 to 50 in 2015. While the routine visual inspection practices lack in-depth insight, the NDE results provide tangible ability to capture and objectively track the progression of deterioration with time. In essence, the issue is bold when the NBI general condition rating remains unchanged within the period from 1997 to 2016 for the case of current surveyed bridge.

To import the NDE results into the modeling framework, Equations 18–20 have been followed and back-calculated to refine the deterioration curve passing through the NDE-specified performance points. Figure 6 presents the refined model obtained based on the NDE test data collected in 2013. The curve is adjusted in such a way as to precisely reflect the performance of the bridge according to the calculated condition ratings obtained from NDE results.

Implementation on LTBP Bridge Portal

As discussed in the introduction to this paper, the LTBP Bridge Portal is a key product of the LTBP Program

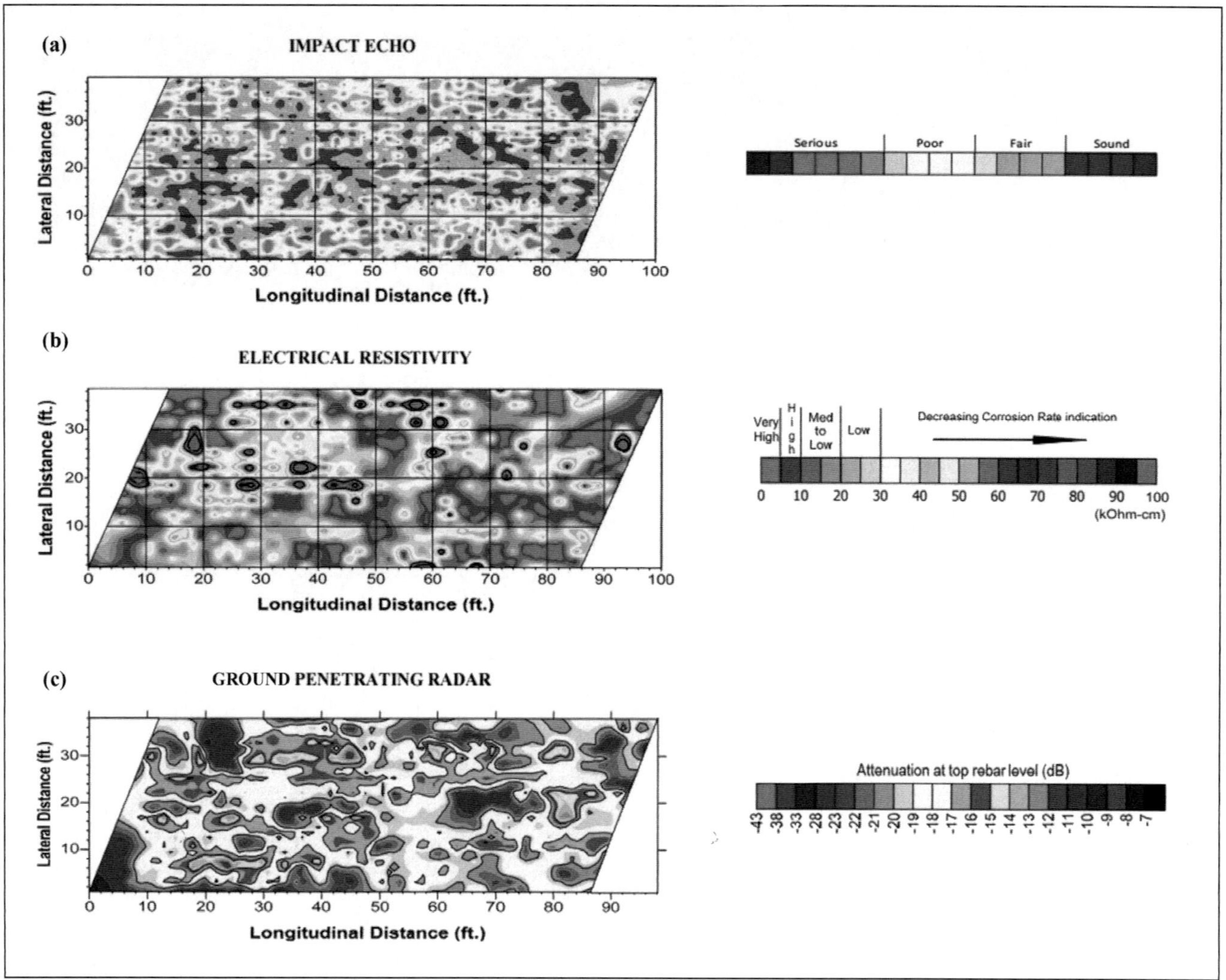

Figure 5. Samples of NDE heat maps conducted under LTBP Program.

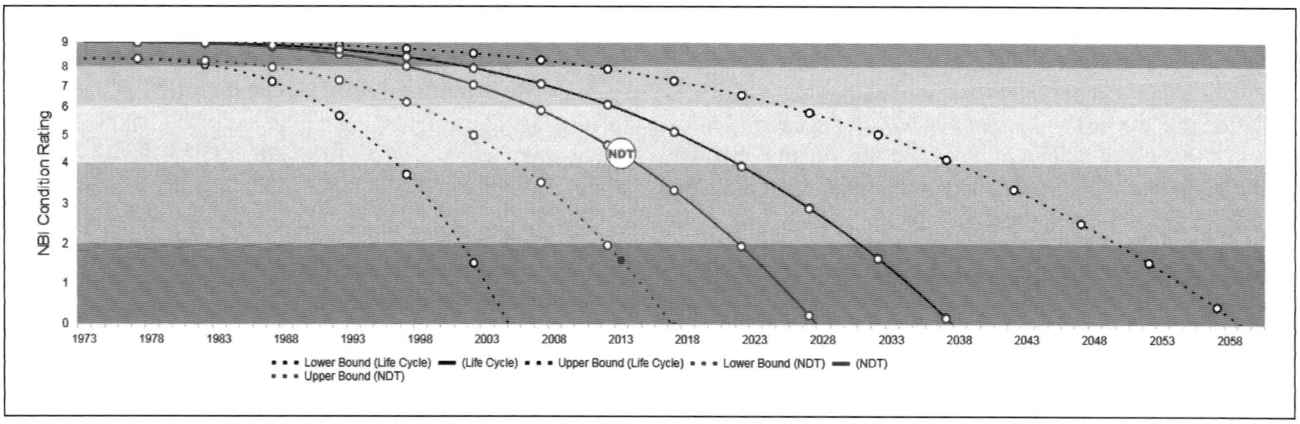

Figure 6. Theoretical global and refined life cycle curves.

containing bridge performance-related data mined from existing sources such as the NBI. The LTBP Bridge Portal offers multiple modules that enhance its capacities in terms of data management and knowledge of bridge performance. These unique Bridge Portal functionalities include deterioration modeling, data management

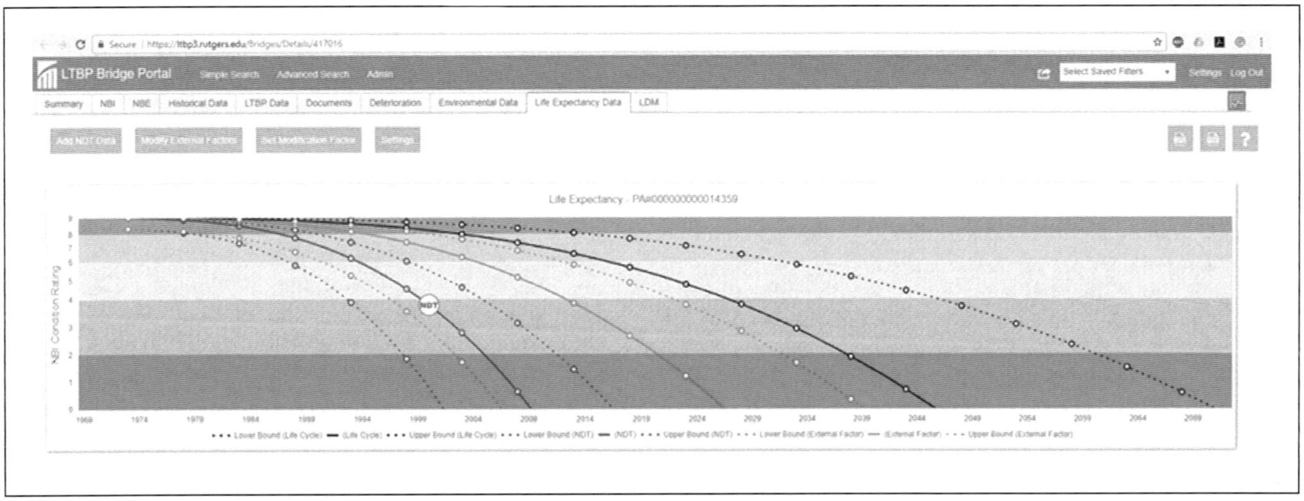

Figure 7. Snapshot view of the "Life Expectancy" interface on LTBP Bridge Portal.

toolboxes, and complementary set of features such as extraction and storage of raw data, search engine, geographical information system (GIS), multi-dimensional charts, and map chart, among others. In particular, it functions as a research and decision-making tool by the implementation of bridge deterioration modeling, based on mined data sources and LTBP-acquired field data. This database platform allows users to investigate bridge performance at many different levels. For a given bridge, the LTBP Bridge Portal browses a separate window which embodies multiple tabs, where the Life Expectancy tab provides the deterioration module developed in the current study (Figure 7).

Conclusion

In the research described in this paper, a detailed description of a step-wise service life expectancy modeling framework has been proposed for the U.S. bridge inventory. A basic model is first developed to reflect the general life expectancy without any specific information about a bridge, which is defined by the 50% (mean), 5% and 95% quantile service life of the entire inventory. An adapted model is then improved through introduction and evaluation of new knowledge gained about the external attributes influencing the structure. Ultimately, the refined model is further developed by incorporating performance-specific information such as more detailed visual inspection and non-destructive testing. An illustrative case using the Mid-Atlantic cluster bridge inventory data is developed to demonstrate the application of the method. The proposed model enables the consideration of advanced bridge performance information such as NDE to be directly routed into life expectancy estimation, eventually reducing uncertainties of prediction. The principle here is that as more and up-to-date information becomes available, that information can be incorporated into the model allowing for more accurate service-life predictions to be made.

For future research, it is recommended to incorporate more detailed information (regarding bridge components) into the modeling scheme. As more data becomes available, predictions of the remaining service life will be more accurate. For example, the AASHTO bridge element data could provide comprehensive information for every single element of the bridge. The campaign of advanced visual inspection, SHM, and non-destructive testing is another example of performance-specific (high resolution) information which could play a significant role in modifying the prognosis of remaining service life. Further sensitivity analysis is also required to assure that the proposed formulations, as well as the calibrated coefficients, are repeatable and reliable.

Acknowledgments

The current study has been funded under contract #DTFH61-08-C-00005, Federal Highway Administration, US Department of Transportation. The assistance of Aditya Chukka in the development of the Bridge Portal's life expectancy interface is greatly appreciated.

Author Contributions

The authors confirm the contribution to the paper as follows: study conception and design: S. Babanajad, H. Wenzel, A. Maher, R. Zobel, F. Moon; analysis and interpretation of results: S. Babanajad, Y. Bai, H. Parvardeh, A. Rezvani, M. Wenzel, H. Wenzel; draft manuscript preparation: S. Babanajad, Y. Bai. All authors reviewed the results and approved the final version of the manuscript.

References

1. Mishalani, R. G., and S. M. Madanat. Computation of Infrastructure Transition Probabilities Using Stochastic Duration Models. *Journal of Infrastructure Systems*, Vol. 8, No. 4, 2002, pp. 139–148. http://doi.org/10.1061/(ASCE)1076-0342(2002)8:4(139).
2. Liu, H. T. *Adaptive Optimization Methods in System-Level Bridge Management*. PhD dissertation. University of California at Berkeley, 2013.
3. Jiang, Y., M. Saito, and K. C. Sinha. Bridge Performance Prediction Model Using the Markov Chain. *Transportation Research Record: Journal of the Transportation Research Board*, 1988. 1180: 25–32.
4. Madanat, S. M., and W. H. Wan Ibrahim. Poisson Regression Models of Infrastructure Transition Probabilities. *Journal of Transportation Engineering*, Vol. 121, No. 3, 1995, pp. 267–272. https://doi.org/10.1061/(ASCE)0733-947X(1995)121:3(267).
5. Bu, G., J. Lee, H. Guan, M. Blumenstein, and Y.-C. Loo. Performance Prediction of Concrete Elements in Bridge Substructures using Integrated Deterioration Method. In *18th CONGRESS OF IABSE*, Seoul, South Korea, International Association for Bridge and Structural Engineering, 2012, pp. 1–8.
6. Tokdemir, O. B., C. Ayvalik, and J. Mohammadi. Prediction of Highway Bridge Performance by Artificial Neural Networks and Genetic Algorithms. *Proc., 17th International Symposium on Automation and Robotics in Construction (ISARC)*, Taipei, Taiwan, International Association for Automation and Robotics in Construction (IAARC), 2000, pp. 1–8.
7. U.S. Department of Transportation, Federal Highway Administration. *Recording and Coding Guide for the Structure Inventory and Appraisal of the Nation's Bridges*. Publication FHWA-PD-96-001. Office of Engineering, Bridge Division, U.S. Department of Transportation, Federal Highway Administration, Washington, D.C., 1995.
8. Miyato, A., and J. J. Ishida. Bridge Maintenance Strategy Based on Life Cycle Cost and Rebuild Cost Stabilization. In *Bridge Maintenance, Safety, Management, Health Monitoring and Informatics* (H.-M Koh and D. M. Frangopol, eds.), Taylor & Francis Group, London, UK, 2008.
9. Wenzel, H., R. Veit-Egerer, and M. Widmann. Life Cycle Analysis Applied to the New Jersey Case Study. In *Industrial Safety and Life Cycle Engineering (IRIS)* (H. Wenzel, ed.), VCE Vienna Consulting Engineers ZT GmbH, Vienna, Austria, 2013, pp. 443–464.
10. Gucunski, N., B. Pailes, J. Kim, H. Azari, and K. Dinh. Capture and Quantification of Deterioration Progression in Concrete Bridge Decks through Periodical NDE Surveys. *Journal of Infrastructure Systems*, Vol. 23, No. 1, B4016005, 2016, pp. 1–11. http://doi.org/10.1061/(ASCE)IS.1943-555X.0000321.
11. Maser, K. R., and A. Rawson. Network Bridge Deck Surveys Using High Speed Radar: Case Studies of 44 Decks (Abridgement). *Transportation Research Record: Journal of the Transportation Research Board*, 2012. 1992: 25–28.
12. Gucunski, N., G. R. Consolazio, and A. Maher. Concrete Bridge Deck Delamination Detection by Integrated Ultrasonic Methods. *International Journal of Materials and Product Technology*, Vol. 26, No. 1–2, 2006, pp. 19–34. https://doi.org/10.1504/IJMPT.2006.008978.

The Standing Committee on Bridge Management (AHD35) peer-reviewed this paper (18-05106).

TRR
JOURNAL OF THE TRANSPORTATION RESEARCH BOARD

Article

Transportation Research Record
2018, Vol. 2672(12) 93–102

DOI: 10.1177/0361198118796003
journals.sagepub.com/home/trr
SAGE

Updating Bridge Deck Condition Transition Probabilities as New Inspection Data Are Collected: Methodology and Empirical Evaluation

Rabi G. Mishalani[1], Abdollah Shafieezadeh[1], and Zequn Li[1]

Abstract

A Bayesian updating method is proposed to estimate a Markov chain based concrete deck deterioration model in a manner that combines condition data collected over two consecutive inspections and the deterioration information available prior to the collection of these data. A dataset of bridge deck condition assessments based on AASHTO condition state definitions collected by a state infrastructure agency spanning two years is used to evaluate the performance of this method. Training and validation datasets are selected from the original dataset where the former is used for estimation and the latter for prediction and evaluation. Single period transition probabilities are estimated using Bayesian updating, where prior deterioration information is combined with the condition data, and maximum likelihood estimation where only the collected condition data over two consecutive inspections are used. The evaluation is based on measuring the degree of similarity between reported condition states and those predicted based on the estimated transition probabilities using the two estimation methods. While updating transition probabilities as new data are collected is found to be advantageous for many cases, this advantage is highly dependent on the extent to which the training dataset is representative of the deterioration nature of the bridge decks for which condition is to be predicted. The less representative the training dataset, the more value is derived from Bayesian updating based predictions where prior deterioration information is considered.

Deterioration is among the primary concerns regarding the structural and functional performance of bridges and their components. Such degradations can potentially lead to various forms of structural deficiency and eventually failure if adequate repair and rehabilitation actions are not performed. Also, degraded bridge decks and the application of time-consuming repairs and rehabilitations lead to increased traffic delays and increased vehicle operating costs. In light of annual budget constraints, infrastructure agencies, such as state Departments of Transportation (DOTs) in the United States., prioritize their bridge maintenance needs. Commonly bridges with higher degradation levels receive higher priority for maintenance. To make trade-offs across bridges and over time, key inputs to the prioritization and decision-making process are bridge condition assessments and predictions.

To support such functionalities, the American Association of State Highway and Transportation Officials (AASHTO) developed the bridge management system AASHTOWare BrM (*1*)—formerly known as Pontis (*2*)—which has been successfully used in numerous practical decision-making settings. It is based on a discrete time Markov chain deterioration model in which the type and extent of deterioration of bridge components are expressed in visual terms. Critical parameters of Markov chain deterioration models are probabilities that describe the likelihoods of bridge components transitioning from one discrete condition state to another. These transition probabilities are either determined based on expert judgement or, ideally, estimated based on field inspection data. Numerous studies have focused on the estimation of transition probabilities for various infrastructure types including bridges, roadway pavements, and storm water and wastewater pipes. Examples include Carnahan et al. (*3*), DeStefano and Grivas (*4*), Jiang et al. (*5*), Madanat et al. (*6*), Madanat and Wan

[1]The Ohio State University, Columbus, OH

Corresponding Author:
Address correspondence to Rabi G. Mishalani: mishalani@osu.edu

Ibrahim (7), Madanat et al. (8), Micevski et al. (9), Mishalani and Madanat (10), Baik et al. (11), Kobayashi et al. (12), Tsuda et al. (13), Kobayashi et al. (14), Li et al. (15), and Chang and Maguire (16).

The AASHTO Bridge Element Inspection Guide Manual (17), subsequently referred to as the AASHTO Manual, establishes a national guideline for performance evaluation of bridge elements and collection of inspection data. In the manual, the state of performance of a bridge element is expressed as a condition state vector. For example, for bridge decks, the focus of this study, four condition states are defined based on the extent of different types of defects, such as cracking, spalls, and delamination. Condition state 1 represents excellent condition, where only hairline cracking may be present, and condition state 4 represents poor condition where extreme defects are present (17). For a given bridge, each entry of the condition state vector is the proportion of the total area of the deck surface that is in the corresponding condition state.

The AASHTO Manual, which is adopted by all state transportation agencies in the United States, requires regular observations and measurements of highway bridges to capture their physical and functional states. State DOTs usually carry out such bridge inspections once every one to two years, providing an opportunity to continually improve the estimates of bridge deterioration transition probabilities. However, typically, and as reflected in (3–16), resulting inspection data are used on a one-time basis to estimate deterioration models whereby subsequent observations are used only as inputs to predict deterioration. While Micevski et al. (9) employed an approach that updates prior information, this information is assumed to be uninformative (i.e., equally likely to take any possible set of values) rather than being derived from other sources. Durango and Madanat (18) addressed the notion of updating infrastructure deterioration estimates using an adaptive control framework. Kobayashi et al. (14) improved the Markovian bridge deterioration model proposed by Tsuda et al. (13) using sampling technique in the Bayesian estimation method. However, these studies are based on computation- and simulation-based analyses rather than empirical ones.

The objective of this study is to adapt a method that uses the bridge inspection data collected on an ongoing basis to update transition probabilities for concrete bridge decks. Doing so is expected to improve the representativeness of these probabilistic deterioration models and the accuracy of their predictions. An experimental design is developed and empirical data are used to demonstrate the transition probability updating method and to assess the value of such an approach in a manner that reveals any involved nuances.

Deterioration Model and Parameter Estimation

Markov Chain Deterioration Model

As noted above, based on the condition state definitions, each entry in the condition state vector represents the proportion of the total deck area that is in that condition state. This definition could be interpreted as the probability of each unit of the deck being in each condition state, whereby a deck is divided into a contiguous set of small 1 ft × 1 ft units forming a grid. A widely-used approach to model such deterioration is the Markov chain (19, 20). Markov chain models are often used to represent the evolution of discrete and finite states of variables at discrete periods (21). A mathematical representation of a Markov chain is $Y^{(t+1)} = \boldsymbol{P} \cdot Y^{(t)}$, where $Y^{(t)}$ and $Y^{(t+1)}$ are the vectors that represent the probability mass functions of the discrete state at time t and time $t+1$, respectively. The transition probability matrix $\boldsymbol{P}$ consists of transition probability elements p_{ij}. Each of these elements represents the probability of transitioning to state j at time $t+1$ given that the condition state is i at time t. The matrix $\boldsymbol{P}$ could be invariant or may vary over time. In this representation, the former is assumed.

This study uses the same Markov chain deterioration model that is adopted by the AASHTOWare BrM Bridge Management System whereby the following assumptions are made: (a) over a single time period the condition state either remains unchanged or drops at most by one (i.e., condition does not improve given that deterioration is being modeled and no multi-step transitions are assumed to occur), and (b) the transition probability matrix $\boldsymbol{P}$ is independent of age. As a result, the Markov chain model used in this study takes the following form:

$$Y^{(t+1)} = \boldsymbol{P} \cdot Y^{(t)} \tag{1a}$$

$$\begin{bmatrix} y_1^{(t+1)} \\ y_2^{(t+1)} \\ y_3^{(t+1)} \\ y_4^{(t+1)} \end{bmatrix} = \begin{bmatrix} p_{11} & 0 & 0 & 0 \\ 1-p_{11} & p_{22} & 0 & 0 \\ 0 & 1-p_{22} & p_{33} & 0 \\ 0 & 0 & 1-p_{33} & p_{44} \end{bmatrix} \cdot \begin{bmatrix} y_1^{(t)} \\ y_2^{(t)} \\ y_3^{(t)} \\ y_4^{(t)} \end{bmatrix} \tag{1b}$$

The only parameters to be determined are p_{11}, p_{22}, p_{33}, and p_{44}. In the case of estimation, $Y^{(t)}$ and $Y^{(t+1)}$ are obtained from data, and in the case of prediction, $Y^{(t)}$ is obtained from data and $Y^{(t+1)}$ is predicted based on Equation 1*a*.

Parameter Estimation Methods

Different statistical estimation methods could be used to estimate the transition probabilities as illustrated in the literature (3–16) whereby, as discussed above, the methods rely on a one-time estimation for a given condition dataset and, therefore, do not take advantage of condition data that stream in on a regular basis to update the estimates of the transition probabilities. In this study, Bayesian estimation is used to determine transition probabilities for concrete bridge decks while taking into account the available estimates determined from previous data (or other datasets, models, or expert judgement), thus, allowing for the incorporation of new inspection data to update the estimates of deterioration models as these data become available. For the purpose of evaluating the proposed Bayesian updating method, maximum likelihood estimation is employed as a reference whereby only inspection data over one period (i.e., two consecutive inspections) are used.

Notation

Given the at most one-step deterioration assumption introduced in the Markov chain model in this study, for each condition state i at time t only two transition probabilities govern the deterioration that takes place between time t and time $t + 1$, namely the transition probabilities p_{ii} and $p_{i\,i+1}$ indicating whether a unit of a deck remains in the same state i or transitions to the next lower state $i + 1$, respectively. Since and $p_{i\,i+1} = 1 - p_{ii}$, only p_{ii} needs to be estimated for each condition state i at time t. The variables $N_i^{(t)}$ and $N_i^{(t+1)}$ represent the area of the deck that is in condition state i at time t and the area that remains in condition state i at time $t + 1$, respectively.

Maximum Likelihood Estimation

The event that a unit of deck remains in the same condition state or not at the time of the next inspection can be described as a Bernoulli process, which has two outcomes, success and failure. Assuming that each 1 ft × 1 ft unit of bridge deck remains in the same condition state independently with the same probability p_{ii}, the maximum likelihood estimate of p_{ii} can be shown to be the following—see (*23*) for the details:

$$\hat{p}_{ii,MLE} = \frac{N_i^{(t+1)}}{N_i^{(t)}} \tag{2}$$

That is, the maximum likelihood estimate of p_{ii} is the ratio of the deck area that remains in state i at time $t +$ 1 to the deck area that was in state i at time t. In this approach, only the inspection data available from two consecutive inspections are used to estimate the transition probabilities and, as a result, the Markov chain deterioration model.

Bayesian Estimation

Another approach to estimate transition probabilities is that of Bayesian estimation, whereby the transition probabilities are assumed to follow certain distributions, the parameters of which are updated when new condition observations are made. In this approach, the estimates of the transition probabilities available before the two most recent sets of inspection data are collected are taken into account in the estimation process.

As in the case of maximum likelihood estimation, the event that a 1 ft × 1 ft unit of the deck remains in the same condition state or not at the time of the next inspection is also described as a Bernoulli process. For such events, the beta distribution is commonly selected as the prior distribution for mathematical convenience whereby the posterior distribution is also derived as a beta distribution through the application of Bayes' law. Using the mode of the posterior distribution as the point estimate of p_{ii}, the Bayesian point estimate of p_{ii} can be shown to be the following—see (*23*) for the details:

$$\hat{p}_{ii,BYE} = \frac{a_{prior} + N_i^{(t+1)} - 1}{a_{prior} + b_{prior} + N_i^{(t)} - 2} \tag{3}$$

Note that when the parameters of the prior distribution a_{prior} and b_{prior} take the values of 1, corresponding to a uniform prior distribution, the Bayesian point estimates of p_{ii} are identical to the maximum likelihood estimates. Also note that when a_{prior} and b_{prior} take the values of 0, the mode and mean of the posterior distribution are approximately equal if $N_i^{(t)}$ and $N_i^{(t+1)}$ are sufficiently large, which is the case for the dataset used in this study.

Data

To evaluate the performance of the Bayesian estimation method in forecasting future condition states of concrete bridge decks with respect to the maximum likelihood estimation method, an empirical investigation is conducted based on available data. In this analysis, the inspection records of a set of bridges for the years 2015 and 2016 are compiled. Once various discrepancies and inconsistencies discussed in more detail below are addressed, a total of 357 bridge deck records are considered. This dataset is then divided into training and validation datasets, which are discussed in more detail in the Analysis Set-up subsection below.

Data Collection and Variable Definition

The data used in this study are gathered from the Ohio Department of Transportation (ODOT) bridge inspection database for the years 2015 and 2016. ODOT

Table 1. Variables in the Inspection Report Database and Their Definitions

Name	Definitions
STATE	STATE is the Federal Information Processing Standards (FIPS) code assigned to each state, and it is 39 for all bridges in Ohio.
STRUCNUM	STRUCNUM is the structure file number to identify bridges; this number is unique for every bridge.
EN	EN is the element number for a certain bridge element. In this project, the bridge element to be analyzed is reinforced concrete deck, and the element number for it is 12. Therefore all records with EN of 12 are selected from the database.
EPN	EPN is the parent element number for bridge elements. In this project, the element to be analyzed is reinforced concrete deck, and the only element that is attached to concrete decks is wearing surface (EN: 510). Therefore, all records with EN of 510 and EPN of 12 are selected from the database. From these data, protection system factors in concrete deck deterioration models can be determined.
TOTALQTY	TOTALQTY is the total quantity of the bridge element. The unit of quantity for bridge decks is square foot.
CS1, CS2, CS3, and CS4	CS1, CS2, CS3, and CS4 are the quantity of a bridge element in condition states 1 to 4, respectively. In this project, they represent the deck area in condition states 1 to 4 based on the condition state definitions provided by AASHTO Bridge Element Inspection Guide Manual (2010).

recently adopted the new four-states based condition definitions and, therefore, the number of years for which such condition data are available in the dataset are limited accordingly. The variables in the inspection report database and their definitions are presented in Table 1.

Data Filtering

Some procedures are applied in this study to organize, compile, and select the original inspection records. This section provides a description of the preparation of the dataset used.

Bridge Element Filter. The focus of this study is reinforced concrete bridge decks. Thus, all inspection records with the element number (EN) of 12, indicating bridge decks, were selected.

Inspection Records Pair-Up. Since not all bridges are inspected every year and the analysis necessitates that condition inspection data have to be available for two consecutive inspections spanning the same period, inspection reports in 2015 and 2016 were paired up based on the structure number. For all the records of reinforced concrete decks in 2015, 914 out of 3,183 decks were inspected in 2016. For these bridges, the inspection records in 2015 and 2016 were extracted.

Bridge Deck Area Discrepancy. The area of bridge decks is found to vary from year to year. Small variation could be a result of measurement errors, while large differences could be indicative of record keeping errors. In this study, records that exhibit large differences were discarded. Considering the empirical cumulative distribution function of the difference in deck area between 2015 and 2016, records with differences from -5 to $+5$ square feet were deemed acceptable (a larger error tolerance would lead to a larger sample size at the expense of larger errors in the data). The remaining records, 458 out of 914, were not considered further.

Table 2. Summary Statistics of Bridge Decks Considered

	Age (years)	AADT (vehicles)	Deck area (ft^2)
Minimum	2	0	2,013
Maximum	60	8,810	139,067
Mean	29.85	2,152.17	17,373.53
Standard Deviation	16.49	1,921.26	19,424.00
Coefficient of Variation	0.5526	0.8927	1.1180

Deterioration Assumption Consistency. The estimation methodologies presented in the Parameter Estimation Methods subsection above are based on the assumptions that there is no improvement in the condition state of bridge decks from year t to $t + 1$ and no multi-step deterioration transition within a unit period takes place (i.e., no condition state transitions from 1 to 3 or 4, and from 2 to 4). The subset of bridges, 357 out of 456, that comply with these criteria considering a tolerance of -5 to $+5$ square feet were, therefore, selected for analysis in this study.

Pertinent summary statistics for the considered bridges are shown in Table 2. Based on the range and the coefficient of variation (the ratio of the standard deviation to the mean), it is clear that the bridge decks reflected in the considered dataset represent a wide range of values regarding age, annual average daily traffic (AADT), and deck area.

Empirical Evaluation

Analysis Set-up

The dataset consisting of 357 bridge deck records described in the Data Filtering subsection is divided into training and validation datasets. The training dataset contains 282 or 287 records (depending on the two scenarios considered as discussed below in this subsection) and is used to determine estimates of transition probabilities using the two methods. The estimated transition probabilities are then applied to the year 2015 inspection records of bridge decks in the validation dataset containing the remaining 75 or 70 records to predict their condition state in the year 2016 using Equation 1. The similarity between the predicted condition states using the estimated transition probabilities as described above and "ground truth" condition states reported by bridge inspectors is used as a measurement to assess the prediction accuracy resulting from the use of the transition probability estimates determined by the Bayesian updating and maximum likelihood estimation methods.

The squared Hellinger distance (HD^2) metric is commonly used to measure the degree of similarity between two probability mass functions. This metric is used in this study to assess the similarity between two condition states, the measured and predicted. The squared Hellinger distance is expressed as the following:

$$HD^2 = \frac{1}{2}\sum\nolimits_{i=1}^{k} \left(\sqrt{p_i} - \sqrt{q_i}\right)^2 \qquad (4)$$

where p_i and q_i are the elements of two probability mass functions for which the degree of similarity is to be assessed, in this case the measured condition state vector in 2016 and the predicted condition state vector in 2016. The upper and lower limits of HD^2 are 1 and 0, respectively. A smaller value indicates a higher degree of similarity between the two probability mass functions.

To apply Equations 1 and 4, the reported bridge deck areas in each condition state for each bridge in a given year must be converted to a condition state vector, which as discussed in the Markov Chain Deterioration Model subsection represents the probability mass function of the condition state. This conversion is achieved by normalizing the bridge deck areas in each state by dividing the area in each state by the total area. In the case of predicted condition states, the predictions based on Equation 1a directly take the form of a probability mass function.

Based on an exploratory analysis, a few bridge decks are identified as having unique characteristics. As a result, the records for these bridge decks are given special treatment in the analysis. As noted above, there are 357 records of bridge decks in the dataset. Among the 357, 70 reflect observed deterioration between 2015 and 2016.

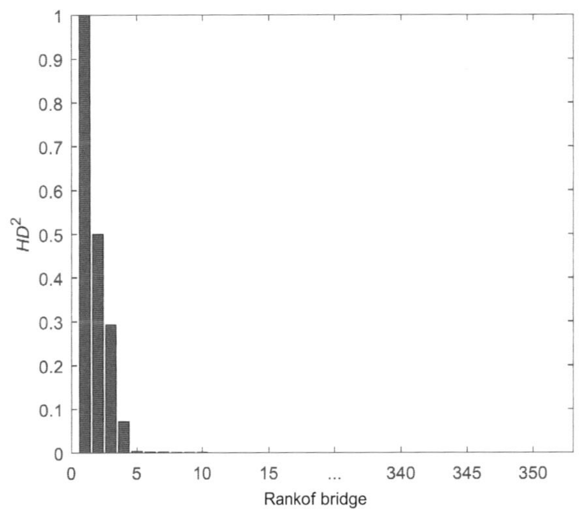

Figure 1. Bar chart of HD^2 between 2015 and 2016 for all bridges.

The extent of this deterioration is also measured by the HD^2 metric, in this case capturing the degree of similarity where p_i and q_i of Equation 4 would represent the elements of the measured condition state vectors in 2015 and 2016, respectively. Figure 1 shows the resulting HD^2 values ordered from largest to smallest along the x-axis. Clearly, four bridge decks exhibit substantial deterioration as reflected in the large HD^2 values with respect to those of the other bridge decks.

In addition, one bridge deck (structure number: 7904983) is found to exhibit a different deterioration pattern with respect to all other bridge decks. More specifically, this bridge deck is the only one that has some deck area in condition state 3 in 2015 where part of this deck area deteriorates to condition state 4 in 2016. The inspection record for this bridge deck is shown in Table 3. Under the one-step transition assumption, the transitions from the condition states in 2015 that lead to the condition states in 2016 are the following: 33 square feet deteriorate from condition state 1 to state 2, 34 square feet deteriorate from condition state 2 to state 3, and 34 square feet deteriorate from condition state 3 to state 4. All the other bridge decks either have no deck area in condition state 3 in 2015 or have some deck area in condition state 3 in 2015 but none of this area deteriorates to state 4 in 2016.

Given that these five bridge decks appear to be distinctly different from all other decks in the dataset, the impact their corresponding records could have on the results may be substantial. Therefore, two scenarios are considered. In Scenario 1 the records of these five bridge decks are included in the validation dataset, resulting in a total of 282 records in the training dataset. In Scenario

Table 3. Inspection Record for Unique Deterioration Pattern Bridge (Structure Number: 7904983, Unit: ft^2)

Year	Total deck area	Area in condition state 1	Area in condition state 2	Area in condition state 3	Area in condition state 4
2015	14,433	12,267	1,877	289	0
2016	14,433	12,234	1,876	289	34

2 these five records are included in the training dataset, resulting in a total of 287 records in this dataset (the total deck area across these five bridges comprises 3.53% of the total area across the 287 bridge decks). The records of all other bridges are randomly assigned to each of the two datasets such that approximately 80% form the training dataset and the rest form the validation dataset. Since the confidence in the estimates depends on the sample size of the training dataset, it is common to assign the majority of the records to the training dataset to ensure a sufficiently large sample size for the estimation. (Each bridge record, excluding those associated with the five distinctly different bridges, is assigned a random number generated to follow a uniform distribution that ranges between zero and one. All the records that are assigned random numbers greater than 0.2 form the training dataset and the rest form the validation dataset.)

To evaluate the performance of the Bayesian updating method with respect to the maximum likelihood estimation method, transition probabilities are determined using the training dataset. Applying the estimated transition probabilities to the condition state vector of bridge decks in the validation dataset for the year 2015, the condition states of these bridges are predicted for the year 2016. The similarity between the predicted and reported condition states for 2016 are then compared across the two methods.

As discussed in the Parameter Estimation Methods subsection, in the Bayesian estimation method, the transition probabilities available before the collection of condition data for 2015 and 2016 are assumed to follow beta distributions defining the priors in the Bayesian framework. In the Parameter Estimation Methods subsection, the beta distribution is described through its parameters a_{prior} and b_{prior}. For the purpose of this analysis, the beta distribution is conveniently parameterized through the mean and sample size, referred to as the prior mean and confidence "parameters." The relationship between this parameterization and that described in the Parameter Estimation Methods subsection is that the confidence (sample size) is the summation of a_{prior} and b_{prior} and the prior mean is the ratio of a_{prior} to the sample size. The prior mean represents the prior belief about the mean value of p_{ii}. The sample size represents the number of samples used to derive p_{ii}, which is a measure of confidence in the prior belief about p_{ii}.

Because the prior transition probabilities are not estimated from data due to the fact that only two years of condition data are available for each bridge deck, in this analysis the confidence is set to vary between 0 and 400,000 square feet. When the confidence is set to 0, the Bayesian point estimate (the mode of the posterior distribution) is equal to the maximum likelihood estimate. The upper limit sample size of 400,000 is selected based on recommendations in Pontis Technical Manual (*22*). This manual suggests that it is reasonable to assume that the prior distribution of transition probabilities is developed based on observations for 10 to 20 bridges. Considering that the average deck area for the considered bridge dataset is about 20,000 square feet, 20 bridge observations yield the upper limit of 400,000 square feet for the assumed confidence.

In general, the mean values of prior transition probabilities could be determined from bridge deck condition data collected in previous years, condition data in other bridge deck datasets (e.g., from other states), models, or expert judgement. Since none of these sources of information are available for this analysis (given that only two years of condition data are available for each bridge deck, and two years of condition data are necessary to estimate one set of transition probabilities), 10,000 mean values of prior transition probabilities (p_{11}, p_{22} and p_{33}) are generated independently from a uniform distribution that ranges between zero and one. Each set of the prior mean transition probabilities is updated using the training dataset through the application of the Bayesian updating method presented in the Parameter Estimation Methods subsection.

Combined with confidence values ranging from 0 to 400,000 in increments of 1,000, 4,000,000 sets of transition matrices are determined based on Bayesian estimation. Applying these transition matrices to 2015 condition state vectors of bridge decks in the validation dataset, the condition state vectors of these bridges are predicted for 2016. The similarity between the predicted and reported condition states for 2016 are then measured using the HD^2 metric.

The HD^2 metric is computed for each bridge deck in the validation dataset for all generated prior mean values of p_{11}, p_{22}, and p_{33} and the levels of confidence in these mean values. Thus, for a given bridge deck and one set of generated mean values, a curve of HD^2 versus confidence can be plotted. Figures 2 and 3 show two examples

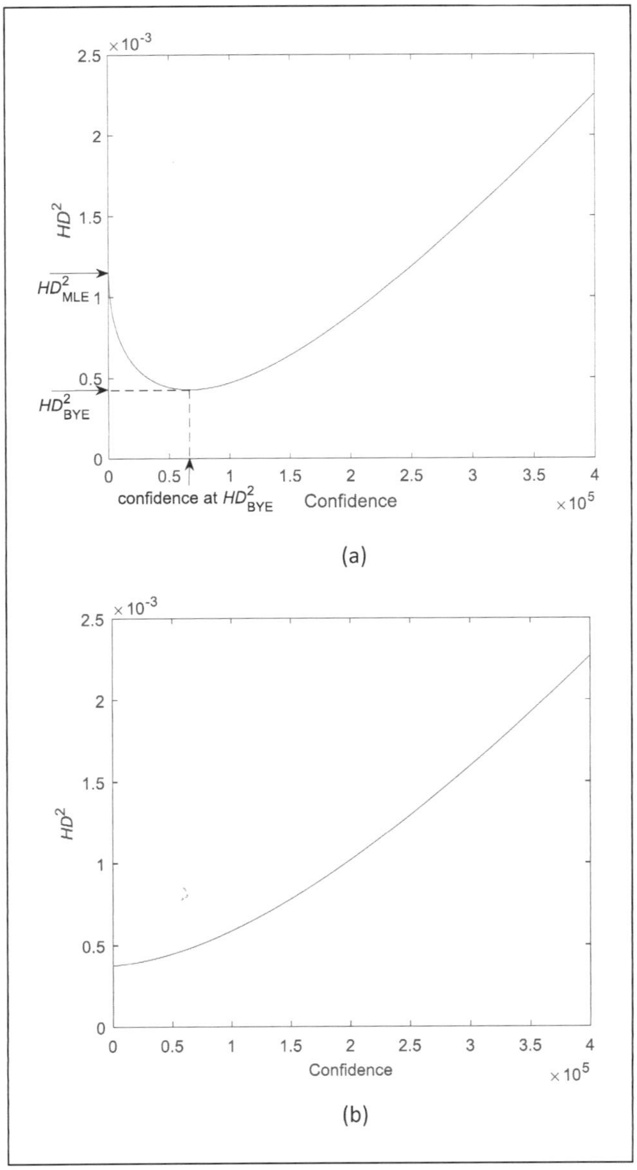

Figure 2. Example HD^2 versus confidence plot where Bayesian estimates at the confidence level associated with the minimum HD^2 value are (*a*) superior to those of maximum likelihood and (*b*) equivalent to those of maximum likelihood.

of such plots. Since, as discussed in the Parameter Estimation Methods subsection, the Bayesian estimates are approximately equal to the maximum likelihood estimates at zero confidence for large values of $N_i^{(t)}$ and $N_i^{(t+1)}$, which are the case for the training datasets, Figure 2*a* indicates a case where a greater-than-zero confidence in the prior mean values leads to Bayesian estimates that produce a superior prediction (lower HD^2) for 2016 than that of the maximum likelihood estimates. Figure 2*b* indicates a case where a confidence greater than zero leads to an inferior prediction (higher HD^2) to that based on the maximum likelihood estimates. That is, the Bayesian and maximum likelihood estimates would reflect the same prediction performance when the confidence in the prior mean values is assumed to be zero for the Bayesian estimation method.

To quantify the improvement in prediction by using Bayesian estimation at the optimal confidence value for a set of prior mean values, the following measure is defined:

$$Reduction = (HD^2_{MLE} - HD^2_{BYE})/HD^2_{MLE} \quad (5)$$

where HD^2_{MLE} is the HD^2 value for the predictions based on the maximum likelihood estimates and HD^2_{BYE} is the HD^2 value based on the Bayesian estimates that correspond the confidence associated with the minimum HD^2 value across all the confidence values considered. The elements of the measure Reduction are depicted in Figure 2*a*. Naturally, for a case like the one shown in Figure 2*b*, the measure Reduction takes a value of zero.

Results

All 4,000,000 sets of the Bayesian updated transition probabilities based on the combinations of prior mean values and confidence levels and the one set of maximum likelihood estimated transition probabilities are applied to the bridge decks of the Scenario 1 and Scenario 2 validation datasets as defined in the Analysis Set-up subsection. Figure 3 shows the number of cases for each bridge where the Bayesian estimates are superior to the maximum likelihood estimates (i.e., Reduction > 0%) under Scenario 1 as shown in Figure 3*a* and under Scenario 2 as shown in Figure 3*b*. Note that in Scenario 1 there are 75 bridge deck records in the validation dataset, while in Scenario 2 there are 70 records.

In Figure 3, bridge ID is defined as the rank among the 75 bridge deck records in the validation dataset of Scenario 1 based on the descending order of the values of HD^2 measuring the similarities of the reported condition states in 2015 and 2016. The counts reported in Figure 3 for each bridge represent the number of sets of prior transition probabilities out of the 10,000 generated ones that lead to Bayesian estimates that produce predictions that are superior to the prediction produced by the maximum likelihood estimates.

Based on Figure 3, the performance of Bayesian updating is clearly different for the two scenarios. Under Scenario 1, where the records of the five distinctly different bridge decks belong to the validation dataset, Figure 3*a* indicates that Bayesian updating is superior to maximum likelihood estimation for a substantially large number of cases associated with a fairly small number of bridge decks, specifically those that exhibit more substantial deterioration between 2015 and 2016.

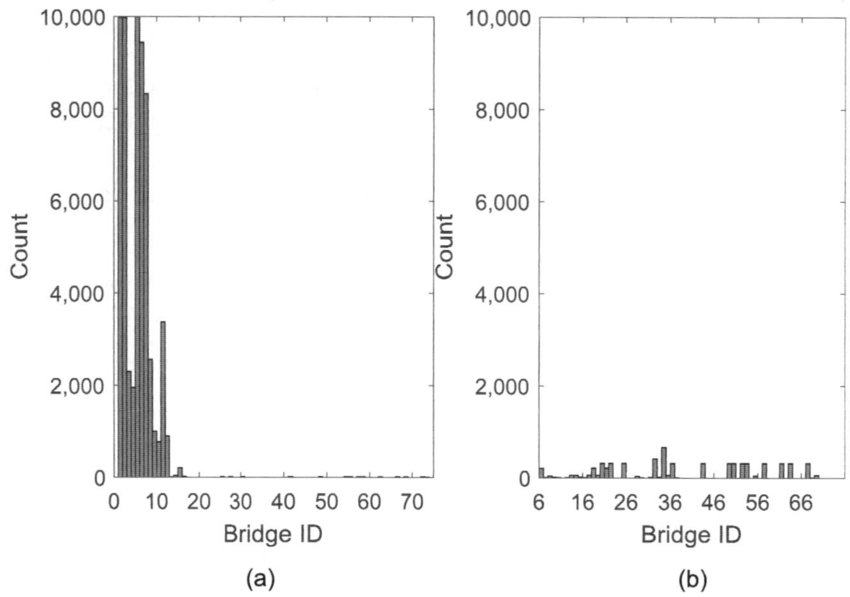

Figure 3. Bar chart by Bridge ID when Reduction > 0%: (*a*) Scenario 1 and (*b*) Scenario 2.

The bridge decks that are associated with the superiority of Bayesian updating over maximum likelihood estimation include the four decks found to exhibit the most deterioration between 2015 and 2016—bridge ID values of 1 through 4 in Figure 3. They also include the one and only deck that exhibits deterioration that involves a transition to state 4 in 2016—bridge ID value of 5 in Figure 3 (the deck with the eighth largest change in deterioration between 2015 and 2016 when considering all bridge decks as depicted in Figure 1). The characteristics of these bridge decks were investigated with the aim of identifying possible variables of similar values among them and, thus, possibly indicating explanations for the substantial deterioration they exhibit. Variables associated with age, loading, environment, and protection systems were considered. However, none were found to possibly explain the high deterioration in a span of one year. Nevertheless, the markedly different results associated with the two scenarios are consistent with the expectation that for decks that experience deterioration that is not well represented in the training dataset (as reflected in Scenario 1), incorporating prior information with the two most recent condition inspections via Bayesian updating is advantageous.

Under Scenario 2, where the records of the five distinctly different bridge decks identified in the Analysis Set-up subsection belong to the training dataset, Figure 3*b* indicates that Bayesian updating is superior to maximum likelihood estimation only for a small number of cases spanning most bridge decks in the validation dataset. This result suggests that due to the unrepresentative impact the five distinctly different decks have on the estimates, only a small set of prior mean values and confidence levels lead to more accurate Bayesian updating based predictions with respect to the maximum likelihood estimation based predictions. That is, only a few combinations of prior mean values and confidence levels lead to Bayesian estimates that sufficiently counter a training dataset corrupted by the presence of the five bridge decks of distinctly different deterioration natures than reflected among the bridge decks for which predictions are determined (the decks in the validation dataset) to improve upon the maximum likelihood estimates.

Summary and Future Research

A Bayesian updating method is proposed to estimate a Markov chain based concrete deck deterioration model in a manner that combines condition data collected over two consecutive inspections and the deterioration information available prior to the collected condition data. A dataset of bridge deck condition assessments collected by a state infrastructure agency spanning two years is used to evaluate the performance of the Bayesian updating method with respect to a maximum likelihood estimation method where only the collected condition data over two consecutive inspections are used.

Training and validation datasets are selected from the original dataset where the former is used for estimation and the latter for evaluation. While Bayesian updating is found to be superior to maximum likelihood estimation for many cases, this superiority is highly dependent on the extent of the representativeness of the dataset used to estimate the transition probabilities (i.e., the training dataset) of the deterioration nature of the bridge decks for which condition is to be predicted (i.e., those that belong to the validation dataset). Specifically, decks whose deterioration behavior is not well represented in the training dataset are the ones that stand to benefit the most from Bayesian updating estimates producing predictions that are superior in accuracy to those produced by maximum likelihood estimates.

Several possible areas for future research are worth pursuing. First, it is important to investigate further the effect of the nature of the datasets on the results and conclusion. A more refined experimental design using datasets spanning several years of condition inspections (as more condition data based on the four-states condition definitions become available) could be used for this purpose. Second, the sensitivities of the results to various assumptions are worth exploring. Such assumptions could then be relaxed if the results are found to be sensitive to them, especially those with high sensitivity. For example, adjacent bridge deck units are not expected to deteriorate independently as is assumed in this study. Also, deterioration could lead to more than a one-level change in condition state, unlike the single-level change assumed in this study. In addition, the transition probabilities may not be age-independent as is assumed in this study. Moreover, confidence in the prior mean values are not likely to be known as is also assumed in considering the Bayesian estimates that correspond to the maximum improvement in the quality of the corresponding predictions with respect to the quality of the predictions based on the maximum likelihood estimates. Furthermore, observed condition values are not error-free as is assumed in the estimation and evaluation aspects of the study. Third, additional variables could be taken into account in the deterioration model such as the effects of age, loading, environment, and protection systems.

Acknowledgments

This research was supported by the Region 5 University Transportation Center funded by the U.S. Department of Transportation, Office of the Assistant Secretary for Research and Technology (OST-R) (Grant No. DTRT12-G-UTC05) with additional financial support provided The Ohio State University. The authors are grateful to Amjad Waheed of the Ohio Department of Transportation for providing the bridge inspection data.

Author Contributions

The authors confirm contribution to the paper as follows: study conception and design: R.G. Mishalani, A. Shafieezadeh, and Z. Li; data collection: R.G. Mishalani, A. Shafieezadeh, and Z. Li; analysis and interpretation of results: R.G. Mishalani, A. Shafieezadeh, and Z. Li; draft manuscript preparation: R.G. Mishalani, A. Shafieezadeh, and Z. Li. All authors reviewed the results and approved the final version of the manuscript.

References

1. Bentley Systems, Inc. *AASHTOWare Bridge Management User Manual Version 5.2.2*. American Association of State Highway and Transportation Officials, Washington, D.C., 2015.
2. Cambridge Systematics, Inc. *Pontis 4.4 Users Manual and Technical Manual*. American Association of State Highway and Transportation Officials, Washington, D.C., 2005.
3. Camahan, J. V., W. J. Davis, M. Y. Shahin, P. L. Keane, and M. I. Wu. Optimal Maintenance Decisions for Pavement Management. *Journal of Transportation Engineering*, Vol. 113, No. 5, 1987, pp. 554–572.
4. DeStefano, P. D., and D. A. Grivas. Method for Estimating Transition Probability in Bridge Deterioration Models. *Journal of Infrastructure Systems*, Vol. 4, No. 2, 1998, pp. 56–62.
5. Jiang, Y., M. Saito, and K. C. Sinha. Bridge Performance Prediction Model Using the Markov Chain. *Transportation Research Record: Journal of the Transportation Research Board*, 1988. 1180: 25–32.
6. Madanat, S., R. Mishalani, and W. H. W. Ibrahim. Estimation of Infrastructure Transition Probabilities from Condition Rating Data. *Journal of Infrastructure Systems*, Vol. 1, No. 2, 1995, pp. 120–125.
7. Madanat, S., and W. H. W. Ibrahim. Poisson Regression Models of Infrastructure Transition Probabilities. *Journal of Transportation Engineering*, Vol. 121, No. 3, 1995, pp. 267–272.
8. Madanat, S. M., M. G. Karlaftis, and P. S. McCarthy. Probabilistic Infrastructure Deterioration Models with Panel Data. *Journal of Infrastructure Systems*, Vol. 3, No. 1, 1997, pp. 4–9.
9. Micevski, T., G. Kuczera, and P. Coombes. Markov Model for Storm Water Pipe Deterioration. *Journal of Infrastructure Systems*, Vol. 8, No. 2, 2002, pp. 49–56.
10. Mishalani, R. G., and S. M. Madanat. Computation of Infrastructure Transition Probabilities Using Stochastic Duration Models. *Journal of Infrastructure Systems*, Vol. 8, No. 4, 2002, pp. 139–148.
11. Baik, H. S., H. S. Jeong, and D. M. Abraham. Estimating Transition Probabilities in Markov Chain-based Deterioration Models for Management of Wastewater Systems. *Journal of Water Resources Planning and Management*, Vol. 132, No. 1, 2006, pp. 15–24.
12. Kobayashi, K., M. Do, and D. Han. Estimation of Markovian Transition Probabilities for Pavement Deterioration Forecasting. *KSCE Journal of Civil Engineering*, Vol. 14, No. 3, 2010, pp. 343–351.
13. Tsuda, Y., K. Kaito, K. Aoki, and K. Kobayashi. Estimating Markovian Deterioration Transition Probabilities for Bridge Deterioration Forecasting. *Structural Engineering/Earthquake Engineering*, Vol. 23, No. 2, 2006, pp. 241–256.

14. Kobayashi, K., K. Kaito, and N. Lethanh. A Bayesian Estimation Method To Improve Deterioration Prediction for Infrastructure System with Markov Chain Model. *International Journal of Architecture, Engineering and Construction*, Vol. 1, No. 1, 2012, pp. 1–13
15. Li, L., L. Sun, and G. Ning. Deterioration Prediction of Urban Bridges on Network Level Using Markov-chain Model. *Mathematical Problems in Engineering*, Vol. 2014, No. 7, 2014, pp. 1–10.
16. Chang, M., and M. Maguire. *Developing Deterioration Models for Wyoming Bridges*. Publication FHWA-WY-16/09F. Federal Highway Administration, U.S. Department of Transportation, 2016.
17. American Association of State Highway and Transportation Officials. *AASHTO Bridge Element Inspection Guide Manual*. American Association of State Highway and Transportation Officials, Washington, D.C., 2010.
18. Durango, P. L., and S. M. Madanat. Optimal Maintenance and Repair Policies in Infrastructure Management Under Uncertain Facility Deterioration Rates: An Adaptive Control Approach. *Transportation Research Part A: Policy and Practice*, Vol. 36, No. 9, 2002, pp. 763–778.
19. Golabi, K., R. B. Kulkarni, and G. B. Way. A Statewide Pavement Management System. *Interfaces*, Vol. 12, No. 6, 1982, pp. 5–21.
20. Mauch, M., and S. Madanat. Semiparametric Hazard Rate Models of Reinforced Concrete Bridge Deck Deterioration. *Journal of Infrastructure Systems*, Vol. 7, No. 2, 2001, pp. 49–57.
21. Ross, S. M. *Stochastic Processes*. John Wiley & Sons, New York, 1983.
22. Golabi, K., P. D. Thompson, and W. A. Hyman. Pontis Technical Manual. Optima Inc. and Cambridge Systematics, Inc., Cambridge, MA, 1993.
23. Li, Z. *Updating Bridge Deck Condition Transition Probabilities as New Inspection Data Are Collected: Methodology and Empirical Evaluation*. MSc. thesis. Department of Civil, Environmental and Geodetic Engineering, The Ohio State University, 2017.

The Standing Committee on Bridge Management (AHD35) peer-reviewed this paper (18-00239).

The views, opinions, findings, and conclusions reflected in this paper are the responsibility of the authors only and do not represent the official policy or position of US DOT, OST-R, OSU, ODOT, or any other person or entity.

Article

TRR
JOURNAL OF THE TRANSPORTATION RESEARCH BOARD

Transportation Research Record
2018, Vol. 2672(12) 103–115

DOI: 10.1177/0361198118793520
journals.sagepub.com/home/trr
SAGE

Senior Community Resilience with a Focus on Critical Transportation Infrastructures: An Accessibility-Based Approach to Healthcare

Richard Twumasi-Boakye[1], John O. Sobanjo[1], Sylvester Kwame Inkoom[1], and Eren Erman Ozguven[1]

Abstract

The importance of bridges to mobility in transportation is well known. However, the identification of bridges that influence the mobility of senior members of communities has not been evaluated. This is imperative because of human frailties associated with aging. In this paper, senior community resilience is assessed through accessibility of seniors to hospitals after bridge damage caused by hurricane events using Pinellas County in the Tampa Bay area as case study. The paper presents: (i) exposure probabilities for hurricane events at bridge locations; (ii) bridge damage state functions and damage state rating assignments using historical data from the National Bridge Inventory database; (iii) identification of bridges at risk of hurricane-induced damage; (iv) bridges identified as serving areas (census districts) with dense population of aging people; and (v) estimated effects of bridge closures on mobility and resilience of the aged population, based on accessibility to hospitals by using congested and free-flow travel times obtained from traffic assignment modeling. Findings showed that: (i) 66 bridges prone to hurricane-induced damage were observed to affect 140 selected aging population areas; (ii) bridge closures resulted in about 15% and 75% increase in free-flow and congested travel times, respectively; (iii) complete loss of accessibility to hospitals for some aging-dense zones; and (iv) resilience indexes of 0.94 and 0.81 were computed for free-flow and congested travel times, respectively. These results highlight significant loss in senior accessibility to hospitals and emphasize the need for policy discussions on the capabilities of highway bridges for efficient senior mobility.

Transportation infrastructures are essential components of intermodal facilities and important to the reliability, accessibility, and resilience of communities. Damage to civil bridge infrastructure poses a major threat to the overall resilience of communities (*1*). With the increasing impact of climate change on the environment, different countries and states may become vulnerable to natural disasters such as hurricanes, tornadoes, and floods.

The recent occurrences of Hurricanes Harvey and Irma (both in 2017) have created a stronger sense of awareness of safety during hurricane events, especially for the aging population. Hurricane Irma caused deaths in Hollywood, Florida of which many victims were the elderly (65 years and over) in the affected communities. From 2004 to 2006, five major hurricanes that hit the state of Florida left catastrophic devastation along their paths. The impacts of these hurricanes included damage to bridges and civil infrastructure in the state, with the most significant being the damage to the I-10 Escambia Bay Bridge. The quantified cost of damage to bridges within the two years were reported as amounting to about $500 million (*2*). Natural and man-made hazards have been postulated to have a synergistic effect in the event of failure. Their impacts have led to road posting and bridge closures for serviceability and structural review due to physical damage and the consequent disruption of the transportation network (*3–5*). Bridge closures have also been determined to have enormous implications on costs to transportation users at the regional level (*6*).

Bridge elements, such as sign structures, movable bridge elements, trusses, and railings, are liable to suffer a high extent of damage in the event of hurricanes (*7*). With the growing concern for climate change (sea level rise and unexpected storms) and its implications, the risk

[1]Department of Civil and Environmental Engineering, Florida State University, Tallahassee, FL

Corresponding Author:
Address correspondence to Richard Twumasi-Boakye: rt12f@my.fsu.edu

of hurricane-induced floods in Florida is high as most of its cities are at low elevations. Bridges located in areas subject to high tides and tropical storms have experienced unseating of decks, undermining of approach slabs, and deterioration of slope protections, channels, culverts, footings, and walls. Probabilities of scouring, vessel collision, and advanced deterioration have been reviewed by past research to ascertain the likelihood of occurrence of hurricanes and other hazards and their respective impacts (*2, 8, 9*).

The concept of community resilience can be explained by the capacity of social units to mitigate hazards, contain their effects after the occurrence, and strategically recover to normal levels of activity (*10–12*). Up to the middle of this century, it is expected that the oldest age categories in the population will grow in both numbers and proportions. This changing age structure of the population will affect both families and society (*13*). This indicates the need for more resilient communities for seniors. Resilient communities are unequivocally vital to the safety of a society with a gradually increasing aging population. Recent literature based on the 2010 Census indicates that Florida had the highest percentage in the U.S.A. of people 65 years and over, representing 17.3% of the total population of the state (*14*).

Several studies have also revealed the aging population's need for shelters during hazard events (*15, 16*), while many others have cited the disproportionate effects of disasters on frail older adults. Records of the effects of Hurricane Katrina indicated that 49% of those who died were over the age of 75 years (*17*). To compound the situation, in the U.S.A. in 2004 about 80% of older adults had at least one chronic condition, such as heart disease, cancer, diabetes, or stroke, while 50% had at least two (*18*). Congestion during evacuation activities means that the majority of nursing homes and assisted-living facilities would have to shelter in place. Asking the elderly to board a bus and making them stay on the road for hours is not the best for their health. This means post-disaster emergency management must enable easy and safe access to healthcare facilities. Road closures or damage to bridges can cause travel delays and increased fatalities among the more frail aging population. Therefore, the need for building and maintaining infrastructure of relevance to the aging population cannot be overstated. Furthermore, resilient communities are essential in ensuring the safety and well-being of individuals living in the community, especially during hazard events. Previous studies in this field include research on the role of interdependencies in community resilience (*19*) and a place-based model for evaluating community resilience (*20*).

The primary focus of this paper is to assess senior community resilience by considering the physical transportation infrastructure within the communities. The paper investigates three issues regarding community resilience, with a focus on bridge infrastructure: (a) the identification of coastal bridges susceptible to hurricane damage; (b) evaluation of the expected damage condition/states of the exposed bridges focusing on critical bridges significant to senior mobility; and (c) the development of performance measures for the assessment of the impact of the closures of selected bridges on senior mobility. Using historical condition data from the National Bridge Inventory (NBI), these effects are evaluated at the network level for the case study region. The above is further explained with a developed resilience index.

Methodology

In order to achieve the goals of this paper, the analysis followed these sequential steps: (i) computing exposure probabilities for categorical hurricane events at bridge locations; (ii) developing and applying damage state functions in allocating damage states to bridges using both historical and NBI data fields; (iii) identifying bridges at risk of hurricane-induced damage; (iv) identifying the bridges affecting areas with a high density of population aged over 65 (aging-dense); and (v) estimating the effects of bridge closures to senior mobility and resilience through accessibility to hospitals based on congested and free-flow travel times obtained from traffic assignment modeling. The framework for selection of the bridges is illustrated in Figure 1.

Computing Exposure Probabilities for Categorical Hurricane Events at Bridge Locations

Coastal bridges have been identified as critical during storm surges and wave loadings resulting from hurricane events. Many studies have therefore focused on the vulnerability of coastal bridges to storm surges and wave loading, and the consequences in terms of agency and user costs (*6, 21–23*).

To forecast the occurrence of hurricanes based on historical records, the number of storm arrivals at an exact coastal location in a single year is modeled as a Poisson distribution. In this paper, more specific attention was given to Category 3 hurricanes due to Florida's coast being prone to such storms and the resulting debilitating effects on physical infrastructure and mobility. Using Hazards United States (HAZUS) software wind data (*24*), the exposure probabilities (Figure 2) were thus estimated. Wind speeds assigned to each census tract were categorized using the well-known Saffir-Simpson Hurricane Wind Scale from the National Hurricane

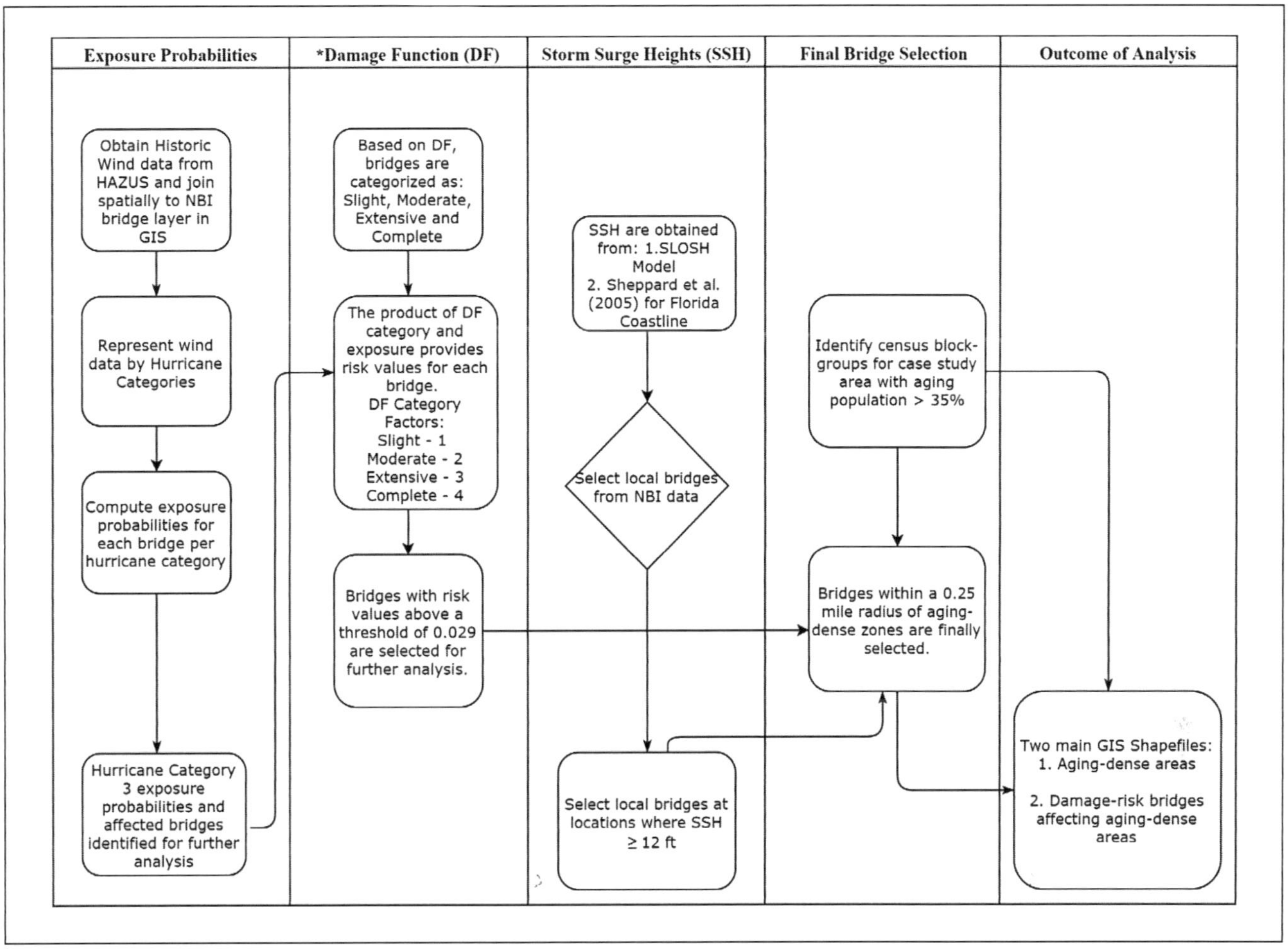

Figure 1. Flow chart showing framework for identifying at-risk bridges critical to aging-dense areas.

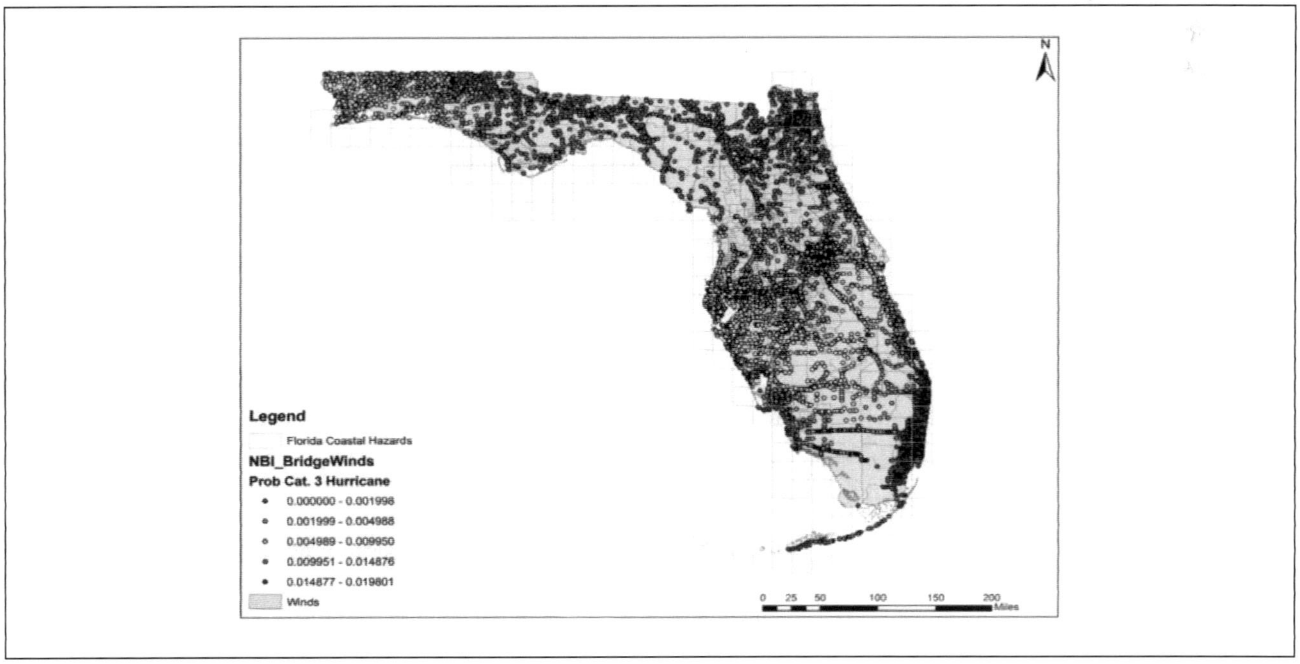

Figure 2. Map indicating probabilities of exposure to Category 3 hurricanes for NBI bridges in the state of Florida.

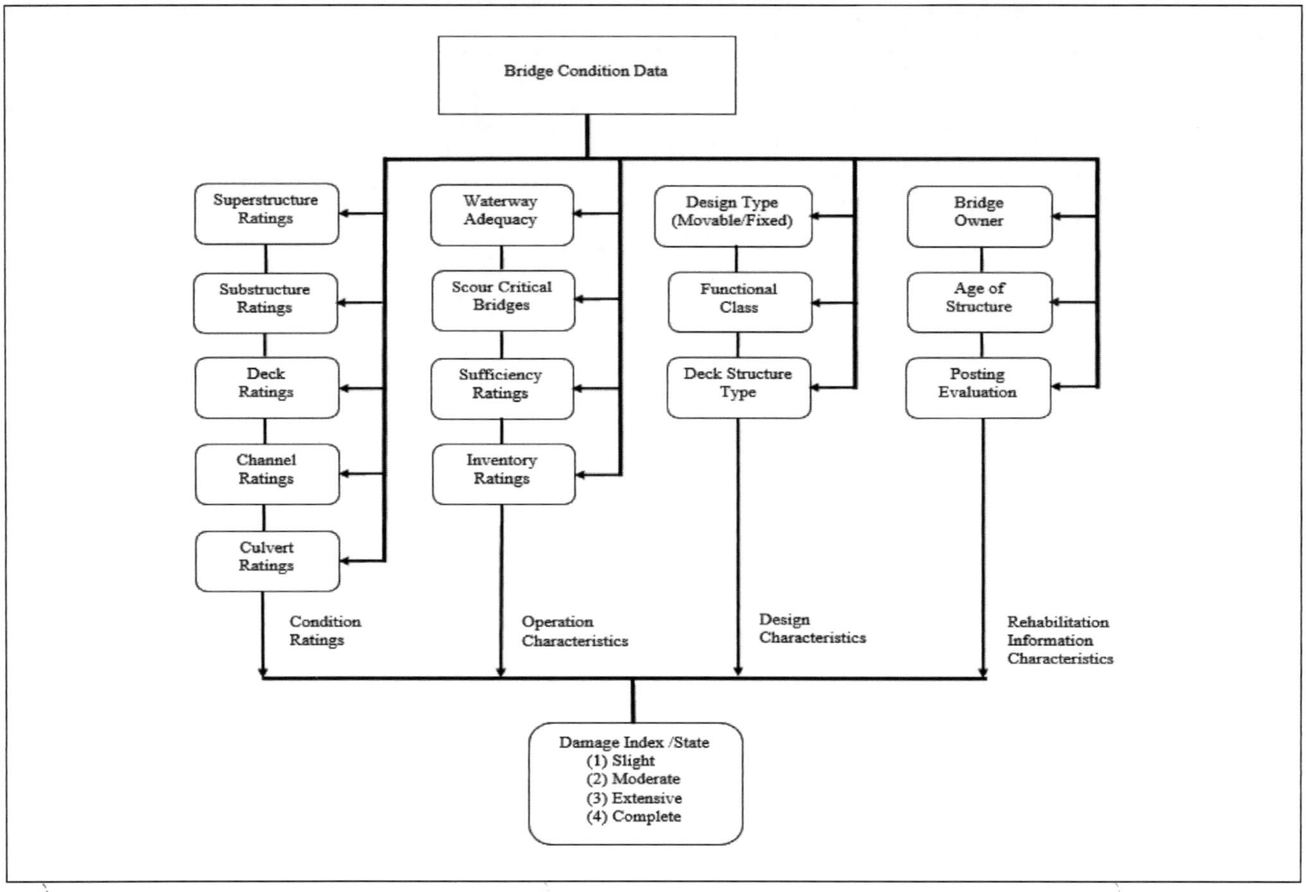

Figure 3. NBI fields selected for computing bridge damage states.

Center. The number of storms arriving at a location in one year is defined as:

$$P_n = \frac{\lambda^n exp(-\lambda)}{n!} \quad (1)$$

$$F_T(t) = P(T \leq t) = 1 - exp[-\lambda t] \quad (2)$$

where

P_n– probability of n number of storms occurring in a year,

λ– mean rate of storms per year, and

$F_T(t)$– cumulative distribution function of an exponential random variable, T, and t is a random variable representing a given period

Developing and Applying Damage State Functions in Allocating Damage States to Bridges using Both Historical and NBI Data Fields

In assessing the performance of bridges in the case study region, pertinent prior studies on the impact of different hurricanes on bridges in different states of the U.S.A. were consulted, to evaluate operational and traffic characteristics. The levels of damage were assessed through probability analysis in addition to engineering expert decision making to predict the expected damages to bridges in the region. Different coded fields (Figure 3) from the NBI database were also utilized for the analysis. The database fields, such as deck ratings, superstructure ratings, substructure ratings, and culvert and channel ratings, are evaluated along with other explanatory variables such as age, location of bridge, type of bridge (fixed or movable), waterway adequacy, and traffic characteristics. From the data, damage is categorized as slight, moderate, extensive, or complete based on the categories previously developed (*2*). Table 1 is a qualitative description of the bridge damage states.

A total of 82 damaged bridges from Florida, Louisiana, Mississippi, and Alabama were used in mapping out the damage state of bridges in the case study region of this research. There were 25% categorized as movable bridges, with the remaining being fixed bridges of different types. Based on an extensive review of damage history, movable bridges are expected to suffer different levels of impact ranging from damage of elements such as the operator facility, failure of gates, signals, and motors due to wind, and storm impacts causing

Table 1. Qualitative Damage State Descriptions Defined by Amending HAZUS for Typical Hurricane-Induced Bridge Damage (2)

Damage state	Hazard type	Description
Slight	Hurricane (Bridges)	Debris (tree logs, boats, etc.) insignificant scour, minor damage to channel, and damage to non-structural elements such as street lights, luminaires, lamps, mounted lights, small signs, and railings. Poses no serious structural problem. Structure may need minor repairs. Minor damage such as loss of sign panels, twisting of luminaires, etc. Poses no serious structural problem. Structure may need minor repairs.
	Hurricane (Sign structures)	Minor damages such as loss of sign panels, twisting of luminaires, etc. Poses no serious structural problem. Structure may need minor repairs.
Moderate	Hurricane (Bridges)	Washouts at embankments/approach slabs and damage to slope protection system. Overtopping due to flood (deck/slab or culvert) and significant scour. Moderate damage including undermining, to abutments, columns, piles, caps, footings, channel, and bulkhead. Moderate damages to fenders, navigational lights, warning gates, traffic signals, operator facilities, electrical conduit, cables, PLCs, transformers, and equipment. Poses serious structural/functional problems. Structure is repairable.
	Hurricane (Sign structures)	Loss of horizontal members, and minor cracks on foundation. Moderate damage to horizontal, vertical members, or foundation. Poses serious structural/functional problems. Structure is repairable.
Extensive	Hurricane (Bridges)	Extensive damage to culvert, deck, superstructure, substructure, and pertinent bridge elements. Structure is repairable. Poses serious structural/functional problems. May require full replacement of structural component(s).
	Hurricane (Sign structures)	Extensive damage to panels, chords, trusses, and foundation. Poses serious structural/functional problems. Structure is repairable. May require full replacement of structural component(s).
Complete	Hurricane (Bridges)	Severe damage to all or critical structural and non-structural components. Structure needs to be completely replaced.
	Hurricane (Sign structures)	Severe damage to all or critical structural components. Structure needs to be completely replaced.

mechanical and electrical faults. On the other hand, fixed bridges are anticipated to fail through unseating of decks, scouring, and failure of slope protection and abutments, as well as pier and column failure due to collision from barges or other vessels.

A number of NBI fields were selected to assess the operational characteristics and ratings of individual bridges. Four key variables were found to be statistically correlated to the damage states of bridges based on the movable bridge data. These fields with correlation coefficient of 30% and higher were the superstructure condition rating, substructure condition rating, posting evaluation, and scour critical condition. For fixed bridges, the key fields which were observed to have high correlation with levels of damage were the deck condition rating, superstructure condition rating, substructure condition rating, inventory rating, deck geometry evaluation, and scour critical condition. To assign the level of damage of bridges some assumptions were made, and steps followed, summarized as follows:

1. Bridge damage states were categorized based on the type of bridge (either fixed or movable).
2. Bridges with unknown foundations are expected to suffer the worst damage (either complete or extensive).
3. Non-waterway bridges as well as bridges which are not scour critical are ranked based on the inventory and operational characteristics (deck, superstructure, and substructure ratings).
4. Bridges without deck, superstructure, and substructure such as culverts and channels are assigned damage states based on other significant NBI fields.
5. Bridges could be assigned two damage states based on certain special characteristics.

Table 2 represents the initial list of NBI fields used in the damage state assessment of bridges.

Discussion for Damage State Analysis. From the analysis, it is observed that a total number of 1162 bridges out of the 1393 fixed bridges are expected to be subjected to slight or moderate damage conditions. The two damage states which amount to 83% of the entire inventory were found to be in consonance with the evaluation made by previous studies (*4, 25*) where many bridges were estimated as at risk of slight or moderate damage levels. Further results indicated that the remaining bridges have equal likelihood of experiencing extensive or moderate damage with each damage state amounting to 8% of the

Table 2. NBI Fields Considered for Bridge Damage State Assessment

NBI field	Operational characteristics
Owner	State, County, Local
Functional class	Interstate / Non-Interstate
Age	Old (>=50yrs) / New(<50yrs)
Year reconstructed	Recent / Previous
Waterway adequacy	Adequate / Inadequate
Scour critical bridges	High (>3) / Low (<=3)/ Unknown Foundation
Type of service	Highway / Railroad, Interchange
Kind of material	Wood /Timber, Steel, Concrete
Type of design	Fixed / Movable
Condition ratings (deck, superstructure, substructure)	High (>4) / Low (<=4)
Channel and channel protection rating	High (>7) / Low (<=7)
Culverts rating	High (>8)/ Low (<=8)
Minimum vertical underclearance	Adequate / Inadequate
Sufficiency rating	High (>50%)/ Low (<=50%)
Status	Functionally Adequate / Inadequate

entire bridge count. It is noteworthy to state that about 87 bridges were classified as at risk of experiencing either complete or extensive damage state because of their peculiar characteristics. A total of 313 culverts were identified from the inventory with about 88% of them suffering slight damage while the rest experience other damage states as follows: 11% moderate; 14% extensive, and 12% complete damage, respectively.

For the 21 recorded movable bridges, about 19% were expected to suffer slight damage, 32% extensive to complete damage levels, with 26% being extensive and 23% with complete damage.

Identifying Bridges at Risk of Hurricane-Induced Damage

Bridges at risk of hurricane-induced damage were identified by combining bridge damage states with exposure probabilities. Storm surge heights (SSH) based on data from a previous Florida study (*25*) and the sea, lake, and overland surge from hurricanes (SLOSH) model (*26*) were also used to identify local bridges at risk to damage. The SLOSH model is a computerized model developed by the National Weather Service (NWS) to estimate SSH and winds resulting from historical, hypothetical, or predicted hurricanes.

Local bridges at locations where SSH were greater than or equal to 12 feet were identified as being at risk of damage. The threshold of 12 feet was chosen since SSH above that level are known to cause inundations, while SLOSH model outputs indicated that SSH for the case study area were mostly in the selected range. Aging-dense zones were selected as census block groups with over 35% of the total population aged 65 years and above. All identified bridges within a quarter-mile radius of these locations were then selected as those having a direct influence on mobility to and from the aging-dense zones. The results are shown in Figure 4.

Computing Resilience

Accessibility is used as a mobility measure in estimating the effects of bridge damage on journeys of the aged population to hospitals. This measure is expressed as the least cost (travel time) between origins (aging-dense zones) and destinations (hospital facilities) prior to and after the hurricane events. The approach is executed through the use of the Environmental Systems Research Institute (ESRI)'s ArcGIS software, applying the closest facility extension. A similar concept was noted as being adopted in the literature for determining geographical access to cancer care (*27*). In our study, travel time data was based on the Tampa Bay Regional Planning Model (TBRPM) provided by the Florida Department of Transportation (FDOT)'s District 7 Metropolitan Planning Organization. The model was exported into ArcGIS and used to obtain origin–destination (OD) matrices for both free-flow travel times (FFTT) and congested travel times (CTT) with the aid of the network analyst extension. Resilience was computed by combining measures of functionality and recovery times for bridge closure events, as illustrated in Figure 5.

Functionality Measure:

$$\gamma_e^N(t) = \frac{Acc_i^T(t)}{Acc_i^T(t\_dis)} \quad (3)$$

where:

$Acc_i^T(t)$– minimum travel time for ith OD prior to hazard

$Acc_i^T(t\_dis)$– minimum travel time for ith OD after hazard

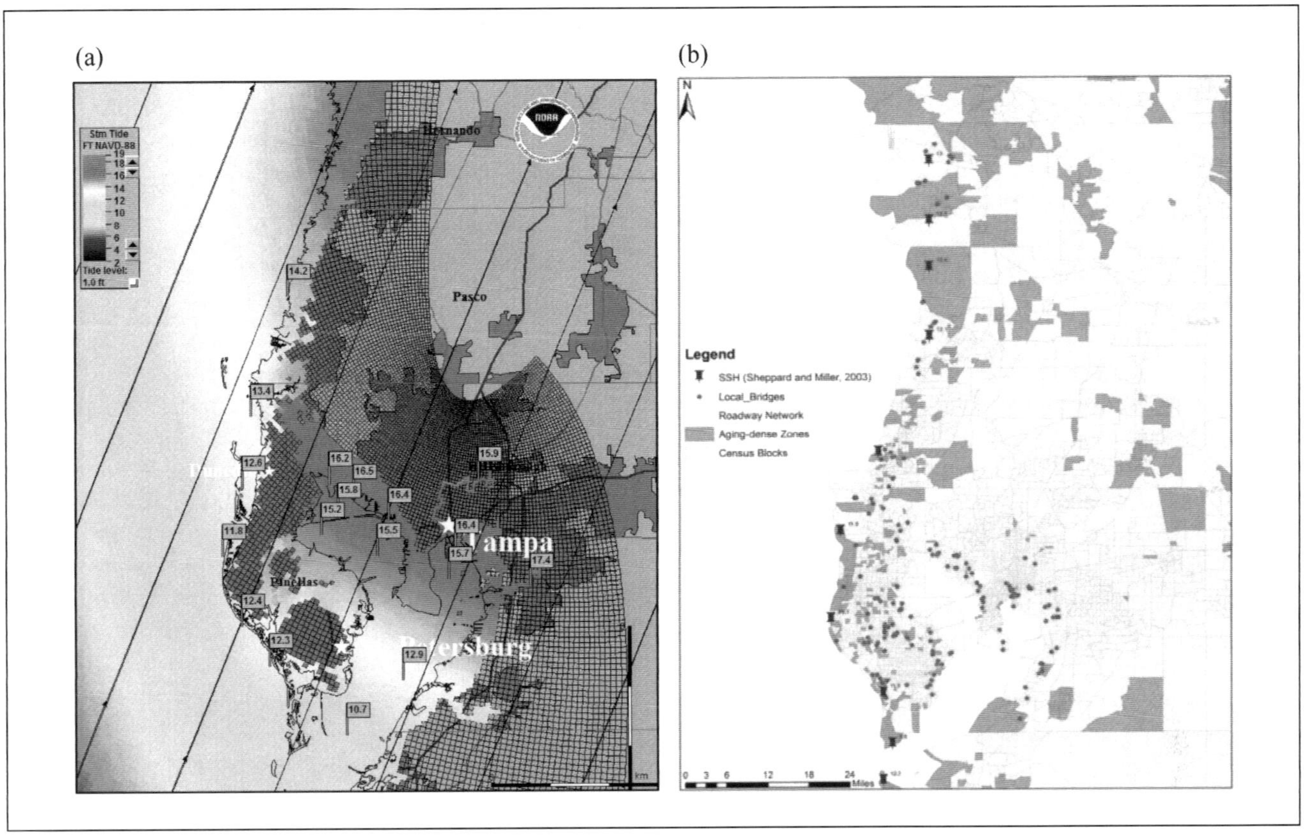

Figure 4. SSH from SLOSH model (*a*) and selected local bridges based on SSH threshold (*b*) for the Tampa Bay area.

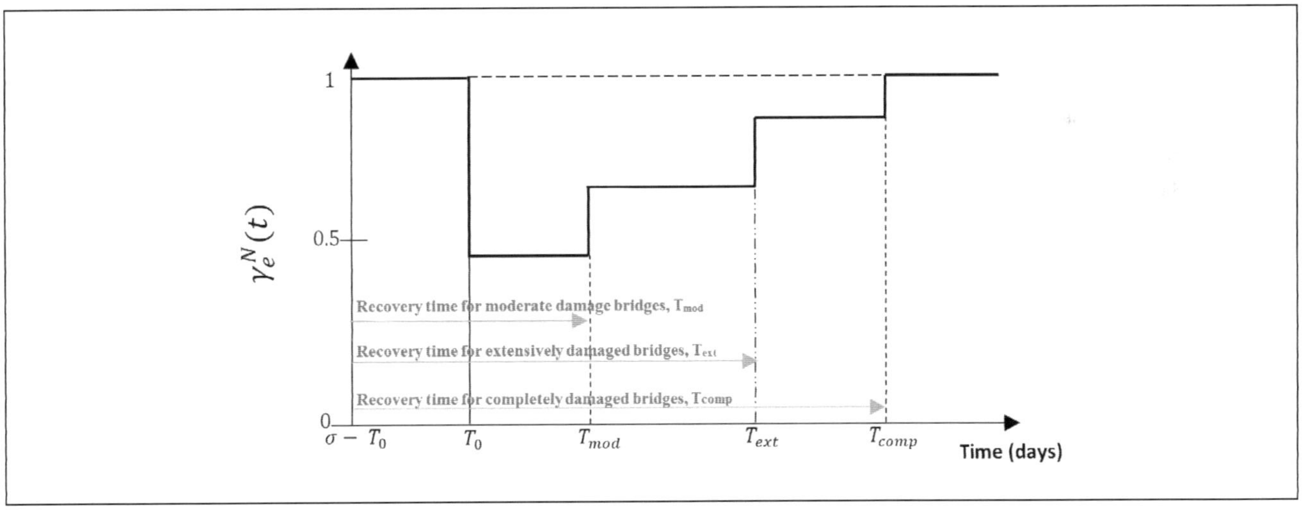

Figure 5. Resilience based on bridge damage states used in this paper.

N– transportation network
Resilience

$$R = 1 - \frac{1}{\bar{T}} \int_{0}^{\bar{T}} (1 - \gamma_e^N(t))dt \qquad (4)$$

where:

$\bar{T}$– mean time to recovery in days

Case Study for Accessibility Analysis

The general area of Tampa Bay, the case study, is prone to hurricane strikes and storm surges. Tampa Bay is a vast natural harbor and estuary which is linked to the Gulf of Mexico on the west central coast of Florida. The specific county for this case study is Pinellas County.

Data Set

Information for hospital facilities, census block groups, and Florida coastal hazards demarcations were obtained from Florida Geographic Data Library. The NBI bridges shapefile was obtained from ESRI, while roadway shapefile and TBRPM were retrieved from the Florida Standard Urban Transportation Model Structure. The hospital shapefile contained attributes for hospital facility locations and capacities (number of beds) for the case study area. The census block groups also contained various demographic details for each block group division within the jurisdiction. The coastal hazards dataset contained cartographic representation of the coastal counties in the state of Florida that are vulnerable to coastal erosion and inundation from sea level rise or storm surge. The database file and its associated layers are utilized by coastal managers to assess hurricane-induced storm surge hazards comprehensively along the coast of Florida.

Results and Discussion

The accessibility analysis for this study included the selection of 140 census block groups that were identified as aging-dense zones, with these zones serving as incident areas. This study also found that 66 bridges critical to the mobility of the populations of these zones are at risk of closure during a Category 3 hurricane. Fifteen designated hospital locations were identified, serving as facilities of interest for accessibility analysis. The scenario adopted in this study assumes that the identified at-risk bridges are damaged during the storm, and hence remain closed after the hurricane event for repair activities which are envisaged to take lengthy periods. Furthermore, as normal activities resume, accessibility of seniors to primary healthcare, expected to be affected, is evaluated. The network analysis methodology used in this study is quite rigorous since resilience is measured based on the accessibility of seniors to healthcare under normal road

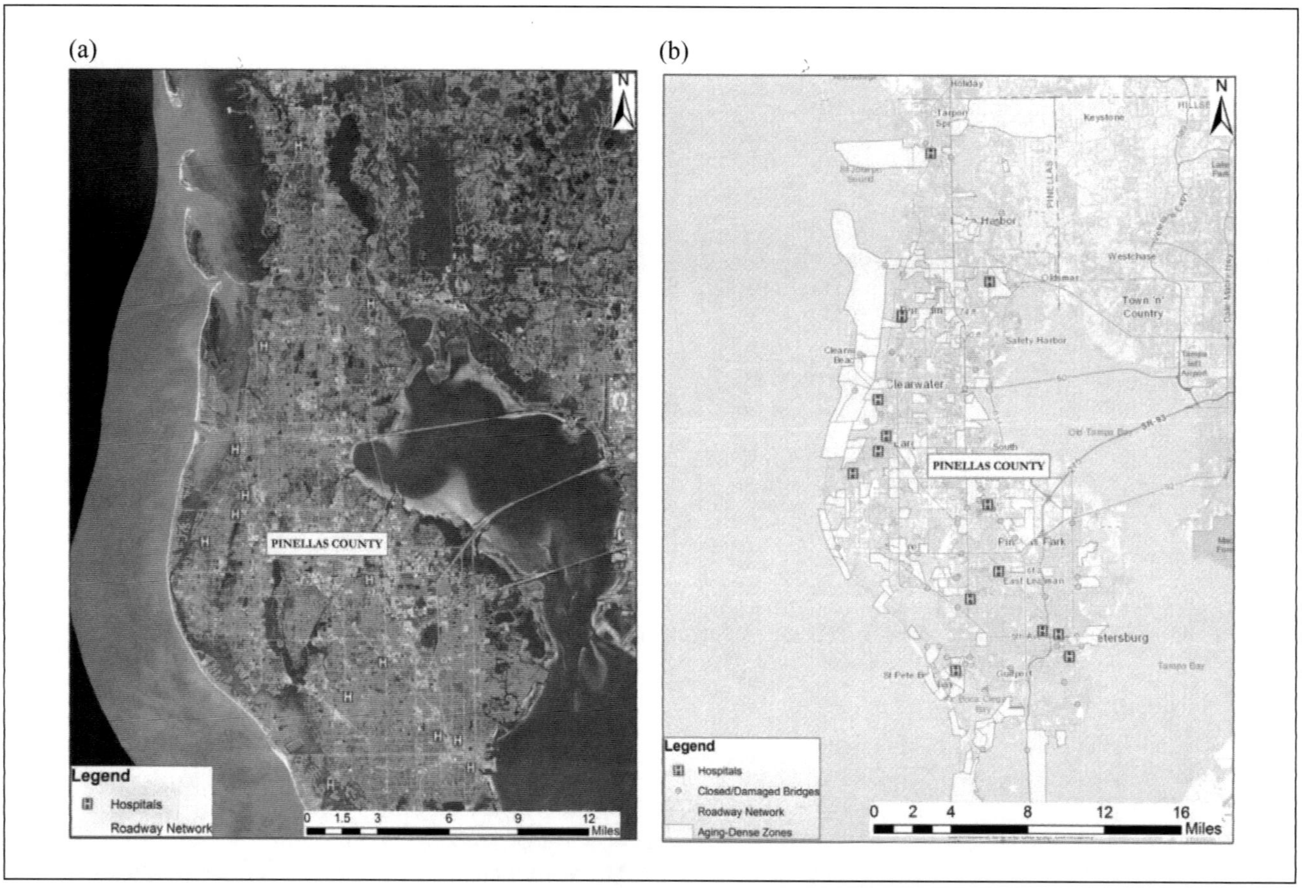

Figure 6. Maps showing Pinellas County with the locations of hospitals (*a*) and bridges at risk of damage near to aging-dense zones (*b*).

network functionality, compared with senior healthcare accessibility during closures of the identified bridges.

In the network analysis, the TBRPM is utilized initially to capture FFTT and CTT for Pinellas County for the base model (without bridge closures). The network is then modified to capture the damaged bridges within the TBRPM environment and re-analyzed until a new equilibrium is reached, hence adequately representing FFTT and CTT during bridge closures. The closest facility analysis tool in network analyst (an extension in ArcGIS) is then used to obtain the minimum times for each aging-dense area to access healthcare prior to and after the hurricane event. Figure 7 depicts the effects of bridge closures on travel times between the aging population zones and the hospitals. There was an observed increase from about 900 to 1100 minutes and from about 1200 to 2100 minutes, for the FFTT and CTT, respectively. This indicates travel time increases of about 15% and 75% for FFTT and CTT, respectively. Furthermore, an additional total travel distance of 52.85 miles was observed for FFTT and CTT.

The mean travel times after bridge closures increased from 6.6 to 7.76 minutes and from 8.43 to 15.1 minutes for FFTT and CTT, respectively. Figure 7*a* represents changes in minimum travel time after bridge closures for each aging-dense zone. Figure 7*b* and *c* are derived from Equation 3, and account for functionality computed as the ratios of minimum travel times for each trip, prior to and after bridge closures. It is observed that while many age-dense zones did not record changes in FFTT as observed in Figure 7*b*, *c* indicates significant changes for CTT resulting from the effects of congestion on travel. This is because post-hazard recovery involves an increase in roadway demand leading to significant impact of bridge closures on network travel time. Such conditions warrant the prioritization and rapidity of bridge restoration activities in order to ensure that the emergent health needs of the aging population are met. Results shown in Figure 8 indicate minimum travel times and routes to various hospitals based on FFTT and CTT for both the base model (Figure 8*a* and *b*) and interrupted network (Figure 8*c* and *d*). The color variations seen indicate the minimum travel times from the centroids to the nearest hospital.

While FFTT accessibility maps (Figure 8*a* and *c*) show some similarities, bridge closures are observed to significantly affect accessibility of the aging population to hospitals. This difference is further evident when comparing CTT accessibility maps (Figure 8*b* and *d*). These results support the findings shown in Figure 7. Additionally, three aging-dense zones were observed as being without access to hospitals after bridge closures. These are seen on the southern boundary of Pinellas County and are highlighted green in Figure 8*c* and *d*.

Resilience indexes for the bridges were based on the functionalities computed from FFTT and CTT using Equation 3, as well as the expected recovery times of the bridges after damage. The damage states for the 66 identified bridges were considered as moderate, extensive, and complete levels; slightly damaged bridges were not taken into account in this study as those bridges are normally not expected to undergo total closure. In computing the resilience index, it is expected that in most cases, moderately-damaged bridges will be restored before those bridges with extensive damage and that the restoration of completely damaged bridges will take the longest time. Computations included the re-evaluation of the traffic assignment model for improvement of network functionality after bridge restoration for each damage state. The resulting resilience index scaled from 0 to 1 is computed based on Equation 4, with 1 representing a perfectly functional network and 0 otherwise. It is expected that six days after all bridges are closed, moderately-damaged bridges will be restored, and this results in functionality improvement from 0.87 to 0.94 considering FFTT, and from 0.57 to 0.83 considering the CTT. Extensively-damaged bridges are expected to open 30 days after the hurricane event, resulting in functionality increase from 0.94 to 0.96, and 0.83 to 0.85, respectively, considering FFTT and CTT. All bridges are expected to be restored 29 days after extensively-damaged bridges are opened. The resilience index for this study was computed as 0.94 and 0.81 for FFTT and CTT respectively, implying significant loss in senior mobility hence the need for mitigation measures.

Conclusion

This paper has presented an accessibility-based approach to healthcare for evaluating senior community resilience with a focus on damage to bridges. The research approach adopted included the identification of bridges which are at high risk of damage as a result of Category 3 hurricane events by computing wind exposure probabilities at each bridge location, assigning damage states to bridges by using national bridge inventory (NBI) attribute fields and historic data, and finally identifying local bridges subjected to high storm surge heights (SSH) during hurricanes. The adopted approach was based on previous studies which identified bridge damages in areas of high-wind exposure probabilities. The essence of this study was to provide an approach for identifying at-risk bridges by utilizing available data sources on hurricane winds, SSH, operational characteristics (from NBI) of previously damaged bridges due to hurricanes, and NBI characteristics of bridges presently located in coastal areas and exposed to hurricanes. Furthermore, the importance of the identified bridges to aging-dense zones

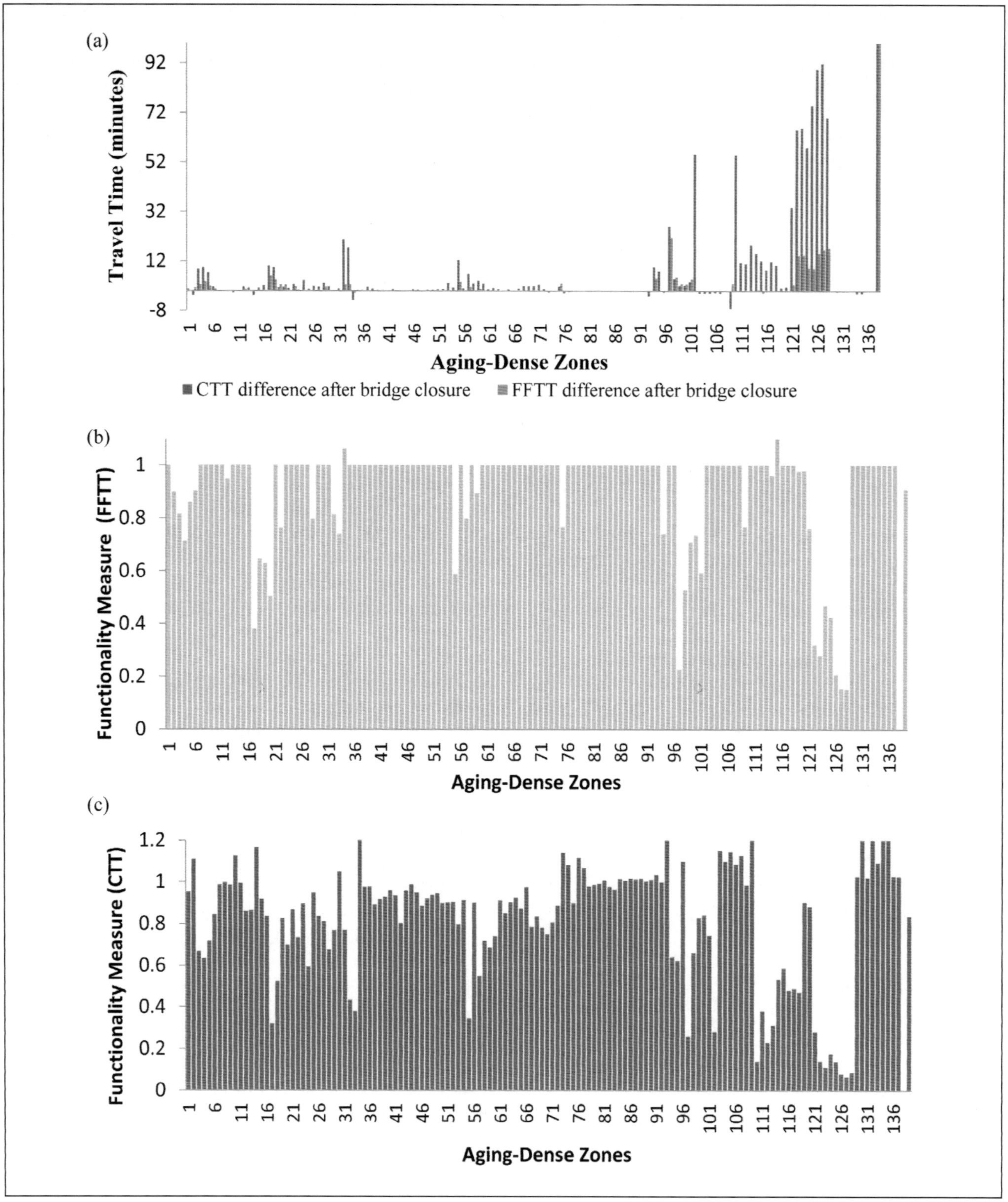

Figure 7. Results indicating differences in FFTT and CTT prior to and after bridge closures (*a*), as well as FFTT and CTT based functionality measures (*b*), (*c*).

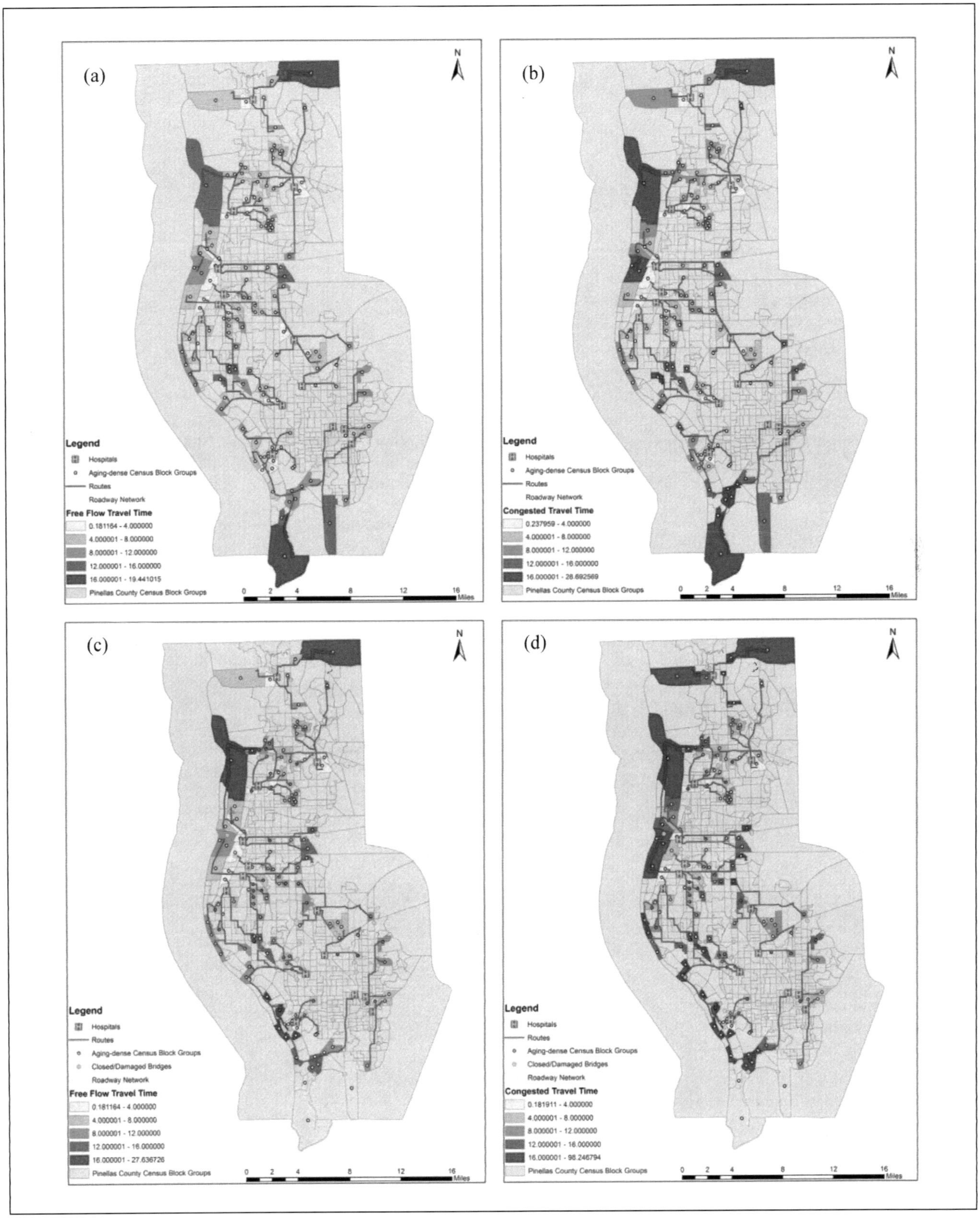

Figure 8. Minimum FFTT and CTT to hospitals for each aging-dense location: (*a*) and (*b*) base network; (*c*) and (*d*) bridge closure network.

was evaluated as well as the effects of bridge closures on access of the aged population to hospitals.

Results indicated that 66 bridges were of specific interest (using proximity analysis) to areas with a high percentage of aging population. Movable bridges were identified as being significantly vulnerable during hurricanes. Accessibility was determined based on closest facility analysis by using 140 identified aging-dense zones and 15 hospitals as origins (incident locations) and destinations (facilities), respectively. Significant increases in minimum travel time to hospitals were observed for both free-flow travel time (FFTT) and congested travel times (CTT). This was more evident for CTT due to congested roadway conditions, yielding a resilience index of 0.81 compared with 0.94 for FFTT. Accessibility of aged population to hospitals is of utmost importance due to human frailties that come with age. In addition, it was observed that some aging-dense zones were more affected than others in this study, therefore, there is the need for health access improvement solutions in the significantly affected areas. Increased financial investment in maintaining and reinforcing both state and locally maintained bridges is requisite for efficient senior mobility. With the population of those aged 65 and above on the increase, this need is timely. Increase in roadway travel times, reduced functionalities, and decreased resilience communicated through this study highlights that senior members of communities are affected by bridge closures.

The authors of this paper recommend that further studies which entail the application of more precise recovery times for affected bridges are undertaken to improve the computation of a more rigid resilience index which will help to communicate the effects of bridge closures on community resilience. Various transportation agencies are also encouraged to develop and maintain a database to document hazard-induced bridge damage and post-disaster recovery activities (especially recovery times). This will contribute to identifying the extent of bridge damage, evaluating recovery efforts, and enhancing mitigation measures and/or rapidity during post-hazard recovery.

Acknowledgments

This project was funded by a grant from the U.S. Department of Transportation and administered by the Center for Accessibility and Safety for an Aging Population (ASAP) at Florida State University.

Author Contributions

The authors confirm contribution to the paper as follows: study conception and design: Richard Twumasi-Boakye, John O. Sobanjo, Eren Erman Ozguven, Sylvester Kwame Inkoom; data collection: Richard Twumasi-Boakye, Sylvester Kwame Inkoom; analysis and interpretation of results: Richard Twumasi-Boakye, John O. Sobanjo, Sylvester Kwame Inkoom; draft manuscript preparation: Richard Twumasi-Boakye, John O. Sobanjo, Eren Erman Ozguven, Sylvester Kwame Inkoom. All authors reviewed the results and approved the final version of the manuscript.

References

1. Zhang, P., and S. Peeta. A Generalized Modeling Framework to Analyze Interdependencies Among Infrastructure Systems. *Transportation Research Part B: Methodological*, Vol. 45, No. 3, 2011, pp. 553–579.
2. Sobanjo, J. O., and P. D. Thompson. Development of Risk Models for Florida's Bridge Management System. Technical Report BDK83 977-11. Florida State University, 2013. https://ntl.bts.gov/lib/47000/47800/47813/FDOT-BDK83-977-11-rpt.pdf. Accessed May 12, 2017.
3. Dalziell, E., & Nicholson, A. Risk and Impact of Natural Hazards on a Road Network. *Journal of Transportation Engineering*, Vol. 127, No. 2, 2001, pp. 159–166.
4. Padgett, J., R. DesRoches, B. Nielson, M. Yashinsky, O. S. Kwon, N. Burdette, and E. Tavera. Bridge Damage and Repair Costs from Hurricane Katrina. *Journal of Bridge Engineering*, Vol. 13, No. 1, 2008, pp. 6–14.
5. Salman, A. M., and Y. Li. Age-dependent Fragility and Life-Cycle Cost Analysis of Wood and Steel Power Distribution Poles Subjected to Hurricanes. *Structure and Infrastructure Engineering*, Vol. 12, No. 8, 2016, pp. 890–903.
6. Twumasi-Boakye, R., and J. O. Sobanjo. Evaluation of Transportation User Costs Based on Simulated Regional Network Models. *Transportation Research Record: Journal of the Transportation Research Board*, 2017. 2612: 121–131.
7. Inkoom, S., J. O. Sobanjo, P. D. Thompson, R. Kerr, and R. Twumasi-Boakye. Bridge Health Index: Study of Element Condition States and Importance Weights. *Transportation Research Record: Journal of the Transportation Research Board*, 2017. 2612: 67–75.
8. Xu, L., and R. E. Brown. A Hurricane Simulation Method for Florida Utility Damage and Risk Assessment. In *Power and Energy Society General Meeting – Conversion and Delivery of Electrical Energy in the 21st Century*, IEEE, 2008, pp. 1–7.
9. Thompson, P., J. Sobanjo, and R. Kerr. Modeling the Risk of Advanced Deterioration in Bridge Management Systems. *Transportation Research Record: Journal of the Transportation Research Board*, 2013. 2360: 52–59.
10. Bruneau, M., S. E. Chang, R. T. Eguchi, G. C. Lee, T. D. O'Rourke, A. M. Reinhorn, M. Shinozuka, K. Tierney, W. A. Wallace, and D. Von Winterfeldt. A Framework To Quantitatively Assess and Enhance the Seismic Resilience of Communities. *Earthquake Spectra*, Vol. 19, No. 4, 2003, pp. 733–752.
11. Bruneau, M., and A. Reinhorn. Exploring the Concept of Seismic Resilience for Acute Care Facilities. *Earthquake Spectra*, Vol. 23, No. 1, 2007, pp. 41–62.
12. Zorn, C. R., and A. Y. Shamseldin. Post-disaster Infrastructure Restoration: A Comparison of Events for Future Planning. *International Journal of Disaster Risk Reduction*, Vol. 13, 2015, pp. 158–166.

13. Ortman, J. M., V. A. Velkoff, and H. Hogan. *An Aging Nation: The Older Population in the United States*. United States Census Bureau, Economics and Statistics Administration, US Department of Commerce, 2014, pp. 25–1140.
14. Howden, L. M., and J. A. Meyer. *Age and Sex Composition: 2010*. 2010 Census Briefs, US Census Bureau, Economics and Statistics Administration, US Department of Commerce. 2010.
15. Gibson, M. J. *We Can Do Better: Lessons Learned for Protecting Older Persons in Disasters*. AARP Public Policy Institute, Washington, D.C., 2006. http://assets.aarp.org/rgcenter/il/better.pdf
16. Fernandez, L. S., D. Byard, C. C. Lin, S. Benson, and J. A. Barbera. Frail Elderly as Disaster Victims: Emergency Management Strategies. *Prehospital and Disaster Medicine*, Vol. 17, No. 2, 2002, pp. 67–74.
17. Brunkard, J., G. Namulanda, and R. Ratard. Hurricane Katrina Deaths, Louisiana, 2005. *Disaster Medicine and Public Health Preparedness*, Vol. 2, No. 4, 2008, pp. 215–223. https://doi.org/10.1097/DMP.0b013e31818aaf55
18. Centers for Disease Control National Center for Health Statistics. *The State of Aging and Health in America*. Centers for Disease Control National Center for Health Statistics, 2004. http://www.cdc.gov/aging/pdf/State_of_Aging_and_Health_in_America_2004.pdf.
19. Cimellaro, G. P., D. Solari, C. S. Renschler, A. M. Reinhorn, and M. Bruneau. Community Resilience Index Integrating Network Interdependencies. In *Structures Congress 2013: Bridging Your Passion with Your Profession* (McHugh, J. ed.), American Society of Civil Engineers, Reston, VA, 2013, pp. 1789–1799.
20. Cutter, S. L., L. Barnes, M. Berry, C. Burton, E. Evans, E. Tate, and J. Webb. A Place-based Model for Understanding Community Resilience to Natural Disasters. *Global Environmental Change*, Vol. 18, No. 4, 2008, pp. 598–606.
21. Maxey, D. Emergency Repair of the I-10 Bridge over Escambia Bay. In *FICE/FDOT Design Conference*, Orlando, Fla, 2006, 43pp.
22. Padgett, J. E., A. Spiller, and C. Arnold. Statistical Analysis of Coastal Bridge Vulnerability Based on Empirical Evidence from Hurricane Katrina. *Structure and Infrastructure Engineering*, Vol. 8, No. 6, 2012, pp. 595–605. http://dx.doi.org/10.1080/15732470902855343
23. Stearns, M., and J. E. Padgett. Impact of 2008 Hurricane Ike on Bridge Infrastructure in the Houston/Galveston Region. *Journal of Performance of Constructed Facilities*, Vol. 26, No. 4, 2011, pp. 441–452. http://dx.doi.org/10.1061/(ASCE)CF.1943-5509.0000213
24. Vickery, P. J., P. F. Skerlj, J. Lin, L. A. Twisdale, Jr, M. A. Young, and F. M. Lavelle. HAZUS-MH Hurricane Model Methodology. II: Damage and Loss Estimation. *Natural Hazards Review*, Vol. 7, No. 2, 2006, pp. 94–103. https://doi.org/10.1061/(ASCE)1527-6988(2006)7:2(94)
25. Sheppard, D. M., and W. Miller, Jr. *Design Storm Surge Hydrographs for the Florida Coast*. FDOT Contract Number BC-354 RWPO 70. University of Florida Contract Number 4910 45-04-920. Department of Civil and Coastal Engineering, University of Florida, Gainesville, 2003. http://www.fdot.gov/research/Completed_Proj/Summary_STR/FDOT_BC354_70_rpt.pdf
26. Jelesnianski, C. P., J. Chen, and W. A. Shaffer. *SLOSH: Sea, Lake, and Overland Surges from Hurricanes*, Vol. 48. US Department of Commerce, National Oceanic and Atmospheric Administration, National Weather Service, 1992.
27. Onega, T., E. J. Duell, X. Shi, D. Wang, E. Demidenko, and D. Goodman. Geographic Access to Cancer Care in the US. *Cancer*, Vol. 112, No. 4, 2008, pp. 909–918.

The Standing Committee on Bridge Management (AHD35) peer-reviewed this paper (18-06705).

Article

TRR
JOURNAL OF THE TRANSPORTATION RESEARCH BOARD

Transportation Research Record
2018, Vol. 2672(12) 116–126

DOI: 10.1177/0361198118797807
journals.sagepub.com/home/trr
SAGE

Systematic Procedures for the Analysis of Agency and User Costs of Bridge Repair Actions

Ehsan Fereshtehnejad[1], Jieun Hur[1], Abdollah Shafieezadeh[1], Mike Brokaw[2], Brad Noll[2], Jared Backs[2], and Amjad Waheed[2]

Abstract
A primary goal of bridge management systems is to identify maintenance, repair, and replacement (MR&R) strategies that maximize benefits and minimize losses, often expressed in terms of cost. A major factor that greatly impacts outcomes of these frameworks, that is, the most appropriate MR&R strategies, is the cost of implementation and associated consequences of performing such actions. Given that the inventory of bridges maintained by a state department of transportation (DOT) is significantly large, this study is aimed at developing a systematic procedure to reliably estimate the costs through effective utilization of DOTs' databases. The considered costs include agency cost of administration, engineering, and mobilization; agency direct cost of performing MR&R actions; agency cost of maintenance of traffic; and user cost incurred from traffic delays, vehicle operation, and excess emissions. The study develops a set of models to estimate the duration of various MR&R work plans. These models are subsequently used to propose analytical formulations and algorithms for the estimation of the above costs. The proposed procedures are employed to estimate the agency and user costs associated with a series of light to extensive repair actions for three bridges in Ohio. Independent calculation of some of these costs by Ohio DOT engineers indicated good agreement with results obtained from the proposed systematic methods. The verified procedures for cost estimation developed in this study enable state DOTs and other entities to reliably estimate implementation costs of actions for their large inventory of bridges and identify the most cost-effective MR&R strategies and work plans.

The need to employ cost-effective strategies for management of bridges has been well recognized by state departments of transportation (DOTs) and other entities. The increase in the number of bridges in need of repair actions and the limited budgets available to state DOTs and other entities are among the primary factors that led to the recognition of this need. Several bridge management systems that employ bridge performance indices and optimization methods have been introduced. One such platform that assists in cost-effective decision-making for bridges is AASHTO BrM (*1, 2*); some other frameworks for optimal decision-making are presented in the literature (*3–10*). Moreover, several bridge performance indices such as the Ohio Bridge Condition Index (*11, 12*), proposed by the authors, and the Bridge Health Index (*13*) have been developed to evaluate the state of structural integrity and functionality of structures. The primary goal of the above frameworks and measures is to identify maintenance, repair, and replacement (MR&R) strategies that result in the least incurred cost for users and agencies. The reliability of suggested optimal policies by these frameworks considerably depends on the accuracy of input information, especially costs associated with implementation of MR&R actions for various bridge members. Given that state DOTs are responsible for a large inventory of bridges, the estimation of these costs should be systematic and based on DOTs' available data. To address this need, the present study develops detailed procedures for objective and reliable estimation of various key costs that are incurred as a consequence of conducting MR&R work plans on bridges.

As mentioned by Fereshtehnejad et al. (*3, 11, 12*), performing MR&R actions on bridge elements incurs several user and agency costs, including agency costs of administration, engineering and mobilization (AEM); the

[1]Department of Civil, Environmental and Geodetic Engineering, The Ohio State University, Columbus, OH
[2]Ohio Department of Transportation, Columbus, OH

Corresponding Author:
Address correspondence to Abdollah Shafieezadeh:
shafieezadeh.1@osu.edu

agency direct cost of performing MR&R actions; agency costs of maintenance of traffic (MOT); and user costs incurred from delay time, vehicle operation, and excess emission (DVE). For AASHTO commonly recognized (CoRe) bridge elements (*14*), several state DOTs have determined element-level MR&R costs. Other required information includes existing condition-states that are suitable for the application of considered actions as well as the likely condition-state of the element after performing such improving actions. These condition-states are often provided in a 1–5 condition-rating system. Some of the states that have collected this information are Colorado (*15*), Michigan (*16*), Delaware (*15*), Minnesota (*17*), Louisiana (*15*), and Florida (*18*). Recently, AASHTO has recommended state DOTs to provide element-level inspections on a condition-state rating system of 1–4 (*19*, *20*). The definition of condition-states in this rating system is different from that in the 1–5 condition-rating system, for which element-level costs are widely available. Thus, for management purposes, in line with AASHTO's new condition-rating system, MR&R costs should be converted to those based on the condition-state rating system of 1–4. For this purpose, this research proposes a simple and practical mapping system based on the definitions of condition-states. Furthermore, specific adjustments are recommended to realistically calculate project-level MR&R costs from element-level ones; this is a feature that is missing in many existing frameworks and measures. A systematic framework is proposed for calculation of the MOT cost, which relies on logical considerations and the unit costs reported by the Ohio DOT for crew, equipment, and police enforcement. A formulation is also proposed for the AEM cost, which uses the overhead factor provided by state DOTs based on their cost histories. Moreover, an analytical step-wise procedure is developed here for the calculation of the expected DVE cost for MR&R work plans of any scale and duration without requiring detailed hourly traffic analyses.

In the following sections, the above cost terms are elaborated and discussed. These costs are then computed for a set of projects on three bridges in Ohio.

Cost Terms and Their Analysis Procedures

This section introduces the cost terms associated with bridge management, and their calculation procedures.

Agency Cost: MR&R Costs

Element-Level MR&R Costs. In general, MR&R costs of an element depend on the material and type of the element, current condition-state of portions of the element that are planned to be repaired, and their target condition-state. As mentioned in the previous section, in many research studies by state DOTs and other researchers, such element-level MR&R costs are presented (e.g., *15–18*). However, the current condition-states corresponding to these costs are based on a 1–5 condition-state rating system, which should be converted to those based on the new 1–4 AASHTO recommended rating system. This task can be accomplished through the application of the conversion tables presented by AASHTO (*19*). Owing to the complexities involved in this process, we also suggest a simple and practical mapping procedure that is based on the general definitions of condition-states and feasible actions that are suggested for these condition-states.

In this procedure, one needs to first identify the material and type of the element. For elements composed of concrete or steel, the following conversion is proposed: condition-states 1, 4, and 5 of the old system are equivalent to condition-states 1, 3, and 4 in the new rating system, and condition-states 2 and 3 of the old system are equivalent to condition-state 2 in the new rating system. It is worthy to note that at least for steel elements, the considered rule is identical to that presented by AASHTO (*19*). An issue in this mapping process is that in some cases there are two different cost values for one action in the 1–5 rating system; one starting from condition-state 2, the other starting from condition-state 3. In these cases, the cost should be averaged and stored as the cost of that action when prior condition-state in the new 1–4 rating system is 2.

For other elements, such as embankments, bearing devices, and drainage systems, the definitions of the 1–5 and 1–4 rating systems are compared and the equivalent current condition in the new 1–4 rating system is identified. For instance, the repair costs of the drainage system that are reported by Hearn (*15*) for the state of Florida are used for MR&R cost calculation of the sample bridges in this article. For this element, cost values for repair actions are provided for four current condition-states: "excellent," "minor deterioration," "moderate deterioration," and "major deterioration." According to the Ohio DOT inspection manual (*21*), condition-states 1–4 for a drainage system are defined as "excellent" with no clog, "satisfactory" with "minor deficiency to drainage system," "poor" with "advanced deterioration," and "critical" with "major deterioration." Thus, there is a one-to-one mapping between the prior condition-states "excellent," "minor deterioration," "moderate deterioration," and "major deterioration," and the new 1–4 AASHTO rating system.

Furthermore, to identify condition-state improvements resulting from taking MR&R actions on elements, the following rules are established:

- Replacing any element improves the condition of that element to condition-state 1.

- For elements that are made from concrete or steel material, all other MR&R actions, such as repairing and major rehabilitation, improve the condition-state to 2 (*20*).
- For elements made from materials other than steel or concrete, the definitions of condition-states 1–4 together with the description of the MR&R action that is applied to the element can be used to determine the condition-state improvements. In these cases, in addition to replacing the unit, a major rehabilitation may also improve the element to condition-state 1. For instance, "having no clog" is defined as condition-state 1 for a drainage system. When the drainage system is in condition-state 2, and one performs flushing, debris are expected to be removed and thus the condition-state of that drainage element will be 1.

Based on the above rules, all actions improve the elements to condition-state 1 or 2. This is also identical to the default assumption in the AASHTO BrM bridge management software (*1*).

Owing to differences between the unit quantity of some elements in the new inspection reports based on the AASHTO rating system and the cost units that are available for MR&R actions for those elements, realistic assumptions for the physical characteristics of those elements may be required for the estimation of MR&R costs. For instance, MR&R costs of pier columns/bents are available in linear foot, whereas the unit quantity of columns in the inspection reports is "each." Considering a realistic value for the width of columns (e.g., 2.5 feet), MR&R costs of pier columns in the unit of "each" can be estimated.

In addition, if MR&R costs particular to state B are employed for state A, the following procedure should be used to adjust the costs:

$$\text{Cost in state A} = \frac{\text{Adjustment factor for the state A}}{\text{Adjustment factor for state B}} \times \text{Cost in state B} \quad (1)$$

where state adjustment factors are provided by the US Army Corps of Engineers (*22*). Notably, since the utilized cost values are reported in previous years (e.g., year n), net present value of each cost term should be calculated, given by:

$$\text{Cost in previous year} = \frac{\text{Annual average CPI for previous year}}{\text{Annual average CPI for year } n} \times \text{Cost in year } n \quad (2)$$

$$\text{Cost in current year} = (1.03) \times \text{Cost in previous year} \quad (3)$$

where CPI is the consumer price index, given in table 24 of the *Consumer Price Index Detailed Reports* (*23*). In addition, 3% is considered as the rate of increase for costs from previous to current year.

Project-Level MR&R Costs. A project-level MR&R work plan may be at component (i.e., approach, deck, superstructure, substructure, culvert, channel, or signs/utilities), bridge, or network level (i.e., a portfolio of bridges). At the component level, the availability of crew and materials for repair actions of constituent elements means the costs associated with mobilization of crew and materials per element are less than those if only one element was to be repaired. A reduction factor, α_C, is considered here to account for these effects in the computation of MR&R costs. With the same analogy, a reduction factor, α_B, is considered for the calculation of bridge-level MR&R costs from component-level costs. Consequently, component- and bridge-level MR&R costs can be calculated as:

$$\text{MR\&R}^C = \alpha_C \times \sum_{e=1}^{M_C} MR\&R_e^C \quad (4)$$

$$\text{MR\&R}^B = \alpha_B \times \sum_{c=1}^{M_B} MR\&R_c^B \quad (5)$$

where $\text{MR\&}R_e^C$ and $\text{MR\&}R_c^B$ are the MR&R cost of element e in component C, and MR&R cost of component c in bridge B, respectively. The terms M_c and M_B stand for the total number of the elements in component C and the total number of components in bridge B, respectively. Agencies can calibrate these reduction factors α_C and α_B based on their cost histories. In this study, α_C and α_C for all components and bridges are taken as 0.8 and 0.9, respectively. It is also worth noting that some state DOTs such as Caltrans (*24*) have provided cost ranges for unit construction cost per deck area of various types of bridges based on their cost histories. These values can be adopted for the calculation of the replacement cost of bridges, which is a substitute for calculating the cost of bridge replacement according to Equation 5.

For network-level MR&R cost calculation, reduction factors can be disregarded, since bridges are mostly apart and thus the MR&R actions on each bridge are performed independently with new laborers and equipment.

Agency Cost: MOT Costs

To assure safety of workers and drivers, and direct traffic smoothly alongside a bridge repair site, agencies protect the working area with the help of crew and equipment, and if necessary utilizing police enforcement. According to the Ohio DOT Office of Estimation in 2016, "three laborers, one arrow board, one truck with attenuator, and one truck/flatbed for barrel placement and removal" to maintain the traffic cost \$260/hour on average (*11*, *12*). Law enforcement cost is also estimated as \$65/hour. To systematically calculate MOT cost of a repair work plan with any duration, the following logical assumptions are considered:

- Following the discussion with personnel from the Ohio DOT Office of Estimating and Road Department, the average number of hours that bridge laborers work is 8 hours/day.
- On weekends no worker is present, and therefore the cost of MOT over that period is reduced to the equipment that directs the traffic. As an estimation, the \$260/hour unit cost for "three laborers, one arrow board, one truck with attenuator, and one truck/flatbed for barrel placement and removal" is reduced by 60% for weekends.
- The \$65/hour cost of law enforcement is not considered for weekends.
- Police enforcement is assumed to be present at working sites where more than 40% of the road on the bridge is closed.

Thus, for an MR&R project at level l (element, component, bridge level), MOT cost is calculated as:

$$\begin{aligned} \mathrm{MOT}^l = &\left(8\times T^l\times \$260 + 8\times T^l\times F^{N_{Cl}}\times \$65 + 16\times T^l\times 0.4\times \$260\right)\\ &+\left(2\times \frac{T^l}{7}\times 24\times 0.4\times \$260\right) \end{aligned} \tag{6}$$

where T^l is the duration of performing the repair work plan, which is at level l. The term $F^{N_{Cl}}$ is a factor taking a value of 1 or 0, indicating the presence/non-presence of police officers. Identifying the number of closed lanes, N_{Cl}, depends on many factors. Without a need for detailed hourly traffic analyses, a method is proposed to assist agencies with optimal decision on the number of closed lanes based on average daily traffic. This also results in more accurate calculation of the MOT cost. Given that the user cost of DVE depends on T^l and N_{Cl}, step-wise algorithms for objective estimation of these two parameters are presented later in the paper, where calculation procedures for DVE costs are elaborated.

Agency Cost: AEM Costs

AEM costs associated with an MR&R project at level l (element, component, bridge level) can be estimated by:

$$\mathrm{AEM}^l = \beta\times\left(\mathrm{MOT}^l + \mathrm{MR\&R}^l\right) \tag{7}$$

where β is the overhead factor. The Ohio DOT Structure Team suggests 0.25 for this factor. In general, this factor depends on the scale, type, and configuration of bridges, among others, and state DOTs may calibrate this factor based on their cost histories.

User Cost: DVE Costs

In performing MR&R actions on members of a bridge, the bridge may be partially or completely closed. This affects the traffic that passes the bridge, since car and truck drivers need to either reduce their speed or decide to drive through detours if available. For each project at level l, DVE cost is calculated using:

$$\begin{aligned} \mathrm{DVE}^l = &\,T^l\times\left(t_{ij}^{D/R} - t_{ij}^{O}\right)\\ &\times\left[(\mathrm{ADT} - \mathrm{ADTT})\times\rho_C + \mathrm{ADTT}\times\rho_T\right] \end{aligned} \tag{8}$$

where t_{ij}^{O} is the average original time that is spent by drivers to drive from point i to point j when no repair actions are performed on the bridge, whereas $t_{ij}^{D/R}$ is the average time to drive from point i to point j when repair projects are conducted on the bridge. Therefore, $\left(t_{ij}^{D/R} - t_{ij}^{O}\right)$ is the extra time spent by users caused by speed reduction and/or traffic rerouting because of repair actions on the bridge. Finally, ρ_C and ρ_T are the unit user costs for cars and trucks caused by delay time, vehicle operation, and excess gas emission, respectively. These values can be found in the literature (*25*, *26*). Following Boccini and Frangopol (*27*), t_{ij}^{O} can be computed by:

$$t_{ij}^{O} = t_{ij}^{F}\times\left[1+\alpha\left(\frac{f_{ij}}{f_{ij}^{c}}\right)^{\beta}\right] \tag{9}$$

In Equation 9, t_{ij}^{F} is the time required to cover path ij at free-flow speed, which passes through the bridge. This can be estimated by the path length and the original speed limit of the path. The term f_{ij} represents the ADT on the highway segment ij. The term f_{ij}^{c} stands for the critical flow (maximum flow capacity) of the bridge, and α and β are also parameters considered as 0.15 and 4, respectively (used by Boccini and Frangopol (*27*)). Moreover, in Equation 9, $t_{ij}^{D/R}$ is calculated according to the formulation presented by Boccini and Frangopol (*27*) as follows:

$$\begin{aligned} t_{ij}^{D/R} = &\,t_{ij}^{R}\times\left[1+\alpha\left(\frac{f_{ij}}{f_{ij}^{c}}\right)^{\beta}\right]\\ &+\sum_{b\epsilon ij} s_{b,ij}\times t_{b,ij}^{d}\times\left[1+\alpha\left(\frac{s_{b,ij}\cdot f_{ij}}{f_{b,ij}^{c}}\right)^{\beta}\right] \end{aligned} \tag{10}$$

where t_{ij}^{R} is the free-flow time for drivers to travel from point i to point j through the bridge that is affected by a repair project. In this case, the speed limit on the bridge is usually reduced compared to its original value, depending on the original speed limit, the protection required for the work zone, and whether workers are present in the work zone (*28*). The term $f_{b,ij}^{c}$ stands for the critical flow (maximum flow capacity) of detour b joining points i and j. In addition, $t_{b,ij}^{d}$ is the time required to pass from point i to j through detour b at free-flow speed. Finally, $s_{b,ij}$ is the fraction of traffic from point i to j that passes through detour b, which on a logical basis can be considered as the ratio of the closed lanes on the bridge to the total number of lanes on that bridge. In this research, an optimization procedure is developed to identify this factor by finding the scenario for the number of closed lanes that minimizes the incurred costs of MOT and DVE for the duration of the repair project. The flowcharts for identifying the optimal number of closed lanes together with the calculation of the associated MOT and DVE costs for repair and bridge replacement work plans are presented in Figures 1 and 2, respectively. As can be seen, estimation of the MOT and DVE costs depends on the duration of the repair/bridge replacement work plan. Correct identification of this parameter is important to arrive at accurate cost calculations. Thus, this study proposes analytical formulations for the estimation of the required time for conducting MR&R work plans.

Element-Level Duration of MR&R Actions. At the element level, formulas are developed that calculate the durations as a function of the quantity of elements receiving MR&R actions, and the type of actions (i.e., repair or replacement), as follows:

- For elements, for which MR&R costs are calculated based on the unit of "each," such as bearing devices, durations of MR&R actions, T^{E}, are calculated as follows:

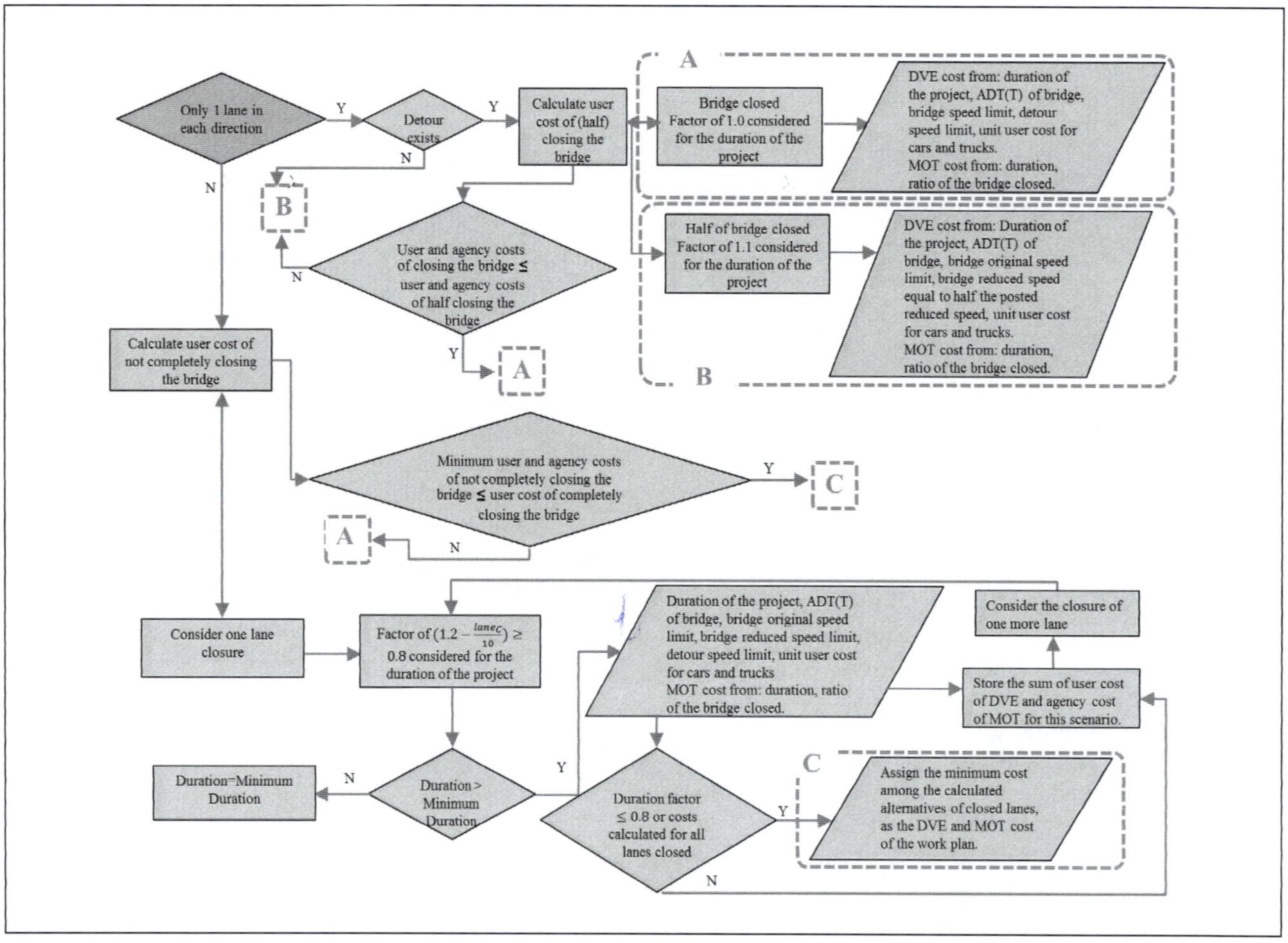

Figure 1. The flowchart for identifying the optimal number of closed lanes, together with the calculation of the associated MOT and DVE costs for a repair work plan.

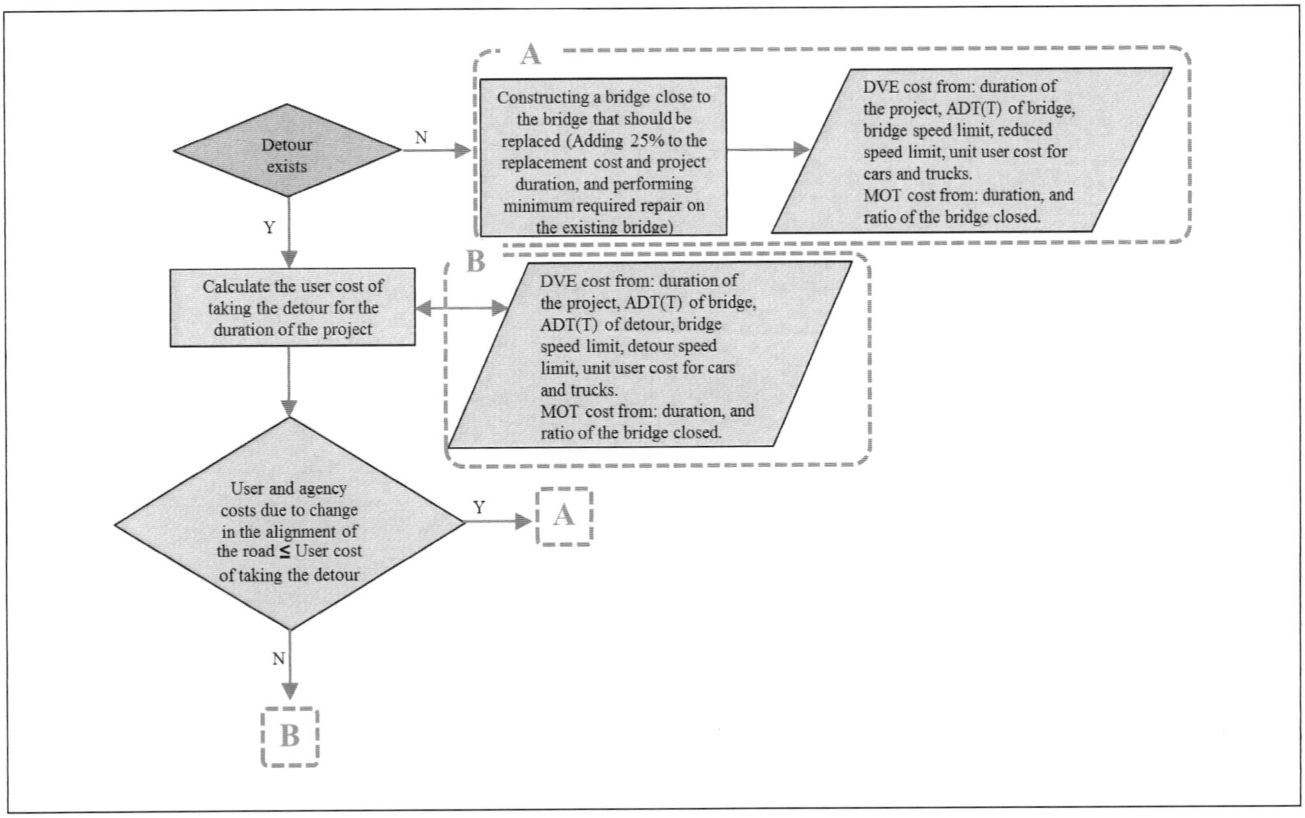

Figure 2. The flowchart for identifying the optimal number of closed lanes, together with the calculation of the associated MOT and DVE costs for a bridge replacement work plan.

$$T^E = RF^E \times \left(N^E \times t_T^E\right) \quad (11)$$

where N^E is the number of element E receiving an MR&R action of type T (e.g., 15 bearing devices receiving repair), t_T^E is the required duration of performing the MR&R action of type T on one element, and RF^E is the reduction factor particular to element E. The latter accounts for the reduced time of the MR&R action on element E when more than one of that element receives MR&R actions. This is because of the reduced time required for mobilization of crew and materials. This term RF^E is an inverse function of the number of elements improved by the MR&R action. As an example, we suggest $\frac{1}{\left(1+\frac{N^E}{100}\times 3\right)} \geqslant \frac{1}{4}$ as the reduction factor for the bearing device element, and $1\,\frac{\text{day}}{\text{each}}$ for the repair action on this element.

- For other elements, T^E is calculated as:

$$T^E = c \times \left(a^E + b^E \times Q^E\right) Q^E > 0 \quad (12)$$

where a^E and b^E are constant values that are particular to element E, c is a factor indicating the type of MR&R action; 0.75 for repair and 1.00 for replacement action, and Q^E is the quantity of element E. For instance, for floor/slab elements, authors suggest 1.05 and 0.005 for a^E and b^E, respectively.

It is worth noting that the existing factors in Equations 11 and 12 should be adjusted based on agencies' information on the duration of MR&R activities and/or engineering elicitation. Furthermore, since the calculated element-level durations are the required *working days* to perform MR&R actions, the total duration of a work plan, including weekdays and weekends, is equal to $T^E + 2 \times \lfloor \frac{T^E}{7} \rfloor$, with $\lfloor\rfloor$ as the floor of the ratio $\frac{T^E}{7}$, which determines the minimum number of weekends that the project faces. The same statement is true for project-level duration of MR&R work plans, which is explained in the next section.

Project-Level Duration of MR&R Actions. Component- and bridge-level work plan durations, T^C and T^B, can be estimated through reduction factors ω^C and ω^B, applied to

Table 1. Specifications of the Three Sample Bridges

Structure file no.	Inventory bridge no.	Bridge type	Year built	No. of spans	No. of lanes	Length (ft)	Average daily traffic	Detour length (mile)	General appraisal
2513927	I.R. 270 over CSX RR & private road	Steel/beam/continuous	1968	3	3	178.0	135,746	1	7
2590271	Drive way over dry run	Prestressed concrete/ box beam/continuous	1992	3	2	110.0	50	99	7
2504316	170 over Hague Avenue	Steel/beam/continuous	1973	2	8	113.8	143,747	1	7

the sum of the required working days for the constituent elements/components. In mathematical terms:

$$T^C = \omega^C \times \sum_{k=1}^{M_C} T_C^e \tag{13}$$

$$T^B = \omega^B \times \sum_{c=1}^{M_B} T_B^c \tag{14}$$

where T_C^e and T_B^c are the required time for performing MR&R action(s) on element e of component C, and on component c of bridge B, respectively.

The foregoing reductions in the project-level MR&R work plans are the result of the reduced time required for the mobilization of crew and material for MR&R actions on constituent members. Based on engineering judgment, we suggest 0.9 and 0.75 for ω^B and ω^C, respectively. However, for the replacement of bridges, owing to the availability of large spaces for the work plan requiring complete closure of the bridge, ω^B is suggested as 0.8. State DOTs and other entities may conduct more objective evaluations to determine these factors.

Cost Calculations of Case Study Bridges

This section presents the application of the proposed cost-calculation procedures for three bridges in Ohio. These three bridges are selected by the Ohio DOT and their specifications are presented in Table 1. All of them have the same ratings of general appraisal of 7. Bridge #2590271 has a unique feature with no detour and low average daily traffic, while Bridges #2504316 and 2504316 have heavy traffic.

Following AASHTO (*19*) and the Ohio DOT (*21*), bridge elements are categorized into three groups: national bridge elements (NBE), bridge management elements (BME), and agency developed elements (ADE). In a collaborative effort with the Ohio DOT, Fereshtehnejad et al. (*11*, *12*) established minimum structural/operational thresholds for the condition-state of these elements, as well as for the defects associated with specific bridge elements (*21*). The condition-rating system follows the most recent rating system suggested by AASHTO (*19*, *20*), which ranges from 1 (good) to 4 (severe). On this basis, an NBE, primary ADE, or defect flag with more than 2% in condition-state 3 or any quantity in condition-state 4, or a BME or non-primary ADE with more than 10% in condition-state 3 and 4, is considered structurally/operationally unsafe. In this case, repair actions should be performed on the quantities in condition-state 3 and 4 to improve them to at least condition-state 2. This level of effort for repair actions is considered as "minimum repair" in this paper. Moreover, Fereshtehnejad et al. (*11*, *12*) suggested another extent for repair actions that improves the condition of a bridge member to its like-new state. For this goal, the quantities of constituent elements of a member in condition-state 3 or 4 should be repaired to be improved to at least condition-state 2, and those quantities in condition-state 2 should be maintained to stay in condition-state 2. This type of repair is referred to as "like-new repair" in this article. Based on the cost formulations and considerations in previous sections, the agency-required budget for the minimum and like-new repair actions for the three sample bridges are estimated as $667,540 and $814,441 for Bridge #2513927, $138,313 and $271,918 for Bridge #2590271, and $20,691 and $527,945 for Bridge #2504316. Notably, these costs do not exceed the agency cost incurred by the replacement of the entire bridge.

Detailed budget information for MR&R actions of each bridge are given in Tables 2–4. The required budget for minimum repairs on Bridge #2513927 exceeds $667,000 since a great portion of elements on this bridge require moderate to extensive repairs. For example:

- 1,400 ft^2 of the approach wearing surface in condition-state 3 needs moderate repair.
- 14 bearings in condition-state 3 and 4 need either extensive repair or replacement.
- 12 diaphragms that have severe deficiencies and are reported in condition-state 4 should be replaced.
- The entire 112 ft of backwalls in condition-state 4 should receive extensive repair. The latter contributes the most to the MR&R cost of the bridge.

Table 2. Disaggregating the Required Budget for Minimum Repairs on Bridge #2513927

Component	Description	Element MR&R cost ($)	Total agency cost ($)	Bridge MR&R cost ($)	Bridge AEM cost ($)	Bridge MOT cost ($)	Bridge DVE cost ($)	Duration (days)
Approach	Repair 1400 ft^2 of the approach slab in condition-state 3	146,564	667,546	454,730	133,510	79,306	297,146	21
Deck	–	–						
Superstructure	Repair 28 ft of the girder in condition-state 3	5,125						
	Replace 12 diaphragm/X-frames in condition-state 4	94,695						
	Repair seven bearing devices in condition-state 3 and replace seven bearing devices in condition-state 4	58,834						
Substructure	Extensive repair the 112 ft of backwalls in condition-state 4	231,760						

Table 3. Disaggregating the Required Budget for Minimum Repairs on Bridge #2590271

Component	Description	Element MR&R cost ($)	Total agency cost ($)	Bridge MR&R cost ($)	Bridge AEM cost ($)	Bridge MOT cost ($)	Bridge DVE cost ($)	Duration (days)
Approach	Repair 258.5 ft^2 of the approach slab in condition-state 3 and 4	28,200	138,313	70,015	27,663	40,636	9,080	10
	Repair entire embankments in condition-state 4	1,860						
Deck	Repair 690 ft^2 of the wearing surface in condition-state 3 and 4	49,594						
	Repair the 10 ft of railings in condition-state 3	1,377						
	Repair the existing two drainage system in condition-state 3	685						
	Repair the 40 ft of expansion joint in condition-state 3	1,000						
Superstructure	–	–						
Substructure	–	–						
Channel	–	–						
Sign	–	–						

Furthermore, this project requires 21 days, incurring about \$73,000 for the MOT cost and \$134,000 for AEM cost. The same analysis is conducted for Bridge #2590271 for the minimum required actions, and for Bridge #2504316 for the project that improves all elements to their like-new state. As expected, the MOT cost of Bridge #2504316 is as high as the MR&R cost since the required time for conducting like-new repairs on this bridge is significant at 66 days. Moreover, the DVE costs of Bridge #2590271 and Bridge #2504316 are estimated as high as the total agency costs because of the heavy traffic that passes these bridges.

Table 4. Disaggregating the Required Budget for Like-New Repairs on Bridge #2504316

Component	Description	Element MR&R cost ($)	Total agency cost ($)	Bridge MR&R cost ($)	Bridge AEM cost ($)	Bridge MOT cost ($)	Bridge DVE cost ($)	Duration (days)
Approach	Maintain 8,520 ft^2 of the approach wearing surface in condition-state 2, e.g., sealing or patching	182,570	527,945	221,280	105,590	201,080	695,438	66
	Maintain 100 ft^2 of the slab in condition-state 2	2,500						
Deck	Maintain 120 ft^2 of the floor/slab in condition-state 2	3,000						
	Maintain 3 ft^2 of the edge of floor/slab in condition-state 2	75						
	Maintain 200 ft^2 of the deck wearing surface in condition-state 2	4,286						
	Maintain 2 ft of the railing system in condition-state 2	78						
	Repair the 95 ft of expansion joint in condition-state 3, and maintain the 300 ft rest of the expansion joint in condition-state 2	2,375						
Superstructure	Maintain 54 bearing devices in condition-state 2, e.g., cleaning, painting, and greasing or rehabilitating them	4,655						
	Maintaining the 2,050 ft of the protective system in condition-state 2	40,307						
Substructure	Maintaining the 40 ft of the abutment walls in condition-state 2	5,791						
	Repairing the 5 ft of defected pier cap in condition-state 3	4,423						
	Repairing 2 ft of backwalls in condition-state 4, and maintaining the 20 ft of backwalls in condition-state 2	8,700						
Sign	–	–						

As a verification for the calculated costs, Ohio DOT engineers independently calculated the agency cost of performing minimum repairs for Bridge #2513927, based on their preliminary bridge work estimates and their engineering judgment. They estimated the cost as \$618,902, which is 7% less than the calculated cost using the proposed procedure (Table 2).

It should be mentioned that the accuracy of the cost estimates considerably relies on the accuracy of the input element-level costs and parameters in cost-calculation formulas. Thus, to enhance the accuracy of cost estimates, agencies and state DOTs may determine those element-level costs and parameters based on their bid histories.

Conclusion

This paper proposes reliable and practical algorithms for the calculation of a comprehensive list of agency and user costs that are incurred as a consequence of performing MR&R actions on bridges. These costs include agency cost of AEM; agency direct cost of performing MR&R actions; agency cost of MOT; and user cost incurred from DVE. The developed procedures for the calculation of these cost terms are systematic, and utilize information that is available to DOTs. These features enable the proposed methods to be applicable to large inventories of bridges.

The recommended procedures are employed for the estimation of the agency and user costs associated with a set of light to extensive repair actions on three bridges in Ohio. Detailed descriptions of the considered element-level MR&R actions are provided and the associated MR&R costs, together with the expected project-level duration, and MR&R, MOT, and DVE costs of such work plans are calculated. Some of these cost terms were determined independently by Ohio DOT engineers. Comparison of these costs with those determined through the proposed procedures showed a good agreement (difference of less than 10%), indicating that the developed formulations and cost-calculation processes are reliable. State DOTs and other entities can benefit from the proposed cost-calculation procedures to identify the most cost-effective repair work plans and leverage the produced information for management of their bridge inventories.

Acknowledgments

This research is supported by the Ohio Department of Transportation through Grant 135240. This support is greatly appreciated. Any opinions, findings, and conclusions or recommendations expressed in this article are those of the authors and do not necessarily reflect the views of the sponsor.
The authors would like to also thank Mr. Tim Keller and Ms. Cynthia Jones from the Ohio Department of Transportation for providing valuable practical insights and effective coordination of this research project.

Author Contributions

The authors confirm contribution to the paper as follows: study conception and design: EF, JH, AS, MB, AW; data collection: BN, JB, AW; analysis and interpretation of results: EF, JH, AS, MB; draft manuscript preparation: EF, JH, AS. All authors reviewed the results and approved the final version of the manuscript.

References

1. Bentley Advancing Infrastructure. *AASHTOWare Bridge Management User Manual*. American Association of State Highway Transportation Officials, Washington, D.C., 2017.
2. Patidar, V., S. Labi, K. C. Sinha, and P. Thompson. *NCHRP Report 590: Multi-Objective Optimization for Bridge Management Systems*. Transportation Research Board of the National Academies, Washington, D.C., 2007.
3. Fereshtehnejad, E., and A. Shafieezadeh. Optimal Retrofit Decision-Making for Bridge Systems Based on Multi-Hazard Lifecycle Cost Analysis. *Proc., 11th International Bridge and Structure Management Conference*, Mesa, AZ, April 26–27, 2017.
4. Fereshtehnejad, E., and A. Shafieezadeh. A Randomized Point-Based Value Iteration POMDP Enhanced with a Counting Process Technique for Optimal Management of Multi-State Multi-Element Systems. *Structural Safety*, Vol. 65, 2017, pp. 113–125.
5. Chang, S. E., and M. Shinozuka. Life-Cycle Cost Analysis with Natural Hazard Risk. *Journal of Infrastructure Systems*, Vol. 2, No. 3, 1996, pp. 118–126.
6. Padgett, J. E., K. Dennemann, and J. Ghosh. Risk-Based Seismic Life-Cycle Cost–Benefit (LCC-B) Analysis for Bridge Retrofit Assessment. *Structural Safety*, Vol. 32, No. 3, 2010, pp. 165–173.
7. Shirole, A. M., and M. J. Loftus. Assessment of Bridge Vulnerability to Hydraulic Failures. *Transportation Research Record: Journal of the Transportation Research Board*, 1992. 1347: 18–24.
8. Daher, B. W. *Use of Sensors in Monitoring Civil Structures*. Master's thesis. Massachusetts Institute of Technology, Cambridge, MA, 2004.
9. Ellis, H., M. Jiang, and R. Corotis. Inspection, Maintenance, and Repair with Partial Observability. *Journal of Infrastructure Systems*, Vol. 1, No. 2, 1995, pp. 92–99.
10. Saydam, D., and D. M. Frangopol. Risk-Based Maintenance Optimization of Deteriorating Bridges. *Journal of Structural Engineering*, Vol. 141, No. 4, 2014, p. 04014120.
11. Fereshtehnejad, E., J. Hur, and A. Shafieezadeh. Performance Measures for the Assessment of the Condition of Bridges: A Critical Review. Presented at 96th Annual Meeting of the Transportation Research Board, Washington, D.C., 2017.
12. Fereshtehnejad, E., J. Hur, A. Shafieezadeh, and M. Brokaw. Ohio Bridge Condition Index Multilevel Cost-Based Performance Index for Bridge Systems. *Transportation Research Record: Journal of the Transportation Research Board*, 2017. 2612: 152–160.
13. Shepard, R. W., and M. B. Johnson. *TR News 215: California Bridge Health Index: A Diagnostic Tool to Maximize Bridge Longevity, Investment*. Transportation Research Board of the National Academies, Washington, D.C., 2001.
14. Jiang, X. *Analysis of Bridge Health Index for the City and County of Denver, Colorado*. PhD dissertation. University of Colorado at Denver, Denver, CO, 2012.
15. Hearn, G. *Deterioration and Cost Information for Bridge Management*. Report No. CDOT-2012-4. Colorado Department of Transportation, Denver, CO, 2012.
16. Hearn, G. *NCHRP Report 668: Framework for a National Database System for Maintenance Actions on Highway Bridges*. Transportation Research Board of the National Academies, Washington, D.C., 2010.
17. Adams, T. M., and E. Juni. *Element Unit and Failure Costs and Functional Improvement Costs for Use in the Mn/DOT Pontis Bridge Management System*. No. MN/RC-2004-05.

Minnesota Department of Transportation, Saint Paul, MN, 2003.
18. Sobanjo, J. O., and P. D. Thompson. *Development of Agency Maintenance, Repair & Rehabilitation (MR&R) Cost Data for Florida's Bridge Management System*, Final Report. Florida Department of Transportation, Tallahassee, FL, 2001.
19. American Association of State Highway and Transportation Officials. *AASHTO Bridge Element Inspection Guide Manual.* 1st ed. American Association of State Highway and Transportation Officials, Washington, D.C., 2010.
20. American Association of State Highway and Transportation Officials. *2015 Interim to AASHTO Guide Manual: Manual for Bridge Element Inspection.* 1st ed. American Association of State Highway and Transportation Officials, Washington, D.C., 2013.
21. Ohio Department of Transportation. *Manual of Bridge Inspection*, revised ed. Ohio Department of Transportation, Columbus, OH, 2014.
22. U.S. Army Corps of Engineers. *Civil Works Construction Cost Index System (CWCCIS).* No. 1110-2-1304. US Army Corps of Engineers, Washington, D.C., 2017.
23. Crawford, M., J. Church, and B. Akin, eds. *CPI Detailed Report.* Office of Prices and Living Conditions, Bureau of Labor Statistics, Washington, D.C., 2017.
24. California Department of Transportation (Caltrans). *Construction Statistics 2014.* California Department of Transportation, Sacramento, CA, 2015.
25. Mallela, J., and S. Sadasivan. *Work Zone Road User Costs Concepts and Applications.* Report No. FHWA-HOP-12-005. Federal Highway Administration, Washington, D.C., 2011.
26. Copeland, L. *User's Manual for QYEWZ-98.* Texas Transportation Institute, The Texas A&M University System, College Station, TX, 1998.
27. Bocchini, P., and D. M. Frangopol. Optimal Resilience- and Cost-Based Postdisaster Intervention Prioritization for Bridges along a Highway Segment. *Journal of Bridge Engineering*, Vol. 17, No. 1, 2010, pp. 117–129.
28. Minnesota Department of Transportation. *Speed Limits in Work Zones Guideline.* Minnesota Department of Transportation, Saint Paul, MN, 2014.

The Standing Committee on Bridge Preservation (AHD37) peer-reviewed this paper (18-05384).

Article

TRR
JOURNAL OF THE TRANSPORTATION RESEARCH BOARD

Transportation Research Record
2018, Vol. 2672(12) 127–136

DOI: 10.1177/0361198118796380
journals.sagepub.com/home/trr
SAGE

Sunlight Degradation of Polymeric Detectable Warning Surface Products

Sukjoon Na[1], Siavash Vahidi[2], Hieu Nguyen[3], Adel ElSafety[4], and Grace Hsuan[2]

Abstract

Detectable warning surface (DWS) is a panel product installed at the edge of curb ramps to warn visually impaired pedestrians about the proximity of the roadway. Most DWS products are made of polymers and are subjected to outdoor weathering. Therefore, sunlight degradation is an important factor that affects their service life. In this study, the effects of sunlight on material degradation in four DWS products made from polyester, neopentylglycol (NPG), polyurethane, and polyolefin were evaluated by exposing DWS test coupons in a xenon weatherometer. The exposure conditions were largely based on ASTM D 2565 for a duration of 3,000 h. Color change induced by radiation was measured using a spectro-colorimeter. The fastest discoloration was measured in DWS made from polyurethane. The largest color change occurred between 500 and 1,000 h for polyurethane and between 1,000 and 3,000 h for other polymers. The changes of surface appearance observed under a digital microscope were consistent with the discoloration. The reinforcing fibers in DWS products made from polyester and polyurethane were revealed after 3,000 h exposure. Decrease in surface wear resistance was obtained in tested coupons that were exposed to the highest irradiance level. The product made from NPG exhibited the greatest decrease in wear resistance, while that made from polyolefin showed the least change.

In order to improve the accessibility of pathways for disabled pedestrians, curb ramps are commonly used around the world (*1*). For example, in the U.S.A., curb ramps are extensively used in response to the requirements of the Americans with Disabilities Act (ADA) (*2*). However, an undesirable consequence is that the gradual transition from the curb level to the street level makes it difficult for visually impaired pedestrians to locate the boundary between the sidewalk and the street. Implementing audible contrast may not be very effective because of the level of noise in the street environment. To address this issue, a warning surface is installed at the edge of the ramp to provide a sensible contrast with the adjacent walking area. Different countries, including Sweden, the UK, the U.S.A., and Japan have considered using warning systems that rely on tactile and visual contrast (*1, 3–5*). Truncated dome detectable warning surface (DWS) has proved to be easy to detect by most vision-impaired pedestrians either by foot or a cane (*6*). Chapter-7 of the 2010 ADA Standards for Accessible Design, published by the United States Department of Justice (*7*), defines DWS as a "standardized surface built in or applied to walking surfaces or other elements to warn of hazard on a circulation path." The chapter outlines tactile and visual requirements for DWS. Tactile requirements consist of specifications regarding the dome size and dome spacing to balance ease of detection and trip risk (*8*). The visual requirement states that "Detectable warning surfaces shall contrast visually with adjacent walking surfaces either light-on-dark, or dark-on-light."

Polymers are the most common material used in the production of DWS since they are easy to shape, cost effective, lightweight, and resistant to corrosion. Manufacturers often produce DWS in different colors so that the required contrast can be achieved in various ambient environments. DWS is used in the outdoor environment, and thus is subjected to weathering, particularly sunlight. When sunlight radiation is absorbed by

[1]Department of Civil and Environmental Engineering, Rowan University, Glassboro, NJ
[2]Department of Civil, Architectural and Environmental Engineering, Drexel University, Philadelphia, PA
[3]Department of Mechanical Engineering and Mechanics, Drexel University, Philadelphia, PA
[4]Department of Civil Engineering, University of North Florida, Jacksonville, FL

Corresponding Author:
Address correspondence to Grace Hsuan: ghsuan@coe.drexel.edu

the polymer, photons that have enough energy break the chemical bonds and cause reactions leading to chain scission and/or crosslinking (*9–12*). These processes can eventually cause deterioration of the mechanical properties, such as wear resistance, of the DWS. Consequently, the abrasion caused by pedestrian traffic can compromise the tactile properties of DWS over time. Sunlight degradation also causes discoloration, which could affect the visual properties of the DWS, resulting in insufficient contrast with the surrounding surface.

Different additives such as color pigments, photo-stabilizers, thermal stabilizers, and filler are likely to be incorporated in DWS during fabrication. Therefore, the results of sunlight degradation studies on polymer resin cannot be extrapolated to the DWS manufactured from that resin. Investigating the effects of sunlight degradation on the actual product provides the most reliable assessment. Even though DWS manufacturers in the U.S. provide product specifications such as compressive, flexural, and tensile strength, chemical stain resistance, water absorption, and color change under accelerated weathering, the sunlight degradation requirements vary from product to product regarding the test method and conditions, as well as the exposure duration. The common property that is used to assess the material changes after exposure is the color. In this paper, the effect of artificial sunlight radiation on the mechanical and visual properties of four DWS products manufactured in the U.S. is evaluated.

Test Products/Materials

Four DWS products were selected according to polymer type such that they represent the polymers commonly used in DWS production. Samples were obtained in actual product sizes. Three colors (black, red, and yellow) were included in each product. Table 1 shows the physical properties and polymer type of the products. Coupons with a width of 1.5 in. and a length of 6 in. were cut from the samples using a band saw and were washed and dried before exposure.

Experiments

Laboratory Weatherometer Exposure

DWS coupons were exposed to simulated daylight inside an ATLAS Ci-4000 weatherometer equipped with a xenon lamp and inner and outer borosilicate filters. The exposure condition was set according to ASTM D2565 "Standard Practice for Xenon-Arc Exposure of Plastics Intended for Outdoor Applications" (*13*). Each 120-minute cycle was composed of 102 min of irradiation followed by 18 min of irradiation and water spray. The temperature was set at 63 $\pm$ 2° C, monitored by an uninsulated black panel. Three exposure conditions with irradiance levels of 41.5, 60, and 80 W/m^2 from 300 to 400 nm were applied to assess the effect of irradiance. The exposure duration for each condition was 3,000 h. The test coupons were retrieved and tested at predetermined time intervals.

Color Change

The effect of radiation on color change was measured using a spectro-colorimeter, X-rite RM200QC. The device uses multiple sensors to determine color parameters (light variable and chromaticity values) required for calculating ΔE (see Equation 1). ΔE represents the distance between two colors in a 3D color space defined by the International Commission on Illumination (CIE); E stands for "Empfindung" in German, meaning "sensation." Three coupons from each sample type were designated for this nondestructive test. Three locations (top, middle, and bottom) were marked on each coupon, and color reading was performed at each location. A total of nine measurements were obtained for each type of sample. The test was performed at 500-hour intervals:

$$\Delta E = \sqrt{(\Delta L^*)^2 + (\Delta a^*)^2 + (\Delta b^*)^2} \quad (1)$$

where;
L^* = lightness variable
a^* and b^* = chromaticity values

Table 1. Properties of the Tested Products

Product	Material	Reinforcement	Sample dimensions (in.)	Colors
P-1	Polyester	Fiberglass	48"× 24"	Black, red, yellow
P-2	Polyester (NPG)	Fiberglass	12"× 12"	
P-3	Polyurethane	None	24"× 24"	
P-4	Polyolefin	Fiberglass	8"× 8"	

Note: NPG = neopentylglycol.

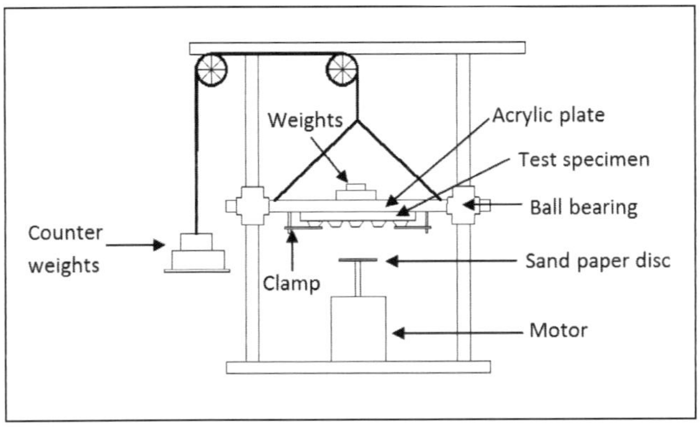

Figure 1. Abrasion test device.

Light Microscopy Examination

A Gaosuo digital microscope was used to examine the surface texture and crack trace on the exposed coupons at different exposure times. Microscopic pictures at a magnification of 100x were taken from two locations (top and bottom) on each color measurement coupon at every 1,000-hour time interval.

Abrasion Test

Sunlight degradation is a surface phenomenon, and thus degradation takes place at the surface and progresses inward. Therefore, surface abrasion test is an appropriate test as opposed to other mechanical tests such as tensile and bending test that measure the bulk property. Due to the dimensions and surface profiles of the DWS test coupons, commercially available abrasion devices are not applicable. A new abrasion test device was therefore designed and built for this study. Figure 1 shows the schematics of this device. The abrasive part consists of a 1-in. diameter sandpaper (120 grit) disk which is driven by a motor at a rotational speed of 1,550 rpm. The test specimen was clamped to an acrylic plate with its exposed surface facing down. Four ball bearings were installed on the corners of the acrylic plate enabling free movement along the steel rods. The weight of the acrylic plate and clamps was balanced by a counter-weight through a pulley system. A 380 g weight was applied on top of the sliding plate to ensure steady contact between the test specimen and the abrasion disk throughout the test duration. The abrasion duration was 60 s and mass loss was measured. A new sandpaper disk was used for each test specimen. Two coupons were retrieved at every 1,000 h of exposure time, and were then cut in half to obtain four specimens.

Results

Color Change

Figure 2*a* shows the ΔE values versus the exposure time for the three colors of the polyester DWS under the low irradiance level. The ΔE value generally increased with exposure time and then levelled off. Overall, the time to reach the maximum ΔE decreased as the irradiance intensity increased. The black color exhibited the smallest change of ΔE, while the yellow showed the largest change. Figure 2*b* shows ΔE values versus the exposure time for yellow coupons at three irradiance levels. The times to reach the maximum ΔE values were at 2,500, 2,000 and 1,500 h for low (41.5 W/m^2), intermediate (60 W/m^2) and high (80 W/m^2) irradiance, respectively.

For the NPG coupons, the changes in ΔE at intermediate and high irradiance are similar to those of the polyester coupons (see Figure 2*a*). However, at the low irradiance, ΔE increased continuously with the exposure

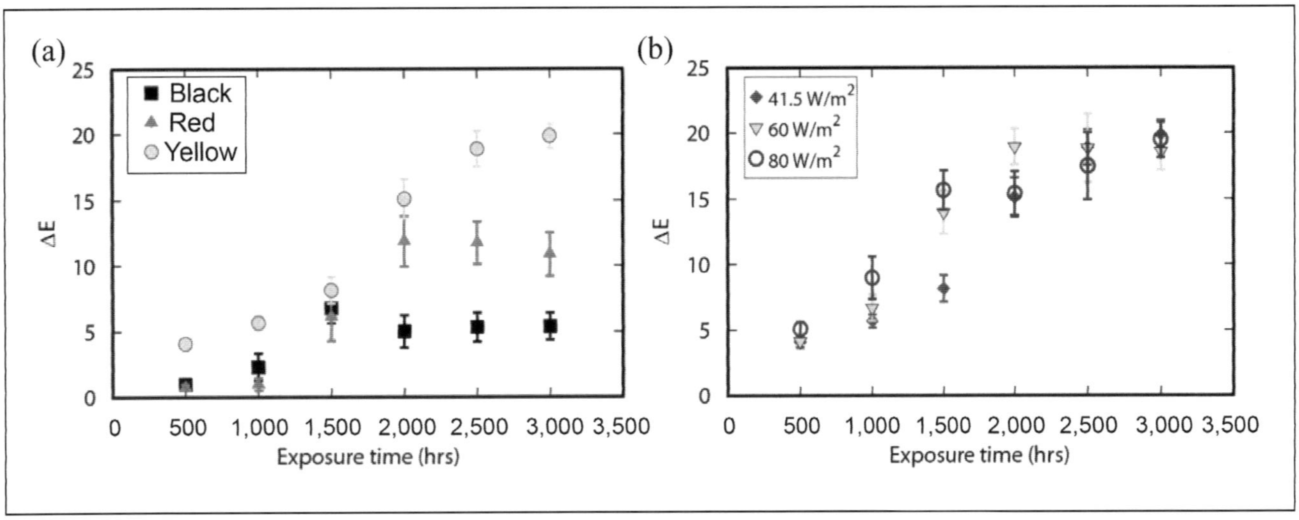

Figure 2. Color change (ΔE) with exposure time for DWS made from polyester: (*a*) different color coupons at low irradiance; (*b*) yellow color coupons at three irradiance levels.

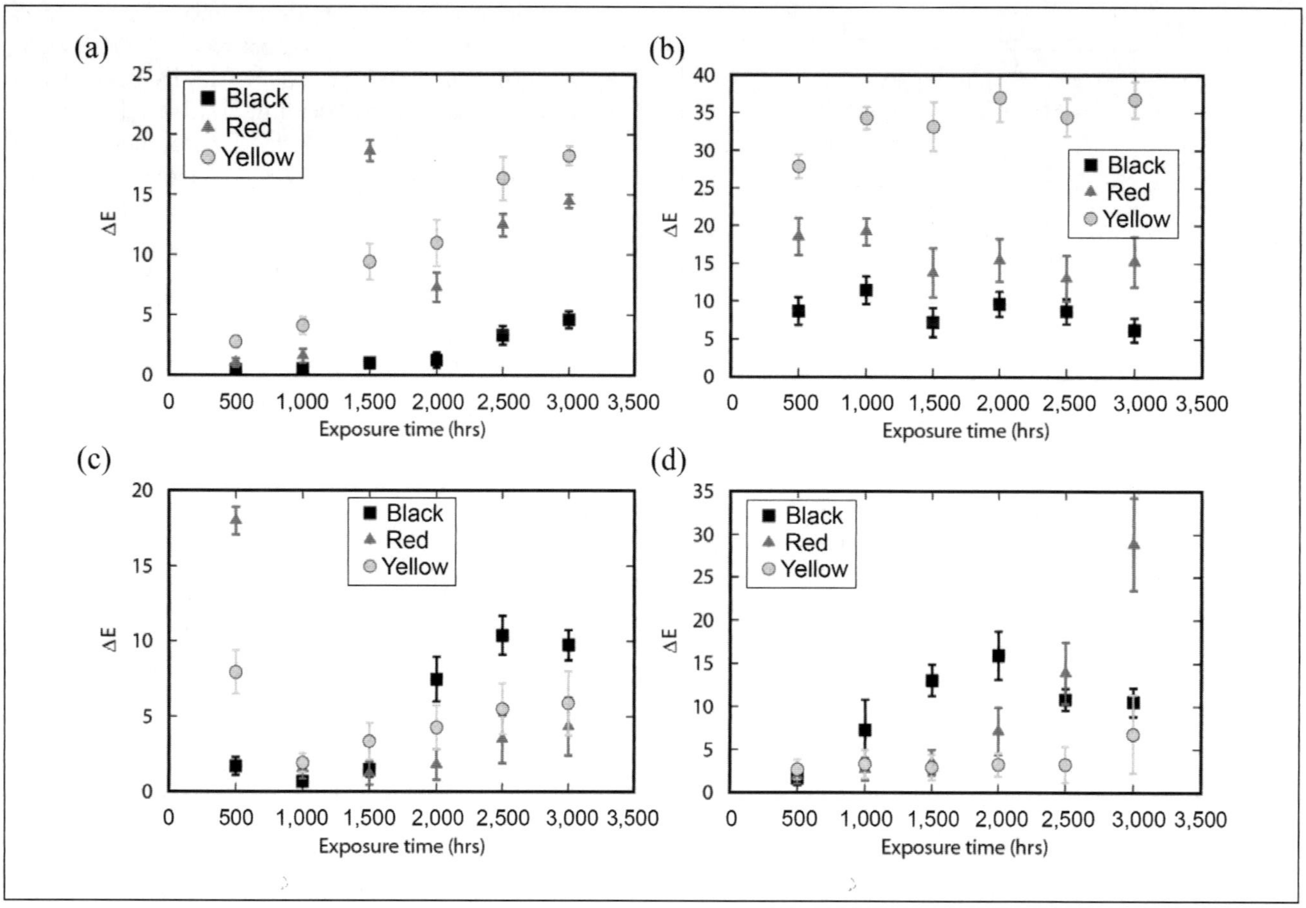

Figure 3. Color change (ΔE) with exposure time: (*a*) NPG at low irradiance; (*b*) polyurethane at intermediate irradiance; (*c*) polyolefin at low irradiance; (*d*) polyolefin at high irradiance.

time (Figure 3*a*). Similar to the polyester coupons, the largest change was measured in the yellow and the least was in the black coupons.

Figure 3*b* shows the change of ΔE with exposure time for polyurethane DWS coupons at intermediate irradiance. For all three colors, a rapid increase in ΔE was measured during the first 500 h and then the ΔE remained relatively constant throughout the rest of the exposure duration. Similar trends were also observed for low and high irradiance conditions. At all three irradiance levels, the largest change was measured in the yellow coupons, while black coupons exhibited the smallest change.

Figure 3*c* and *d* show the ΔE change with exposure time for the polyolefin coupons. For the black coupons, the ΔE value increased with exposure time and then leveled off or decreased slightly after 2,500 and 2,000 h for low and high irradiance levels, respectively. On the other hand, a continuous increase in ΔE with time was observed for red and yellow coupons. Unlike the other three products, the black color exhibited the largest change at low and intermediate irradiance levels (Figure 3*c*), whereas the red color showed the most ΔE changes after 3,000 h of exposure at high irradiance (Figure 3*d*).

Figure 4 shows the time to reach the maximum ΔE in terms of color and irradiance level for the four DWS products. Polyurethane exhibited the quickest color changes followed by polyester and then NPG. At the low level of irradiance (Figure 4*a*), polyurethane reached the maximum ΔE after 1,000 h of exposure while NPG did not reach the maximum even after 3,000 h, the duration of the test program. Overall, the time to reach the maximum ΔE decreased as the irradiance increased, with few exceptions.

Light Microscopy Examination

Figures 5 to 8 depict the surface morphology that corresponds to the largest ΔE change in each product. For the yellow polyester coupon in Figure 5, as exposure time increased, the texture on the surface faded and the contrast of circular patterns decreased. For the yellow NPG coupon, the coral-shaped pattern on the surface (Figure 6*a*) gradually faded as exposure time increased (Figure 6*b*). Furthermore, surface degradation led to the exposure of imbedded reinforcing fibers. Figure 7 reveals the appearance of the yellow polyurethane coupon before

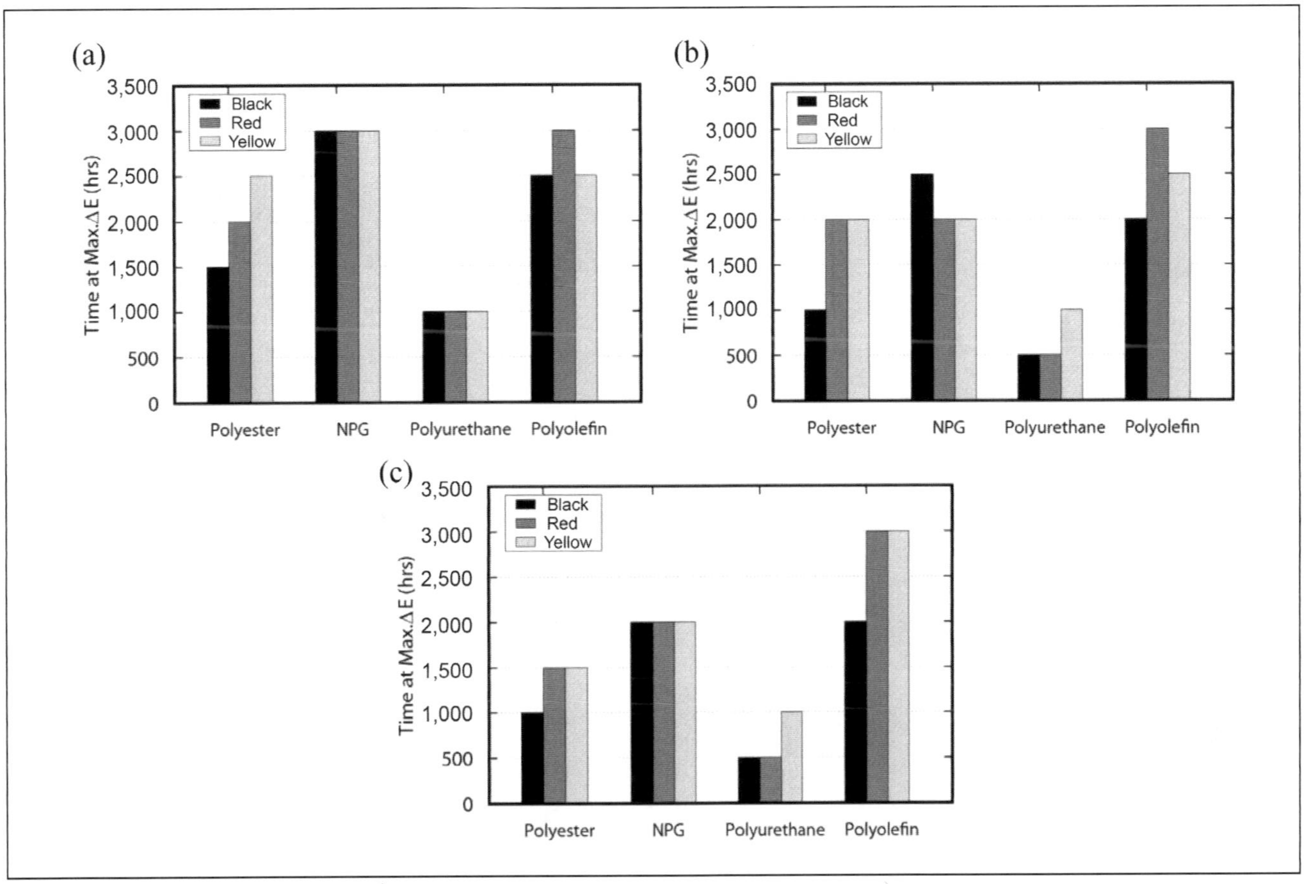

Figure 4. Exposure time to reach the maximum ΔE in three irradiance levels: (*a*) low; (*b*) intermediate; (*c*) high irradiance.

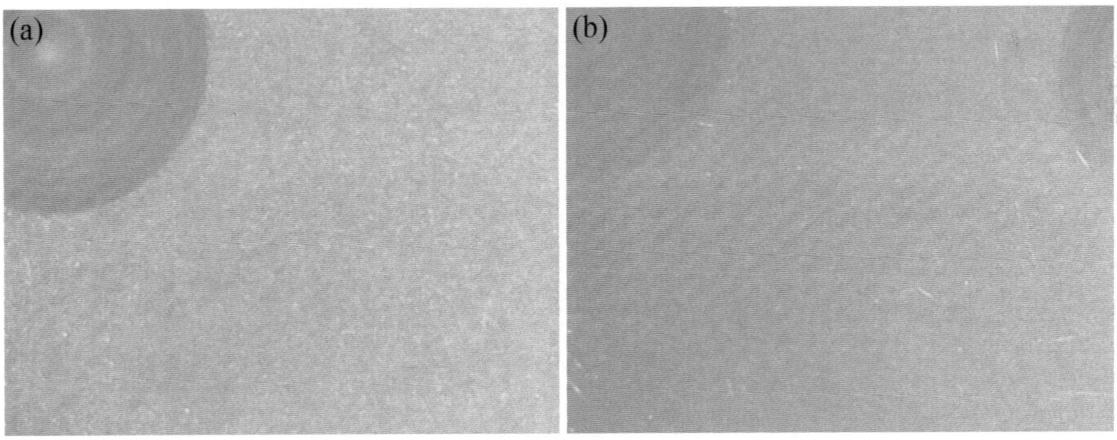

Figure 5. Comparing surface appearance of polyester in high irradiance: (*a*) original; (*b*) 3,000 h.

and after 3,000 h of exposure at high irradiance. The circular spots shown in Figure 7*a* drastically faded after 500 h and became invisible after 1,000 h, while the reinforcing fibers started to appear after 500 h at an increasing rate. Figure 8 shows the difference in the appearance of the red polyolefin coupon before exposure and after 3,000 h of exposure at high irradiance. A white chalky material was observed on the surface, as can be seen in Figure 8*b*. Similar observations were reported by other researchers. They suggested that polymer fragments have

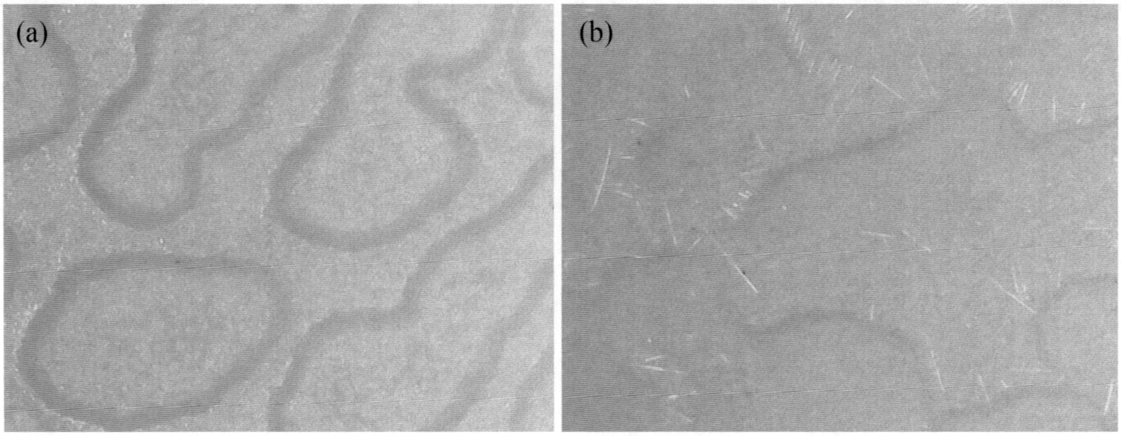

Figure 6. Comparing surface appearance of polyester (NPG) in high irradiance: (*a*) original; (*b*) 3,000 h.

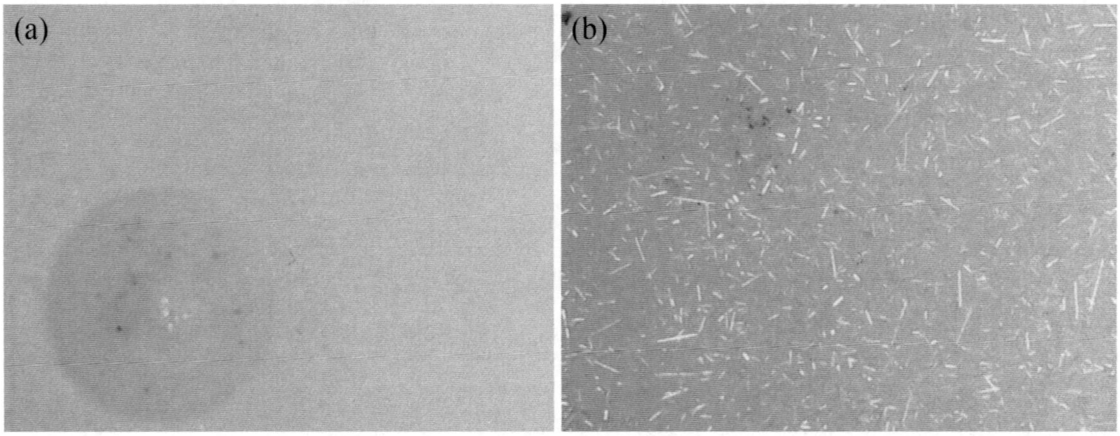

Figure 7. Comparing surface appearance of polyurethane in high irradiance: (*a*) original; (*b*) 3,000 h.

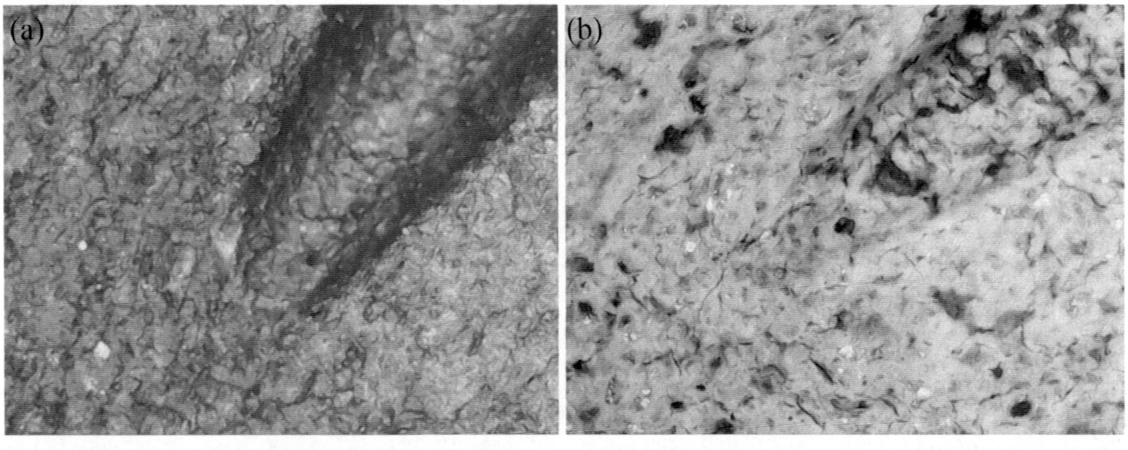

Figure 8. Comparing surface appearance of polyolefin in high irradiance: (*a*) original; (*b*) 3,000 h.

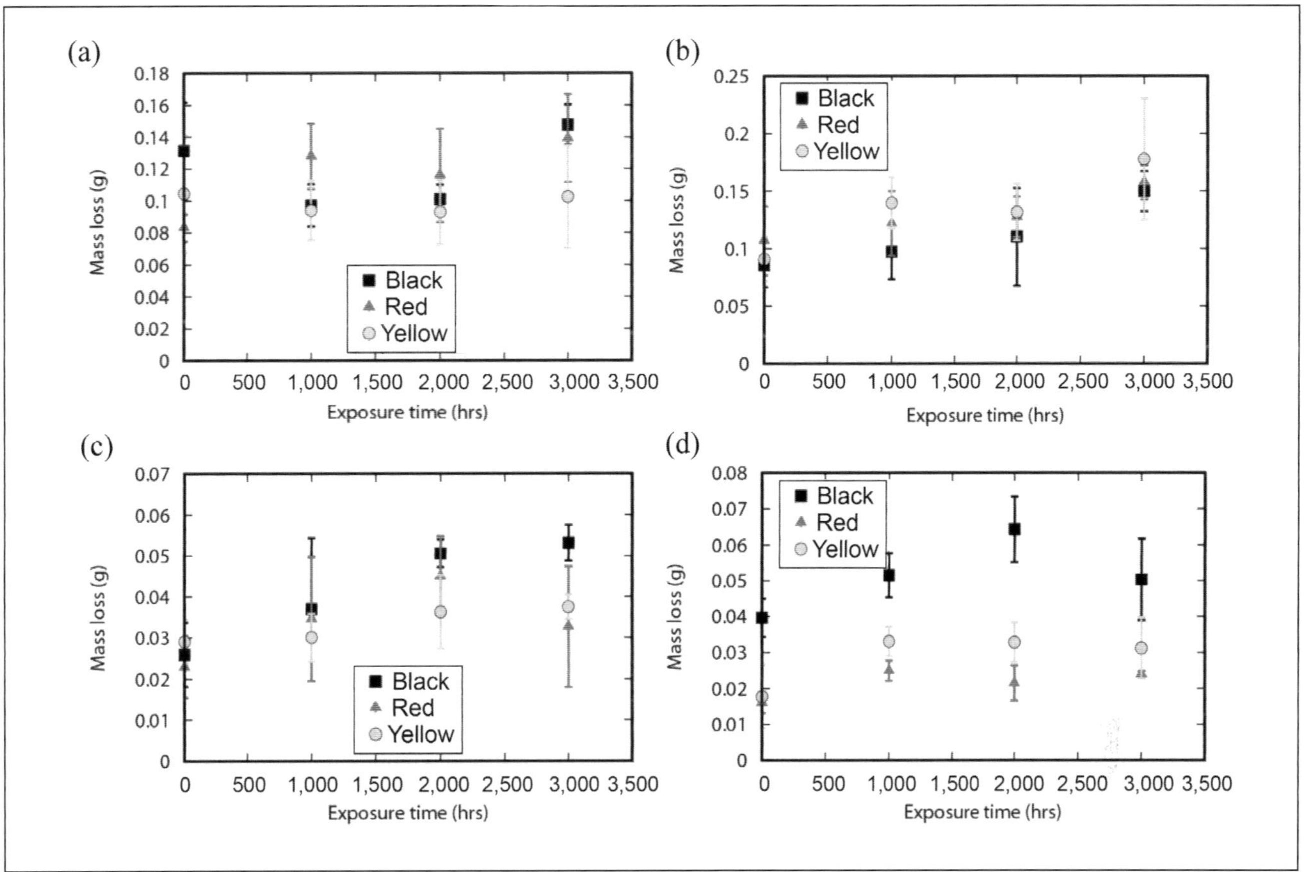

Figure 9. Mass loss versus exposure time at high irradiance: (*a*) polyester; (*b*) NPG; (*c*) polyurethane; (*d*) polyolefin.

been washed out from the surface by water, exposing pigments and fillers which appeared as a chalky material on the surface (*14–17*). The significant change of ΔE (Figure 3*d*) is probably attributed to this chalky effect.

Abrasion Test

For polyester coupons, the mass loss at low and intermediate irradiance was less than the variability of the test. On the other hand, a small increase in mass loss was observed after 3,000 h exposure at the high irradiance condition regardless of the color, as can be seen in Figure 9*a*. The result suggests that a high level of irradiance can lead to the reduction in surface wear resistance for polyester products while low and intermediate levels had minimal effect.

The abrasion test results for NPG coupons are presented in Figure 9*b*. High levels of irradiance led to a steady increase of the mass loss with the exposure time, while low and intermediate irradiance exhibited minimal effect on the resistance. A relatively higher mass loss was measured in the yellow coupons whereas the least change was obtained in red.

Figure 9*c* shows the abrasion test results for the polyurethane coupons at high irradiance. The mass loss generally increased with the exposure time, but the amount is significantly less than polyester and NPG coupons. For all three colors, an increase in irradiance level led to a greater mass loss. Unlike NPG in Figure 9*b*, the black colored coupons showed the greatest mass loss over the testing time while the yellow coupons had the smallest change of the mass.

Figure 9*d* exhibits the results of the polyolefin coupons. The amount of mass loss is in the similar range as polyurethane (Figure 9*c*). The greatest mass loss was obtained under the high irradiance condition. The black coupons had the largest mass loss while the red had the smallest.

Figure 10 shows the mass loss comparison of four products under three irradiance levels. The surface wear resistance decreased at high irradiance while no significant changes were measured at the low and intermediate levels, except for the black coupons. Both the polyurethane and polyolefin products showed significantly less loss of weight (i.e., higher wear resistance) than the polyester and NPG products. This difference may be caused by the stiffness of the product. In brittle materials such as polyester and NPG, this abrasion action may lead to the fracture of the surface and detachment of small fragments (*18*). However, in more ductile materials such as

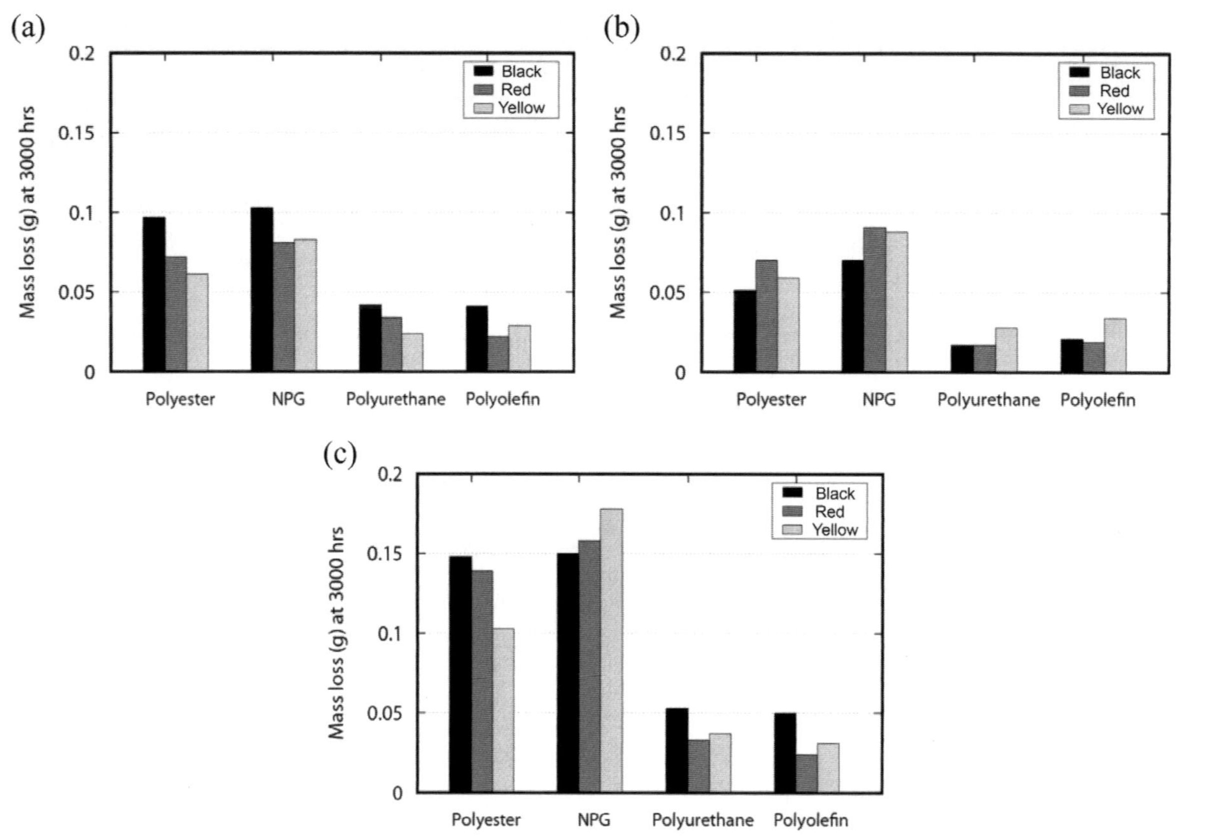

Figure 10. Mass loss at 3,000 h of exposure at three irradiance levels: (*a*) low; (*b*) intermediate; (*c*) high.

polyurethane and polyolefin, the surface tends to deform plastically and dampen the abrasion action, resulting in less mass loss during the abrasion. Overall, among the four tested polymers, the largest mass loss was measured in NPG.

Summary and Discussion

In this study, the effect of sunlight degradation on the visual and mechanical surface properties of four types of polymeric DWS products was examined. Exposure to xenon irradiance resulted in changes in appearance, particularly the color. The degree of discoloration, determined by the ΔE value, varied with polymer types, color, irradiance level, and exposure time. Table 2 summarizes the test results of the four DWS products. At both low and high irradiance levels, the yellow polyurethane exhibited the largest change of color, whereas the black polyester and black NPG showed the least change. On the other hand, relatively small color change was measured in the yellow polyolefin, whereas a significant increase in ΔE of red samples was noticed between low and high irradiance levels.

In Table 2, the specific change in surface morphology for each polymer type is described. The variation of surface morphology is consistent with the color change. All tested DWS products initially possess textures which help to improve the detection for the disabled pedestrians. These textures were degraded and faded as the exposure time increased. Particularly for the polyurethane products, the reinforcing fibers could be seen after 500 to 1,000 h, and simultaneously surface textures were disappearing. For the polyolefin products, the red samples exhibited the most significant discoloration under the high irradiance level due to the chalky effect on the exposed surface. Although white chalky material was also observed on black and yellow samples, its effect on the ΔE value was not as pronounced as that on red samples.

Surface wear resistance is an important mechanical property to assess the sunlight degradation of the DWS products. Although the abrasion device used in this study is a prototype that requires further improvement, as reflected by the error bars associated with each data point, the consistent trends among the tests confirm surface degradation occurred during the 3,000 h of exposure.

Table 2. Summary of the Test Results

Material	Color	Color change (Max. ΔE)/ exposure time (hours)		Surface morphologies	Abrasion (Mass loss [g] at 3,000 h)	
		Low irradiance	High irradiance		Low irradiance	High irradiance
Polyester	Black	5/1,500	5/1,000	• Texture faded	0.10	0.15
	Red	10/2,000	13/1,500	• Contrast decreased	0.07	0.14
	Yellow	20/2,500	20/1,500		0.06	0.10
Polyester NPG	Black	5/3,000	5/2,000	• Texture faded	0.10	0.15
	Red	15/3,000	15/2,000	• Reinforcing fibers exposed	0.08	0.16
	Yellow	18/3,000	19/2,000		0.08	0.18
Polyurethane	Black	14/1,000	9/500	• Texture faded	0.04	0.05
	Red	10/1,000	16/500	• Reinforcing fibers exposed	0.03	0.03
	Yellow	30/1,000	40/1,000		0.02	0.04
Polyolefin	Black	10/2,500	10/2,000	• Chalky appearance	0.04	0.05
	Red	4.5/3,000	29/3,000		0.02	0.02
	Yellow	5.5/2,500	6/3,000		0.03	0.03

The loss of mass increased (i.e., surface wear resistance decreased) with exposure time at high levels of irradiance, while the loss of mass was relatively small at the low and intermediate levels. Comparing low and high irradiance, mass loss of polyester increased by 52% to 93% while that of NPG increased by 45% to 115%. Polyurethane and polyolefin showed a relatively small increase. Mass loss of polyurethane and polyolefin increased by up to 56%, and by 6% to 21%, respectively.

The correlation between abrasion resistance and change in surface color and morphology is not so apparent. The reinforcing fibers appearing on the degraded surfaces of NPG increased mass loss in abrasion, while the same phenomenon in polyurethane did not have the same effect. This may suggest that the stiffness of the polymer matrix plays a more significant role than the reinforcing fibers in the abrasion test. In the case of the polyolefin product, the chalky material which appeared after 2,000 h of exposure in the red and yellow samples was acting as a protecting layer, minimizing the mass loss in the abrasion test.

Conclusion

In this study, sunlight degradation of four DWS products was assessed by exposing them to simulated sunlight radiation inside a laboratory weatherometer and performing tests on the degraded coupons. Color change is a simple non-destructive test method which has been included in the product specification by some manufacturers; however, this test did not show a consistent correlation to the abrasion test results in this study. Despite the potential limitation of a lack of correlation between color change and the abrasion test results, this study found the black NPG DWS to be the most durable against discoloration over the testing duration whereas the red polyolefin DWS has the highest wear resistance among the tested materials.

Acknowledgments

The funding of this study is provided by the Florida Department of Transportation. The project is managed by Dr Chase Knight at the State Materials Office in Gainesville, FL.

Author Contributions

The authors confirm contribution to the paper as follows: study conception and design: G. Hsuan; data collection: S. Na, S. Vahidi, H. Nguyen; analysis and interpretation of results: S. Na, S. Vahidi, G. Hsuan; draft manuscript preparation: S. Na, S. Vahidi, G. Hsuan. All authors reviewed the results and approved the final version of the manuscript.

References

1. Ståhl, A., M. Almen, and M. Wemme. *Orientation using Guidance Surfaces: Blind Tests of Tactility in Surfaces with Different Materials and Structures*. Publication No. 2004: 158E. Swedish Roads Administration, Borlange, Sweden, 2004.
2. U. S. Congress. Americans with Disabilities Act of 1990. Public Law 101-336, 101st Congress. 104 STAT. July 26, 1990. https://www.gpo.gov/fdsys/pkg/STATUTE-104/pdf/STATUTE-104-Pg327.pdf.
3. DETR. *Guidance on the Use of Tactile Paving Surfaces*. Department of the Environment, Transport and the Regions, London, UK, 1998.
4. Horne, D. A. *Information: ADAAG Detectable Warnings*. U.S. Department of Transportation, Washington, D.C., 2002.
5. Fujinami, K., N. Mizukami, H. Ohno, H. Suzuki, A. Shinomiya, O. Sueda, and M. Tauchi. Tactile Ground Surface Indicator Widening and Its Effect on Users' Detection

Abilities. *Quarterly Report of RTRI*, Vol. 46, No. 1, 2005, pp. 40–45.
6. Bentzen, B. L., J. M. Barlow, and L. S. Tabor. *Detectable Warnings: Synthesis of U.S. and International Practice*. U.S. Access Board, Washington, D.C., 2000.
7. U.S. Department of Justice. ADA Standards for Accessible Design. Department of Justice, Washington, D.C., 2010. https://www.ada.gov/regs2010/2010ADAStandards/2010A DAStandards.pdf.
8. Øvstedal, L. R., I. M. Lid, and T. Lindland. How To Evaluate the Effectiveness of a Tactile Surface Indicator System? *International Congress Series*, Vol. 1282, 2005, pp. 1051–1055.
9. Jeon, H. J., and M. N. Kim. Degradation of Linear Low Density Polyethylene (LLDPE) Exposed to UV-Irradiation. *European Polymer Journal*, Vol. 52, 2014, pp. 146–153.
10. Yashchuk, O., F. S. Portillo, and E. B. Hermida. Degradation of Polyethylene Film Samples Containing Oxo-Degradable Additives. *Procedia: Materials Science*, Vol. 1, 2012, pp. 439–445.
11. Yang, X. F., J. Li, S. G. Croll, D. E. Tallman, and G. P. Bierwagen. Degradation of Low Gloss Polyurethane Aircraft Coatings under UV and Prohesion Alternating Exposures. *Polymer Degradation and Stability*, Vol. 80, No. 1, 2003, pp. 51–58.
12. Gil-Castell, O., J. D. Badia, R. Teruel-Juanes, I. Rodriguez, F. Meseguer, and A. Ribes-Greus. Novel Silicon Microparticles to Improve Sunlight Stability of Raw Polypropylene. *European Polymer Journal*, Vol. 70, 2015, pp. 247–261.
13. ASTM International. *Standard Practice for Xenon-Arc Exposure of Plastics Intended for Outdoor Applications*. ASTM D2565-16. ASTM International, West Conshohocken, PA, 2016.
14. Santos, D., M. R. Costa, and M. T. Santos. Performance of Polyester and Modified Polyester Coil Coatings Exposed in Different Environments with High UV Radiation. *Progress in Organic Coatings*, Vol. 58, No. 4, 2007, pp. 296–302.
15. Jones, M. S. *Effects of UV Radiation on Building Materials*. Building Research Association of New Zealand (BRANZ) Judgeford, New Zealand, 2002. https://www.niwa.co.nz/sites/niwa.co.nz/files/import/attachments/Jones.pdf.
16. Seldén, R., B. Nyström, and R. Långström. UV Aging of Poly(propylene)/Wood-Fiber Composites. *Polymer Composites*, Vol. 25, No. 5, 2004, pp. 543–553.
17. Fabiyi, J. S., A. G. McDonald, M. P. Wolcott, and P. R. Griffiths. Wood Plastic Composites Weathering: Visual Appearance and Chemical Changes. *Polymer Degradation and Stability*, Vol. 93, No. 8, 2008, pp. 1405–1414.
18. Ashrafizadeh, H., A. McDonald, and P. Mertiny. Erosive and Abrasive Wear Resistance of Polyurethane Liners. In *Aspects of Polyurethanes*, InTech, Rijeka, Croatia, 2017. https://www.intechopen.com/books/aspects-of-polyurethanes/erosive-and-abrasive-wear-resistance-of-polyurethane-liners.

The Standing Committee on Polymer Concretes, Adhesives, and Sealers (AHD40) peer-reviewed this paper (18-03232).

The opinions, findings, and conclusions expressed in this paper are those of the author(s) and not those of the Florida Department of Transportation or the U.S. Department of Transportation.

TRR
JOURNAL OF THE TRANSPORTATION RESEARCH BOARD

Article

Transportation Research Record
2018, Vol. 2672(12) 137–147
© National Academy of Sciences: Transportation Research Board 2018
Article reuse guidelines: sagepub.com/journals-permissions
DOI: 10.1177/0361198118758035
journals.sagepub.com/home/trr
SAGE

Parametric Ordinal Logistic Regression and Non-Parametric Decision Tree Approaches for Assessing the Impact of Weather Conditions on Driver Speed Selection Using Naturalistic Driving Data

Ali Ghasemzadeh[1], Britton E. Hammit[1,2], Mohamed M. Ahmed[1], and Rhonda Kae Young[3]

Abstract

The impact of adverse weather conditions on transportation operation and safety is the focus of many studies; however, comprehensive research detailing the differences in driving behavior and performance during adverse conditions is limited. Many previous studies utilized aggregate traffic and weather data (e.g., average speed, headway, and global weather information) to formulate conclusions about the impact of weather on network operation and safety; however, research into specific factors associated with driver performance and behavior are notably absent. A novel approach, presented in this paper, fills this gap by considering disaggregate trajectory-level data available through the SHRP2 Naturalistic Driving Study and Roadway Information Database. Parametric ordinal logistic regression and non-parametric classification tree modeling were utilized to better understand speed selection behavior in adverse weather conditions. The results indicate that the most important factors impacting driver speed selection are weather conditions, traffic conditions, and the posted speed limit. Moreover, it was found that drivers are more likely to significantly reduce their speed in snowy weather conditions, as compared with other adverse weather conditions (such as rain and fog). The purpose of this study was to gather insights into driver speed preferences in different weather conditions, such that efficient logic can be introduced for a realistic variable speed limit system—aimed at maximizing speed compliance and reducing speed variations. This study provides valuable information related to drivers' interaction with real-time changes in roadway and weather conditions, leading to a better understanding of the effectiveness of operational countermeasures.

Adverse weather conditions are known to be one of the main contributing factors affecting traffic operation and safety. Inclement weather conditions impede drivers' ability to perceive and react to their environment, and this decrease in driver performance has a dramatic impact on network-wide operations and the predictability of traffic flow. Studies have shown that drivers may reduce their speed, maintain a larger headway, and drive more carefully in adverse conditions to compensate for reduced visibility and slippery road conditions (*1*). A previous study investigated the impact of rain on freeway capacity, revealing that wet pavement and decreased visibility affects drivers' speed selection and roadway capacity (*2*). Although this study focused on the impact of heavy rain, other studies have shown that light rain can also affect travel speed and roadway capacity (*3,4*). A study by Kyte et al. showed that both light rain and snow can reduce speed up to 50% (*5*). In addition, the study identified a 9 km/h speed reduction during wind speeds >48 km/h; however, the impact of reduced visibility on speed reduction was found to be marginal. Another study using the same test sites revealed that snow caused a speed reduction of 18 km/h (*6*). Ibrahim and Hall investigated the difference in traffic conditions during light rain, heavy rain, and snow compared with matching trips in clear weather conditions using the data from two rainy, two snowy, and six clear weather days. Results indicated a 3–5% speed reduction during light rain and snow, a

[1]Department of Civil and Architectural Engineering, University of Wyoming, Laramie, WY
[2]FHWA Turner Fairbank Highway Research Center, McLean, VA
[3]Department of Civil Engineering, Gonzaga University, Spokane, WA

Corresponding Author:
Address correspondence to Ali Ghasemzadeh: aghasemz@uwyo.edu

14–15% speed reduction because of heavy rain, and a 30–40% speed reduction caused by heavy snow (*4*).

According to the Fatality Analysis Reporting System, >5,800 fatal crashes between 2005 and 2014 occurred during adverse weather conditions. Previous studies have attempted to quantify the impact of weather conditions on car crashes; however, the results are not globally consistent (*7*). Based on a National Highway Traffic Safety Administration report, adverse weather conditions contributed to 15% of fatal, 19% of injury, and 23% of property-damage-only crashes (*8*). Other studies in Canada and the UK revealed that weather-related factors contributed to approximately 30 and 20% of crashes in those locations, respectively (*9,10*). Although the magnitude of impact of adverse conditions on weather-related crash risk estimates varies, the overall trend from these studies indicates a significant increase in crash risk during adverse weather conditions.

While the safety and operational effects of adverse weather on transportation networks have been extensively researched, specific considerations related to driver behavior and performance is noticeably absent from these studies. To address this gap, this research leverages trajectory-level vehicle data in adverse and clear weather conditions collected by the second Strategic Highway Research Program (SHRP2). SHRP2 created a comprehensive naturalistic driving study (NDS) database, which comprises data from >3,400 participants spanning a time period of 3 years (2010–2013). The advantage of using naturalistic driving data is that the data are not collected in a controlled environment, such as driving simulators, instrumented research vehicle tests, or self-reporting questionnaires; therefore, the collected data are more representative of natural driving and contain a wide variety of environmental and traffic conditions.

This study is intended to develop insights into driver preferences for speed, while correlating these preferences and behavior with environmental and traffic conditions that can be used to establish an effective variable speed limit (VSL) system. This research evaluates additional aspects of driver behavior and performance that can be used to develop effective countermeasures to reduce the impact of adverse weather on freeway operations.

Data Preparation

The SHRP2 safety data comprise two complementary databases: the NDS database and the Roadway Information Database (RID). As part of the SHRP2 NDS, >3,400 participants were recruited and their driving behavior was recorded continuously during their participation, which ranged from a few months to a couple of years. The RID contains inventory data related to the six NDS states (Florida, Indiana, New York, North Carolina, Pennsylvania, and Washington) and includes additional data on crash histories, traffic patterns, weather conditions, work zones, and ongoing safety campaigns. The NDS and RID can be linked to provide a rich data source associating driving behavior with roadway and environmental characteristics (*11*). The following sections describe the initial acquisition of SHRP2 NDS trips, the data reduction procedures established for efficient data processing, and the subset of data selected for the speed selection modeling presented in this paper.

Data Acquisition

Characteristics of weather conditions for each NDS trip are not readily known; therefore, the Wyoming research team identified three unique and complementary data acquisition methodologies to query the large NDS database and extract weather-related trips:

I. **Windshield Wipers**: The first method queried NDS trips based on the reported windshield wiper status, which is directly available in the NDS vehicle data. This method was successful in acquiring trips occurring during various levels of precipitation, but did not enable the collection of other trips of interest (such as fog conditions or snow-covered roads).

II. **Weather Stations**: The second method leveraged the weather station data available through the National Climate Data Center (NCDC). Weather stations located throughout the six NDS sites were identified and significant weather events were flagged. A five nautical mile radius was created and a spatial-temporal query of the NDS database was requested to extract trips passing through the radius occurring during the selected time frame.

III. **NDS Crash Reports**: The third method utilized the weather-related crashes. Crash reports were reviewed and crash locations with weather conditions of interest were essentially considered "weather stations." The same process as described in Method II was used to identify NDS trips within a 5 nautical miles radius occurring within the time period of the weather-related crash.

A representation of routes flagged in the spatial-temporal query for trips occurring within a 5 nautical mile radius of a weather station or weather-related crash is shown in Figure 1. In this image, the weather station is located directly off Interstate 275; however, trips traversing Interstate 75 could be flagged if they pass through the represented 5 nautical mile radius.

Using the three data acquisition methods, the research team received >33,000 trips, one-third potentially occurring in adverse weather conditions and the remaining representing matched trips (same driver, same vehicle, and same route) in clear conditions. The identified NDS trips involved 1,523 drivers between 16 years and 99 years of age, with the majority of the drivers in young age group: 16–29 years old. Gender representation was balanced in most age groups,

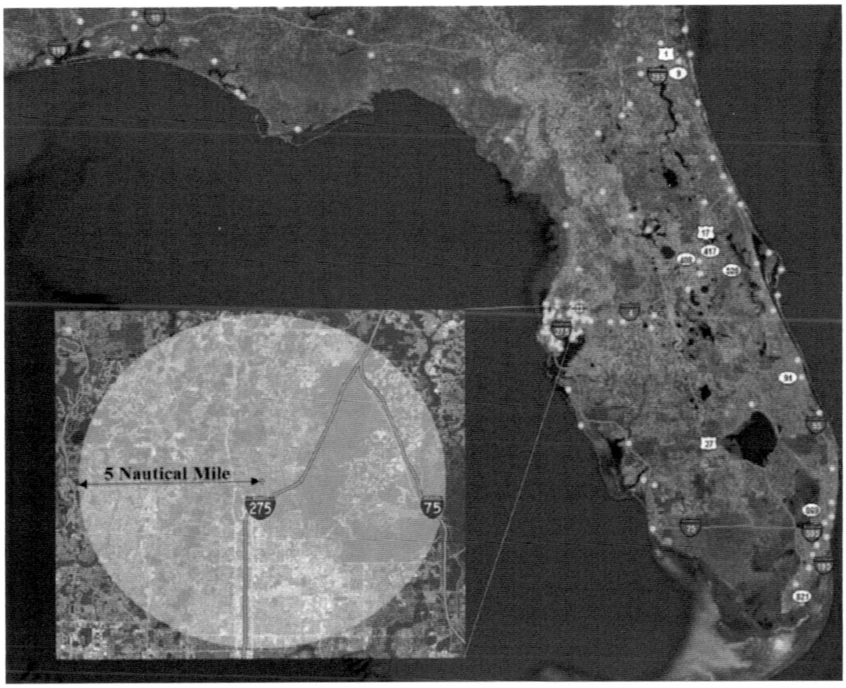

Figure 1. Location of weather stations in Florida and a 5 nautical mile coverage area for a sample weather station.

with the exception of a slight overrepresentation of female drivers between 20 years and 24 years old. The total duration of the received trip data represents over 11,205 h of driving.

Data Reduction

Extensive manual video observation and preliminary processing was conducted to screen the received NDS trips and filter out trips flagged as "weather-related" but which actually occurred in clear weather or dry surface conditions. This preliminary screening process was required because the data acquisition query protocol included a wide temporal coverage of adverse weather events over an entire day, which allowed for the collection of trips that would not have been otherwise identified. Of the received trips, 4,094 (37%) freeway trips were verified to have adverse weather or road surface conditions.

The research team developed the Wyoming NDS Data Analysis Tool (DAT), which enables a semi-automatic process for reducing the time series and video data. The DAT reduces the dimensionality of the data by extracting relevant time series variables for the identification of driving behavior characteristics in inclement weather conditions. Next, each trip is reduced to 1-min segments of time series data to increase the usability during analyses and enable manual video observation. Manual video observation was required to verify the roadway type, weather conditions, surface conditions, visibility, and traffic conditions. The research team established discrete categories and options for each manually observed condition (e.g., traffic conditions were reported according to perceived level of service: A, B, C, D, E, F), and detailed descriptions and images were used to train video reviewers and limit bias in the observation. The video reviewers leveraged the benefits of Wyoming's Visualization and Visibility Identification Tool for efficiently reporting conditions for each trip (*12*).

Trips Used in Speed Selection Study

In this study, a total of 212 trips in adverse weather conditions (22 trips in fog; 102 trips in rain; and 88 trips in snow—plus 424 matching clear weather trips) were randomly selected from the extensive Wyoming NDS database. The selected NDS trips involve 145 drivers between 16 years and 89 years old, with the majority of drivers in the young age group (16–29 years old). Gender was mainly balanced among age groups, except for a slight overrepresentation of female drivers between 20 and 24, which follows the same distribution that is reported by Virginia Tech Transportation Institute (VTTI) for all SHRP2 NDS trips.

A total of 14,923 1-min segments—equivalent to nearly 249 hours and 18,453 km (rain: 3,582 km; snow: 1,615 km; and fog: 954 km) of driving, plus their matching trips in clear weather conditions—were processed. The speed limit data provided in the RID was used to merge speed limits with each 1-min segment. Once non-freeway segments were removed, 10,606 1-min segments were used to model driver speed selection.

Methodology

To identify the impact of weather conditions on driver speed selection, two models using both parametric and non-parametric methods were developed. Parametric models, such as probit and logistic regression models, provide the relationship between a response variable and predictors. However, parametric models have some limitations—they cannot provide a high level of prediction accuracy because there are many embedded assumptions (*13*). Another complication in using parametric models is their inability to automatically handle missing values (*14*). These shortcomings cannot be addressed using common parametric models such as logistic and probit models (*13,15*). Despite their limitations, parametric logistic/probit models are effective in interpreting the marginal effects of various risk factors (*16,17*). On the other hand, there are several key advantages of using non-parametric models, including the ability to provide high prediction accuracy, handle missing values automatically, and process many explanatory variables in a timely manner—all of which might be extremely beneficial for assessing traffic operation and safety in real time with weather and traffic data directly ingested into the model (*14*). However, the trade-off is that their classification results cannot be explicitly interpreted (*13*).

In this study both ordinal logistic regression (parametric) and classification and regression tree (non-parametric) methods are used to analyze the impact of various factors (e.g., weather and roadway conditions) on speed selection.

Ordinal Logistic Regression

Logistic regression is a common model used in traffic safety and operation studies. Logistic regression allows the formulation of predictive models on a probabilistic basis. Similar to other regression analyses, it predicts the value of a dependent variable from one or more explanatory variable(s). Logistic regression can be applied to a binary, nominal, or ordinal dependent variable. Logistic regression (Equation 1) can also be utilized to rank the relative importance of the response variables (*18,19*).

$$Logit[P(x)] = \log(\frac{P(x)}{1-P(x)}) = \alpha + \beta x \quad (1)$$

Equation 1 shows a logistic regression model with x representing the independent variable and $P(x)$ indicating the probability of success for a binary response variable y—considering explanatory variable x. α represents the response probability when explanatory variables are at the reference level (or when $x = 0$) (*19*); β represents the regression coefficients. As mentioned earlier, logistic regression can be conducted using an ordinal response variable. The ordinal logistic regression (OLR) equation is shown in Equation 2 (*18*).

$$Ln(\Pi_j) = \alpha_j - (\beta_1 x_1 + \beta_2 x_2 + \beta_3 x_3 + \ldots) \quad (2)$$

An ordinal logistic regression (ordered logit) model was applied for this analysis because of the ordinal nature of speed selection that would not be accounted for in multinomial analyses.

Classification and Regression Tree Model (CART)

Decision tree modeling can be used for both continuous and nominal dependent variables. Utilizing a decision tree to classify a nominal dependent variable is called a classification tree (*20,21*). Classification can be defined as a procedure for predicting the "class" of an object—considering the object's features (*22*). Classification models are built from a training dataset in which trends of explanatory and response variables are identified and used to predict the value of the target variable for different datasets (*23*). The two main components of decision trees are the "root node" and the "leaf node." The "root node" is the node located at the top of the tree, which contains all ingested data, and the "leaf node" refers to the termination node, which has the lowest impurity.

The root node is divided into two child nodes, based on the independent variable (splitter) that creates the best homogeneity. This procedure of partitioning the target variable recursively is repeated until all of the data in each node reach their highest homogeneity. At that point, tree growth stops, and the node(s) that do not have any branches become the "leaf node(s)." Each path from the top of the tree (root node) to the bottom/termination of the tree (leaf node) can be considered a rule. Following this sequence, the data in each child node are purer (more homogenous) than the data in the upper parent node (*24*).

To identify possible splits among all variables, a splitting criterion is generated. The splitting criterion is the main design component of a decision tree (*25*). In a decision tree learning algorithm, the splitting criterion's role is to measure the quality of each possible split among all variables. Two common tests used to generate splitting criteria are: (1) chi-square and (2) Gini reduction. In this study, the Gini splitting criterion is used to select the variable and split pattern to best partition the node. Gini impurity indicates the data purity; in other words, it shows the probability of incorrect classification for a randomly chosen record from the specific node in the data subset.

Variable importance measure (VIM) is one of the main outputs of the classification tree model, showing the most important factors affecting the target variable (*26*). In this study, the most significant factors affecting driver speed selection—considering adverse weather conditions—were identified using the VIM.

Preliminary Analysis

A preliminary analysis was conducted to better understand the differences between driver speed behavior in clear and

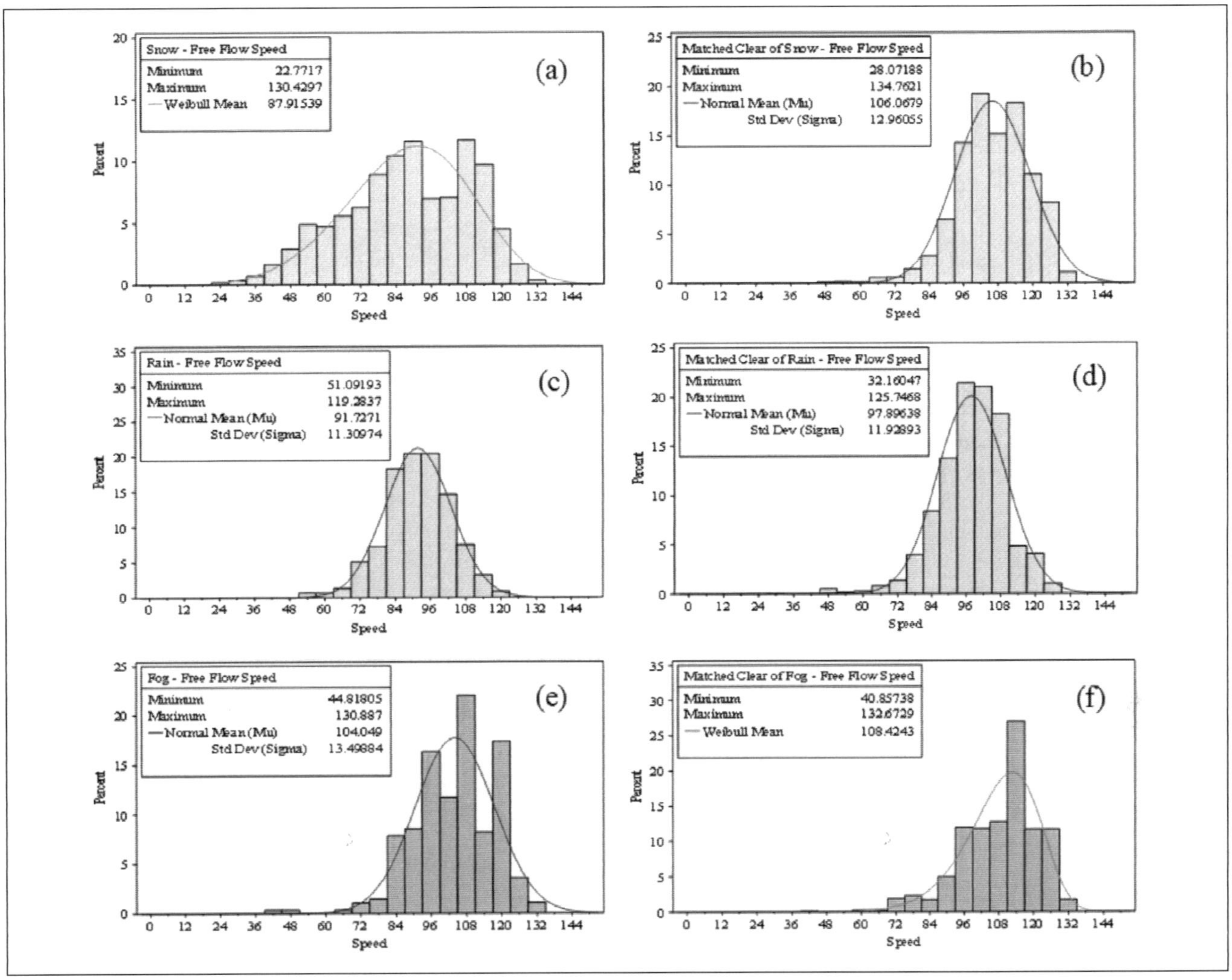

Figure 2. Observed and fitted distributions for speeds during adverse and clear weather under free-flow traffic. (*a*) Speed distribution in snow and (*b*) speed distribution in matched clear weather. (*c*) Speed distribution in rain and (*d*) speed distribution in matched clear weather. (e) Speed distribution in fog and (*f*) speed distribution in matched clear weather.

adverse weather conditions using the matched trips. The distribution and variation of speeds between clear and adverse weather were investigated in free-flow traffic conditions. Investigating the speed distribution in free-flow conditions is important for VSL applications, as the speed is not affected by the interaction with traffic.

Six possible scenarios were considered and compared as shown in Figure 2, including trips in snow, rain, fog, and their matched trips in clear weather conditions. The advantage of using this matching technique is that environmental, traffic, and roadway conditions are controlled as the weather-related trips are matched with trips in clear conditions that have the same driver, traversed the same route, and are in similar traffic conditions.

Results from the preliminary analysis revealed that free-flow speed followed a Weibull distribution in snow and normal distribution in rain and fog. A T-test revealed that the average speed in snow, rain, and fog was significantly lower than clear weather conditions; specifically, the average speed was found to be 18.35 km/h (snow), 6.17 km/h (rain), and 4.25 km/h (fog) lower than the correlated speed in clear conditions, indicating that drivers exhibit greater speed reduction in snow conditions than in rain and fog.

Modeling Speed Selection and Discussion of Key Factors

The ordered logit and classification tree models were calibrated using all reduced data; representing a dataset of 10,606 1-min segments of NDS trips occurring in various weather and traffic conditions (matching was not required for this analysis). The speed selection was considered as a dependent variable with four speed intervals. These intervals were selected based on the median of the speed selection above or below the speed limit. The four-quantile intervals were defined to be:

1. Speed reduction >14%,
2. Speed reduction between 0% and 14%,
3. Speed increase between 0% and 10%, and
4. Speed increase >10%.

These intervals were selected to establish equal sample sizes throughout each speed selection category because modeling with an unbalanced dataset may lead to a skewed distribution of predicted values and result in low model prediction accuracies (*13*). Moreover, it is practical for decision makers in the traffic management center to adjust the VSL based on the speed that a larger number of drivers are following, as opposed to using unrealistic speed, which may be illogical to many drivers.

Therefore, speed selection is the dependent variable for both models. The remaining explanatory variables consist of information extracted from questionnaires, including driver demographics (e.g., age, marital status, gender, and education), driver experience, roadway factors, environmental conditions, and traffic conditions. A full description of the variables used in the OLR and CART modeling procedures for driver percent speed selection in different weather conditions is shown in Table 1.

Ordinal Logistic Model

To confirm the suitability and fitness of the model, the log likelihood ratio was used. The multi-collinearity was assessed by calculating the variance inflation factor (VIF) for each explanatory variable, which indicates how much the variance of an estimated regression coefficient increases if the predictors are correlated. A VIF between 5 and 10 shows high correlation between predictors, and a VIF >10 indicates that the regression coefficients are poorly estimated because of multicollinearity (*27*). The explanatory variables introduced to the model produced VIF values between 1.03 and 1.40, excluding any concerning multi-collinearity. Only statistically significant variables were retained in the final model.

Table 2 provides the results of the Ordered Logit Model. Adverse weather conditions (snow, rain, and fog) were found to have a significant effect on speed selection, indicating that drivers exhibit a greater speed reduction during adverse weather conditions. In addition, this finding shows the negative effect of adverse weather conditions on drivers' performance. Specifically, results show that the odds of drivers reducing their speed were 9.29, 1.55, and 1.29 times higher for drivers traveling in snow, rain, and fog conditions, respectively, in comparison with drivers who were driving in clear weather conditions.

Findings related to visibility indicate that drivers are more likely to reduce their speed when their visibility is impaired. This finding correlates with findings from previous studies that show visibility is one of the most important factors affecting driver speed (*6,28*). The derived model indicates that the odds of a driver reducing their speed were 1.75 times greater in adverse visibility conditions, compared with good visibility conditions.

As expected, traffic conditions had a significant negative effect on speed selection. The odds of a driver experiencing speed reduction were 3.6 times greater for drivers who were driving in higher traffic density compared with those traveling in free-flow conditions (level of service A and B). This finding is in agreement with previous studies (*29,30*). Further consideration of driver demographics (i.e., gender) indicated that female drivers were 1.09 times more likely to reduce their speed than their male counterparts.

An analysis of driver demographic information, using the developed speed selection model, reveals that drivers older than 40 years old were 1.23 times more likely to reduce their speed in comparison with younger drivers. In addition, when considering miles traveled in the previous year as an indicator for driver experience, it was determined to be a significant factor in the developed speed selection model. The results indicated that drivers who drove >10,000 miles in the previous year were less likely to reduce their speed in comparison with drivers who drove <10,000 miles.

Classification Tree Model

Three data subsets were considered in developing the classification tree model (i.e., training, validation, and testing datasets). Moreover, 60% of the data was allocated to the training dataset, 30% to the validation, and 10% to the testing datasets. More specifically, among the 10,606 observations used for developing the classification tree model, 6,363 observations were assigned to the training subset, 3,181 to the validation subset and 1,060 observations to the testing subset. Figure 3 shows the decision tree diagram for drivers' speed selection in different weather conditions produced from the training dataset. In each node box, the node number and the percentage of data in each category are provided. One beneficial characteristic of a decision tree, compared with other modeling methods, is that it gives decision makers rules to address "if-then" questions efficiently.

The misclassification rate, based on the training and validation datasets, indicated that the best tree could be obtained with 15 terminal nodes. In other words, with 15 terminal nodes, the misclassification rate for the model reaches a minimum value of 0.42 and remains fairly steady with the increases of nodes beyond 15. Node 3, on the right side, shows the data related to driving in snowy conditions. On the right branch of the tree, there are four terminal nodes (nodes 7, 13, 24, and 25). In three of these terminal nodes, the drivers were predicted to reduce their speed by >14% (Class Label 1), which implies that if a driver is traveling in snowy conditions, he/she is more likely reduce their speed, regardless of other factors.

Table 1. Description of Variables

Variable	Description	Type	Source	Definition	Assigned Code
Speed behavior	Speed selection in various weather conditions (the difference between the 1-min average driver speed and speed limit divided by speed limit)	Ordinal	Naturalistic driving time series data and Roadway Information Database	>14% reduction	4
				0–14% reduction	3
				0–10% increase	2
				>10% increase	1
Weather conditions	Predominant weather conditions in 1-min video observation	Categorical	Video observation	Clear	1
				Heavy rain	2
				Snow	3
				Fog	4
Visibility	Predominant visibility conditions in 1-min video observation	Categorical	Video observation	Not affected	1
				Affected	2
Traffic condition	Predominant traffic conditions in 1-min video observation	Binary	Video observation	A	1
				B	
				C	2
				D	
				E	
				F	
Gender	The gender the participant identifies with	Binary	Electronic online questionnaire administered during participant in-processing	Male	1
				Female	2
Age	The age group corresponding to the driver's birthdate	Categorical	Electronic online questionnaire administered during participant in-processing	Younger than 40	1
				Older than 40	2
Education	The participant's highest completed level of education	Categorical	Electronic online questionnaire administered during participant in-processing	Some high school	1
				High school diploma or GED	
				Some education beyond high school but no degree	2
				College degree	
				Some graduate or professional school, but no advanced degree (e.g., JDS, MS or PhD)	3
				Advanced degree (e.g., JDS, MS or PhD)	
Marital status	The participant's marital status	Categorical	Electronic online questionnaire administered during participant in-processing	Single	1
				Divorced	2
				Widow(er)	3
				Unmarried partners	4
				Married	5
Driver mileage last year details	The approximate number of miles the participant drove last year	Categorical	Electronic online questionnaire administered during participant in-processing	<10,000	1
				10,000–20,000	2
				>20,000	3

As a function of traffic conditions, node 3 is split into node 6 and terminal node 7; terminal node 7 shows that when a driver travels in any level of traffic congestion (not in free-flow conditions) with snow-covered road surface conditions, there is an 86% probability that the driver will reduce their speed by >14%. Node 6 is further split into node 12 and terminal node 13 based on visibility conditions. Node 13 shows that 56% of drivers are likely to reduce their speed by >14% in snowy surface conditions, free-flow traffic, and reduced visibility. Node 12 is split into node 24 and 25, based on

Table 2. Estimation of Ordered Logit Model for Speed Selection in Different Weather Conditions

Analysis of Maximum Likelihood Estimates

Parameter		DF	Estimate	Standard Error	Wald Chi-Square	Pr > ChiSq	Odds Ratio	Confidence Interval	
Intercept	4	1	−2.57	0.09	800.41	<0.0001	–	–	–
Intercept	3	1	−1.3	0.09	218.23	<0.0001	–	–	–
Intercept	2	1	0.32	0.09	13.93	0.0002	–	–	–
Weather cond.	Rain	1	0.44	0.09	25.35	<0.0001	1.55	1.31	1.83
Weather cond.	Snow	1	2.23	0.06	1,612.52	<0.0001	9.29	8.33	10.36
Weather cond.	Fog	1	0.26	0.09	7.61	0.0058	1.29	1.08	1.55
Visibility	Affected	1	0.56	0.09	35.24	<0.0001	1.75	1.45	2.1
Traffic cond.	C-F	1	1.28	0.04	995.02	<0.0001	3.6	3.32	3.89
Gender	Female	1	0.09	0.04	5	0.0254	1.09	1.01	1.18
Age	>40	1	0.2	0.05	18.24	<0.0001	1.23	1.12	1.35
Marital status	Divorced	1	0.81	0.09	86.57	<0.0001	2.25	1.9	2.67
Marital status	Widow(er)	1	1.2	0.11	121.33	<0.0001	3.31	2.68	4.1
Marital status	Unmarried partners	1	−0.94	0.1	88.74	<0.0001	0.39	0.32	0.48
Marital status	Married	1	0.34	0.05	45.09	<0.0001	1.4	1.27	1.55
Mileage last year	10,000–20,000	1	−0.5	0.05	122.3	<0.0001	0.61	0.56	0.66
Mileage last year	>20,000	1	−0.58	0.06	92.33	<0.0001	0.56	0.5	0.63
Model fit statistics:									
AIC						25,561.928			
SC						25,692.773			
Log likelihood at convergence						−12,762.964			
McFadden pseudo R^2						0.1198			
Number of observations						10,606			

driver mileage (from the previous year). Lastly, node 24 shows that if a driver drove <10,000 miles last year, they were 59% more likely to reduce their speed >14%—indicating that less experienced drivers are more likely to reduce their speed in the aforementioned weather, traffic, and visibility conditions. The other rules in the developed classification tree model can be interpreted in the same manner.

Table 3 provides the relative variable importance, which is one of the most important classification tree outputs.

As can be seen in Table 3, weather conditions turned out to be the most important variable affecting driver speed selection based on the developed classification tree model. This finding is consistent with the previous study demonstrating the effect of weather conditions on driver speed selection (*30*). Traffic conditions were the second most important variable affecting driver speed selection, followed by posted speed limit, education level, age, mileage traveled last year, and the visibility conditions.

Model Efficiency. The area under the Receiver Operating Characteristic (ROC) curve is used to compare the two developed models. The ROC value can be between 0.5 (for a poorly fitted model) and 1.0 (for a perfectly fitted model) (*31*). The ROC indices for both developed models are significantly >0.5 (ROC Indexes were 0.773 and 0.770 for the classification tree and OLR respectively); therefore, results indicate that both models can assess driver speed selection behavior well, with the classification tree slightly outperforming the OLR model.

Conclusions

Multiple data sources—including the SHRP2 NDS, SHRP2 RID, and NCDC—were used in three complementary methodologies to identify weather-related trips from the SHRP2 NDS database. The Wyoming NDS DAT was used by the research team to efficiently process the trip data into homogeneous 1-min segments of data with known weather, traffic, and roadway conditions. The processed data were used as input into speed selection models generated in this study.

Both parametric logistic regression and non-parametric classification tree models were developed to better understand driver speed selection in different weather conditions, that is, snow, rain, and fog. Each modeling technique has its advantages and disadvantages. Although the classification tree model can easily explain the complex interactions between several explanatory variables, it is difficult to fully describe the complicated effects of contributing factors owing to non-linearity and the interaction effects in the logistic regression (*13*). On the other hand, using parametric logistic

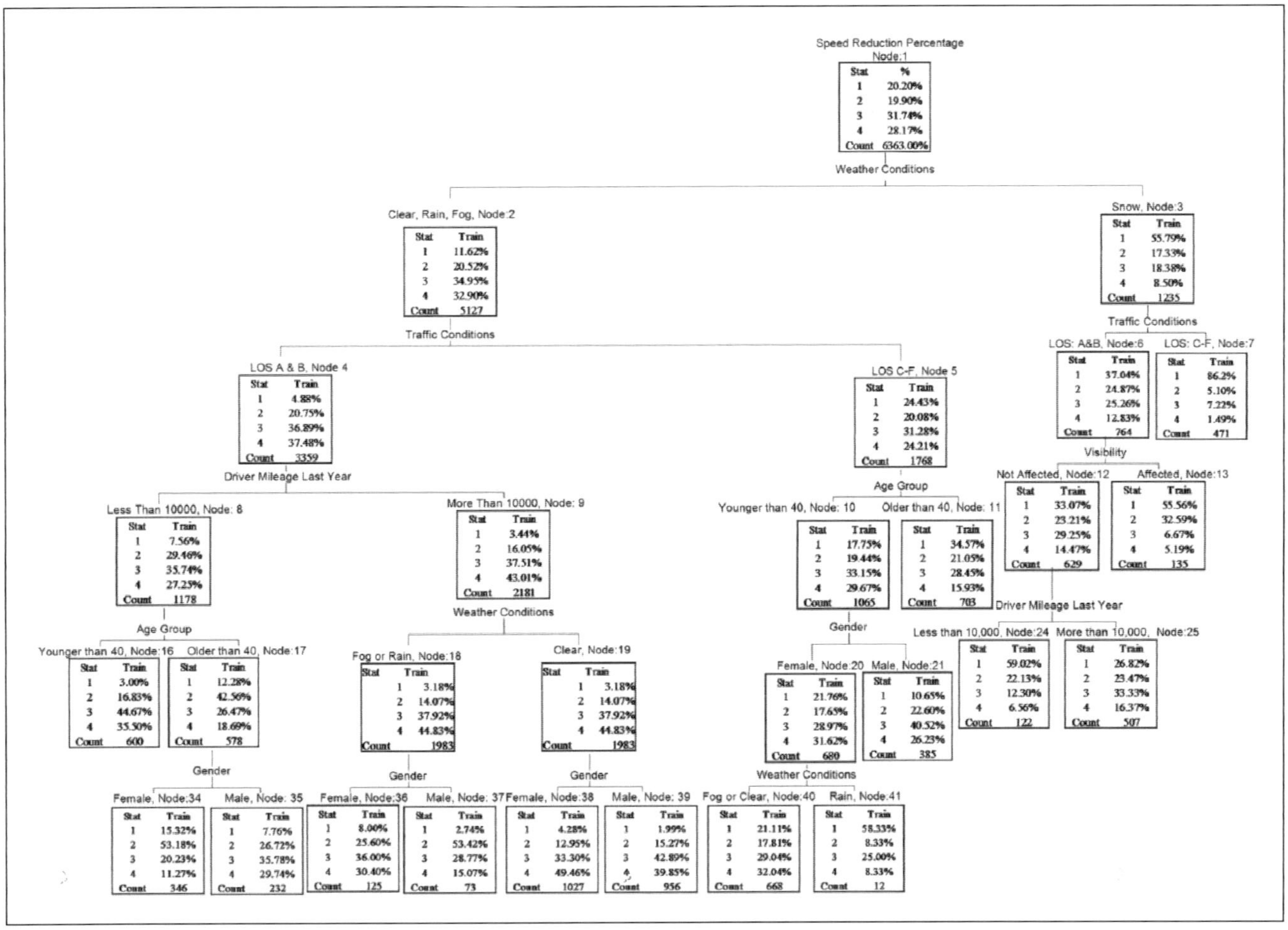

Figure 3. Classification tree diagram for speed selection model.

Table 3. Relative Importance of Variables

Variable	Score
Weather cond.	100
Traffic cond.	74.09
Speed limit	64.82
Education	50.49
Age	44.63
Mileage traveled last year	44.36
Visibility	21.08

regression is beneficial in interpreting the marginal effect of risk factors. Therefore, it is justified to use both models to take the advantage of the benefits and compensate for the shortcomings of each method. Combined, the use of parametric and non-parametric speed selection models provides a deeper understanding of speed selection behavior in adverse weather conditions. The focus of this paper was not to show that one model is superior to the other one; it attempts instead to show how the two proposed complementary parametric and non-parametric approaches can help researchers provide better insights into the factors that impact speed selection in adverse weather conditions.

The speed selection models revealed that among various adverse weather conditions, drivers are more likely to reduce their speed in snowy weather conditions. Specifically, the odds of drivers reducing their speed were 9.29 times higher in snowy weather conditions, followed by rain and fog with 1.55 and 1.29 times, respectively (compared with clear conditions). In addition, variable importance analysis using the classification tree method revealed that weather conditions, traffic conditions, and the posted speed limit are the three most important variables affecting driver speed selection behavior.

Selecting the appropriate driving speed for prevailing conditions is considered one of the most important driving tasks on high-speed facilities. Because of the previously limited understanding of the interaction between driver behavior/performance and weather conditions, the continuation of this research aims to establish a Connected Human-in-the-Loop VSL system, which is aligned with the SHRP2 Task Force's focus areas. An important component of the driver–weather interaction is the characterization of traffic flow, because heterogeneity in driver behavior exists between adverse weather conditions and traffic flow conditions, meaning that driving behavior is different for different levels of congestion and weather conditions. Modeling variation in driver behavior with adverse weather conditions and traffic flow states is crucial to assigning effective VSLs, as these algorithms must consider the impact of both weather and traffic conditions when suggesting the safest and most efficient speed.

An additional benefit from these developed models may be introduced in connected vehicle applications, where the VSL system could be expanded to incorporate mobile vehicle data as an input and to export VSL data to on-board units (OBUs). The OBUs could then provide speed advisories, regulatory speeds, or other related advisories to the driver. Messages, such as "turn off cruise control," could be sent in real time to more effectively regulate driving speed and preserve a safe flow of traffic. If unusual traffic patterns are detected or inclement weather events are forecasted or experienced, these geospatial locations could be flagged for implementation of an appropriate and timely mitigation strategy to reduce the impact of the adverse weather condition.

Acknowledgments

This work was conducted under the second Strategic Highway Research Program (SHRP2), which is administrated by the Transportation Research Board (TRB) of the National Academies of Sciences, Engineering, and Medicine, and it was sponsored by the Federal Highway Administration (FHWA) in cooperation with American Association of State Highway and Transportation Officials (AASHTO).

References

1. Rahman, A., and N. E. Lownes. Analysis of Rainfall Impacts on Platooned Vehicle Spacing and Speed. *Transportation Research Part F: Traffic Psychology and Behaviour*, Vol. 15, No. 4, 2012, pp. 395–403.
2. Kwon, T., L. Fu, and C. Jiang. Effect of Winter Weather and Road Surface Conditions on Macroscopic Traffic Parameters. *Transportation Research Record: Journal of the Transportation Research Board*, 2013. 2329: 54–62.
3. Maze, T., M. Agarwai, and G. Burchett. Whether Weather Matters to Traffic Demand, Traffic Safety, and Traffic Operations and Flow. *Transportation Research Record: Journal of the Transportation Research Board*, 2006. 1948: 170–176.
4. Ibrahim, A. T., and F. L. Hall. Effect of Adverse Weather Conditions on Speed-Flow-Occupancy Relationships. *Transportation Research Record: Journal of the Transportation Research Board*, 1994. 1457: 184–191.
5. Kyte, M., Z. Khatib, P. Shannon, and F. Kitchener. Effect of Weather on Free-Flow Speed. *Transportation Research Record: Journal of the Transportation Research Board*, 2001. 1776: 60–68.
6. Liang, W. L., M. Kyte, F. Kitchener, and P. Shannon. Effect of Enviromental Factors on Driver Speed: A Case Study. *Transportation Research Record: Journal of the Transportation Research Board*, 1998. 1635: 155–161.
7. Qiu, L., and W. A. Nixon. Effects of Adverse Weather on Traffic Crashes: Systematic Review and Meta-Analysis. *Transportation Research Record: Journal of the Transportation Research Board*, 2008. 2055: 139–146.
8. Hamilton, B. A. *Ten-Year Averages from 2005 to 2014 Based on NHTSA Data*. US Department of Transportation- Federal Highway Administration.
9. Hambly, D., J. Andrey, B. Mills, and C. Fletcher. Projected Implications of Climate Change for Road Safety in Greater Vancouver, Canada. *Climatic Change*, Vol. 116, No. 3–4, 2013, pp. 613–629.
10. Vogt, A., and J. Bared. Accident Models for Two-Lane Rural Segments and Intersections. *Transportation Research Record: Journal of the Transportation Research Board*, 1998. 1635: 18–29.
11. Hankey, J. M., M. A. Perez, and J. A. McClafferty. Description of the SHRP2 Naturalistic Database and the Crash, Near-Crash, and Baseline Data Sets. *Task Report, The Strategic Highway Research Program 2 Transportation Research Board of The National Academies*, Vol. 53, No. 9, 2013, pp. 1689–1699.
12. Ahmed, M. M., A. Ghasemzadeh, H. Eldeeb, S. Gaweesh, J. Clapp, K. Ksaibati, and R. Young. *Driver Performance and Behavior in Adverse Weather Conditions: An Investigation Using the SHRP2 Naturalistic Driving Study Data—Phase* 1, No. FHWA-WY-16/08F. 2015.
13. Weng, J., Q. Meng, and D. Z. W. Wang. Tree-Based Logistic Regression Approach for Work Zone Casualty Risk Assessment. *Risk Analysis*, Vol. 33, No. 3, 2013, pp. 493–504.
14. Ahmed, M., and M. Abdel-Aty. A Data Fusion Framework for Real-Time Risk Assessment on Freeways. *Transportation Research Part C: Emerging Technologies*, Vol. 26, 2013, pp. 203–213.
15. Das, S., and X. Sun. Investigating the Pattern of Traffic Crashes under Rainy Weather by Association Rules in Data Mining. Presented at 93rd Annual Meeting of the Transportation Research Board, Washington D.C., 2014.
16. Abdel-Aty, M., and H. Abdelwahab. Modeling Rear-End Collisions Including the Role of Driver's Visibility and Light Truck Vehicles Using a Nested Logit Structure. *Accident Analysis & Prevention*, Vol. 36, No. 3, 2004, pp. 447–456.
17. Lord, D., S. R. Geedipally, and S. D. Guikema. Extension of the Application of Conway-Maxwell-Poisson Models: Analyzing Traffic Crash Data Exhibiting Underdispersion. *Risk Analysis*, Vol. 30, No. 8, 2010, pp. 1268–1276.
18. Chen, H., Q. Chen, L. Chen, and G. Zhang. Analysis of Risk Factors Affecting Driver Injury and Crash Injury with Drivers Under the Influence of Alcohol (DUI) and Non-DUI. *Traffic Injury Prevention*, Vol. 17, No. 8, 2016, pp. 796–802.
19. Agresti, A. *An Introduction to Categorical Data Analysis*, 2nd ed., New York: Wiley, 2007.
20. Ghasemzadeh, A., M. M. Ahmed, and T. R. Board. A Probit-Decision Tree Approach to Analyze Effects of Adverse Weather Conditions on Work Zone Crash Severity Using Second Strategic Highway Research Program Roadway Information Dataset. Presented at 96th Annual Meeting of the Transportation Research Board, Washington D.C., 2017, pp. 1–17.
21. Ghasemzadeh, A., and M. M. Ahmed. A Tree-Based Ordered Probit Approach to Identify Factors Affecting Work Zone Weather-Related Crashes Severity in North Carolina Using the Highway Safety Information System Dataset. Presented at

96th Annual Meeting of the Transportation Research Board, Washington D.C., 2017.
22. Han, J., and M. Kamber. *Data Mining: Concepts and Techniques*. University of Illinois, Urbana-Champaign, 2011.
23. Kashani, A. T., and A. S. Mohaymany. Analysis of the Traffic Injury Severity on Two-Lane, Two-Way Rural Roads Based on Classification Tree Models. *Safety Science*, Vol. 49, No. 10, 2011, pp. 1314–1320.
24. Weng, J., and Q. Meng. Decision Tree-Based Model for Estimation of Work Zone Capacity. *Transportation Research Record: Journal of the Transportation Research Board*, 2011. 2257: 40–50.
25. Kovalerchuk, B., and E. Vityaev. *Data Mining in Finance: Advances in Relational and Hybrid Methods*. Springer Science & Business Media, New York, 2000.
26. Breiman, L., J. Friedman, C. J. Stone, and R. A. Olshen. *Classification and Regression Trees*. CRC Press, Boca Raton, 1984.
27. Marquaridt, D. W. Generalized Inverses, Ridge Regression, Biased Linear Estimation, and Nonlinear Estimation. *Technometrics*, Vol. 12, No. 3, 1970, pp. 591–612.
28. Kang, J., R. Ni, and G. Andersen. Effects of Reduced Visibility from Fog on Car-Following Performance. *Transportation Research Record: Journal of the Transportation Research Board*, 2008. 2069: 9–15.
29. Ghasemzadeh, A., and M. M. Ahmed. Driver's Lane Keeping Ability in Heavy Rain: Preliminary Investigation using the SHRP2 Naturalistic Driving Study Data. *Transportation Research Record: Journal of the Transportation Research Board*, 2017. 2663: 99–108.
30. Ahmed, M. M., and A. Ghasemzadeh. Exploring the Impacts of Adverse Weather Conditions on Speed and Headway Behaviors Using the SHRP2 Naturalistic Driving Study Data. Presented at 96th Annual Meeting of the Transportation Research Board, Washington D.C., 2017.
31. Beshah, T., and S. Hill. Mining Road Traffic Accident Data to Improve Safety: Role of Road-Related Factors on Accident Severity in Ethiopia. *AAAI Spring Symposium: Artificial Intelligence for Development*, 2010.

The Standing Committee on Surface Transportation Weather (AH010) peer-reviewed this paper (18-01096).

Article

TRR
JOURNAL OF THE TRANSPORTATION RESEARCH BOARD

Transportation Research Record
2018, Vol. 2672(12) 148–157

DOI: 10.1177/0361198118758012
journals.sagepub.com/home/trr
SAGE

Development of a Pavement Marking Life Cycle Cost Tool

Adam M. Pike[1] and Bharadwaj Bommanayakanahalli[1]

Abstract

Pavement marking materials can be broadly classified into two types: durable and nondurable. The durable markings have a longer service life compared with the nondurable and, thus, need to be replaced less frequently. The material cost of the durable markings is higher compared with the nondurable markings. However, the durable markings have several benefits compared with the nondurable markings. Durable markings require less frequent striping resulting in less traffic delays faced due to marking installation and fewer accidents due to fewer work zone closures. The administrative costs associated with contracting and monitoring the work is also lower for durable markings due to less frequent contracting. The primary purpose of this project is to explore the cost differences between durable and nondurable markings, and how various factors can impact the life cycle cost of different marking materials. Researchers developed a life cycle cost calculator that considers the various costs associated with pavement markings to assist in comparing total costs of different marking materials. All costs associated with the implementation of pavement marking projects are considered, including material and installation costs, traffic delay costs and crash costs due to marking installations, and administrative costs to cover contracting and management of the pavement marking assets. This life cycle cost calculator would be another tool that decision makers could use to help choose the most cost-effective pavement marking for the selected input condition.

Pavement markings need to be visible during the day, and at night where visibility may be more critical due to the reduction in other cues to aid the driver in navigating the roadway. Properly maintained pavement markings can potentially reduce crashes that occur at night (mainly run off the road and head-on crashes), especially in areas where the roadway alignment changes (*1*).

Pavement markings should be replaced on a regular basis to maintain adequate retroreflectivity levels for the road users. The service life of the markings depends on several factors but a primary difference is due to the type of the marking material. Pavement marking materials can be broadly classified into two types: durable and nondurable. Durable markings have a longer service life compared with the nondurable and, thus, need to be replaced less frequently. The material cost of the durable markings is higher compared with nondurable markings. However, the durable markings have several benefits compared with the nondurable markings. Durable markings require less frequent striping resulting in less traffic delays faced due to marking installation and fewer accidents due to fewer work zone closures.

The administrative costs associated with contracting and monitoring the work is also lower for durable markings due to less frequent contracting.

The primary purpose of this project is to explore the cost differences between durable and nondurable markings, and how various factors can impact the life cycle cost of different marking materials. Researchers developed a life cycle cost calculator that considers the various costs associated with pavement markings to assist in comparing total costs of different marking materials. All costs associated with the implementation of pavement marking projects are being considered, including material and installation costs, traffic delay costs and crash costs due to marking installations, and administrative costs to cover contracting and management of the pavement marking assets. This life cycle cost calculator would be another tool that decision makers could use to help choose the most cost-effective pavement marking for the selected input condition. This paper is a part of a larger study developing the marking cost tool. Additional information on the literature review, tool development, and tool use can be found in the larger not yet published report. The report and the tool will be made available on the Texas A&M Transportation Institute website once the work is complete.

[1]Signs and Markings Program, Texas A&M Transportation Institute, College Station, TX

Corresponding Author:
Address correspondence to Adam M. Pike: a-pike@tti.tamu.edu

Table 1. Summary Pavement Marking Installation Cost Ranges ($/ft)

Pavement marking material	Literature review range	Bid price range 4 in. wide	Bid price range 6 in. wide
Paint	0.02–0.20	0.05–0.22	0.08–0.53
Thermoplastic	0.08–1.18	0.11–0.91	0.16–1.08
Epoxy	0.26–0.45	0.30–1.32	0.54–0.69
Tape	1.00–2.65	1.94–3.78	2.08–5.62
Polyurea	0.43–0.90	0.56–1.32	0.80
Methyl methacrylate	0.25–1.75	1.25	0.79–0.80
High build paint	0.12–0.30	–	–

Note: Cells with only a – indicate that bid price information was not found.

Background and Methodology

This study considers the various costs associated with pavement markings which would affect the total life cycle costs of the markings. This section presents the results of a literature review and review of state bid prices to gather information on the different cost factors and how they apply to the tool. A description of other factors which affect the life cycle cost of pavement markings such as traffic delay costs, work zone crash costs, and service life of markings are also explored in this section. The costs associated with the markings can be classified as follows:

- Primary Costs
  - Material and Construction Costs
  - Grooving Costs (when grooving is performed)
  - Special Requirements (when there is need for wet retro requirements, warranty, etc.)
  - Marking Removal Costs (when removal is done before applying the new markings)
- Secondary Costs
  - Administrative Costs
  - Agency Costs
  - Societal Costs
    1. Delay Costs
    2. Crash Costs

Primary Costs: Material and Construction Costs

The primary costs associated with the pavement markings are the material costs and the installation costs. In this study the following marking material types are considered:

- Paint
- Thermoplastic
- Epoxy
- Tape
- Polyurea
- Methyl Methacrylate

Sources of Data for the Primary Costs. Costs were summarized after looking at two areas for information, a literature review and state bid prices. The costs from literature were gathered from sources mostly from the past 20 years (*2–25*). The bid prices of 10 states were summarized in this study. These states were California, Oregon, Washington, Tennessee, Florida, Michigan, Pennsylvania, Texas, Virginia, and Minnesota. These states were selected due to their geographic distribution, and the availability of their data. Several other states were also considered but their data was not considered adequate. The bid prices have been taken from the DOT websites and are mostly annual averages, with the most recent available data being selected. A summary of the expected ranges of pavement marking installation costs for standard marking types is presented in Table 1.

The average 4- and 6-in. bid prices within individual states showed a range of price increases and in some cases price decreases. For an individual project there should be a price increase when selecting a wider 6-in. marking compared with the standard 4-in. wide marking due to applying 50% more material. The cost increase should be less than 50%, because the labor costs do not increase by 50%. It is safe to assume between a 15% and 45% increase in marking cost when going from a 4- to a 6-in. wide marking.

Table 1 summarizes the expected costs and provides ranges for the various types of markings. Other factors that can potentially influence the primary costs include the color of the marking, width of the marking, thickness of the marking, type of line marked (solid, skip, etc.), contract size (prices are very sensitive to contract size), regional differences, special requirements (wet night visibility, retroreflectivity requirements, etc.), and installation requirements such as surface preparation, removal of preexisting markings, and so forth. The above costs should be considered when estimating pavement marking installation unit costs

Additional Pavement Marking Installation Costs. The application of a new marking or restriping an old marking may require surface preparation or removal of the preexisting marking. In some cases, pavement surface preparation is bid as a separate line item, but is often factored into the unit cost of the marking itself. In many cases the removal of preexisting markings is bid as a separate line item. When restriping a roadway with a new pavement marking material it is critical to ensure compatibility of the restripe material with the preexisting material if removal is not taking place. If the markings are incompatible, then removal should take place. Compatibility of materials, removal requirements, and surface preparation needs should follow state guidelines and manufacturer recommendations. There are several ways to remove pavement markings such as grinding, flailing, sand blasting, water blasting, and so on. NCHRP Report 759 discusses the effective removal of the pavement markings (*27*). A summary of the removal costs is available in the NCHRP report.

In areas with winter maintenance activities, markings are often installed in grooves (inlaid/recessed). The process of placing the markings in grooves is expected to increase the service life of the markings. It is important to account for this additional cost when considering life cycle cost of the marking. The cost associated with grooving has been found in literature as well as in the bid prices. Michigan Department of Transportation (DOT) had an average bid price of \$0.55/ft for recessing longitudinal pavement markings. A report for the FHWA (*28*) included costs associated with grooving pavement markings obtained through a literature review. The report cites the work of Lagergren et al. (*29*) which reports a cost of \$1.05/ft for a 100-mil groove and a cost of \$0.95/ft for a 60-mil groove. The report also cites the work of Hawkins et al. (*30*), which reports a cost of between \$0.40 and \$1.40/ft for the grooves.

Secondary Costs: Delay, Crash, and Administrative Costs

The installation of the pavement markings results in several secondary costs. These costs are often not considered when evaluating the cost-effectiveness of markings. In some situations, the secondary costs may be minor, but in other situations the secondary costs could play a significant factor in determining the most cost-effective marking.

Cost of Delay Due to the Pavement Marking Operations. Pavement marking operations generally involve the slowdown of traffic behind the application equipment and associated traffic control or complete lane closures. Both kinds of work zone operations cause delays, particularly if the marking operations are on two-lane two-way roads, high-volume roads, or work is being conducted during peak hours. The total delay depends upon the traffic at the time of the marking operations, number of lanes on the project roadway, speed limit on the roadway, and the driver behavior in the work zone. It is difficult to account for all these factors and obtain a good estimate of the delay. The cost associated with the delay is obtained by multiplying the delay and estimated average value of time of the road users.

The computation of delay and the associated cost is a major challenge in this study since it depends on multiple parameters and it is difficult to accurately account for all of them. There has been research that has studied the delay due to complete lane closure but there has been very little work that has studied the delay impacts of mobile work zones. A paper by Weng and Meng (*31*) summarized the work done by researchers in evaluating the capacity and delay in work zones. It is important to know the capacity of the work zone since that plays a significant role in the delay. Their paper cites the work of Weng and Meng (*32*) that identified 16 factors that affect work zone capacity. The approaches used by the researchers can be broadly classified into three categories: parametric, nonparametric, and simulation approaches.

A FHWA report on road user costs (*33*) summarizes the traffic analysis tools used by state DOTs to compute road user costs caused by work zones.

Methodology to Evaluate the Delay Costs in This Study

Work Zones with Lane Closures. For evaluating the delay due to the lane closures the research team utilized the procedure by Fu and Wilmot (*14*). Fu and Wilmot tried to estimate the user costs due to marking operations that involve lane closures. Even their procedure is not perfect since the delay calculated is for a specific set of roadway conditions. However, they have developed a set of regression equations that simplifies the process. The study assumes two operating schedules in urban areas during daytime off-peak (from 9 a.m. to 3 p.m.) and nighttime (from 8 p.m. to 6 a.m.). The study assumes the operations occur during just the daytime off-peak between 8 a.m. to 2 p.m. for the rural areas. The study also mentions several parameters to calculate the user costs. In addition, the authors also developed regression equations to calculate the user costs (*14*). A summary of the parameters and equations can be found in their paper. The cost tool developed in this study adjusted the delay costs to 2017 values.

Mobile Work Zones. The most feasible way to evaluate the delays and costs for mobile work zones is through the use of microscopic simulation software. Cottrell and Hanson (*3*) estimated traffic delays caused by mobile pavement marking work zone operations. They used the simulation software CORSIM and simulated two-lane, four-lane, and six-lane road sections with low, moderate, and heavy traffic conditions. They assumed that the marking truck moves at a speed of 7 mph in the rightmost lane. They also assumed that 2% of the traffic was trucks. The simulation also used a 1-mile stretch to calculate the delays. The authors pointed out that if long queues develop behind the marking truck, the marking crew stops the operation and releases the queue. This scenario is likely on two-lane roads and to analyze this they included a second scenarios of two-lane roads with 0.5-mile section. The cost has been calculated assuming a value of time of \$16.10 for cars and \$29.24 for trucks.

For development of the cost tool it was decided to use the delay values that Cottrell and Hanson developed but use

Table 2. Delay Costs for the Mobile Work Zone

Scenario	Delay (veh-h/mile)	Delay cost ($/mile/pass)	Total delay cost ($/mile)	Delay cost per foot**
2 lanes 400 vph	3.94	$89.07	$148.74	$0.028
2 lanes 400 vph 0.5 mi	2.3	$51.99	$86.83	$0.016
2 lanes 1,000 vph	16.3	$368.48	$615.36	$0.117
2 lanes 1,000 vph 0.5 mi	13.7	$309.70	$517.20	$0.098
2 lanes 2,000 vph	35	$791.21	$1,321.32	$0.250
4 lanes 800 vph	0.05	$1.13	$1.89	$0.000
4 lanes 2,000 vph	1.6	$36.17	$60.40	$0.011
4 lanes 4,000 vph	22.5	$508.64	$849.42	$0.161
6 lanes 1,500 vph	0	$0.00	$0.00	$0.000
6 lanes 3,000 vph	0.6	$13.56	$22.65	$0.004
6 lanes 6,000 vph	30.3	$684.96	$1,143.89	$0.217

the latest value of time and occupancy values to calculate the cost associated with the delay. A value of time of $22.40/h is used for passenger cars and a value of time of $32.70/h is used as per the TxDOT guidelines (*34*). Similarly, an average vehicle occupancy of 1.67 is used as per the results of the 2009 NHTS survey (*35*). An example of the cost associated with the delay is provided in Table 2.

Shortcomings Work Zone Delay EvaluationMethodology

Mobile Work Zone Operations. A significant shortcoming of the work is that the mobile work zone delay is calculated using the same speed of the moving truck (7 mph) for all the marking types. The application speed is expected to be different for the different marking types. There is no way to account for the type of marking while evaluating the delay costs. This is a shortcoming in this study that was outside the scope of the original research. It may be difficult to evaluate the *k*-factor (% of AADT occurring during a particular hour) during the hours of marking operations. The tool does not accurately use the traffic during a particular hour to calculate the costs. It uses the next highest volume for which the delay is available. Thus, the tool always overestimates the delay. The tool cannot evaluate the delay costs when the marking operations are done on high-volume roads during the peak hours.

Recommendations for Practitioners. As mentioned earlier, many states have their own tools developed for evaluating the road user costs. Whenever possible the practitioner should use the tool specified by the state to evaluate the road user costs.

Crash Costs Due to Marking Operations (Work Zones). Work zones activities are expected to increase the probability of a crash. There has been very little research done in this area on crash frequency at pavement marking work zone operations. The total crash cost associated with the markings can be obtained by multiplying the total number of expected crashes by severity with the associated cost of each severity.

The evaluation of the cost of crashes associated with marking operations is very challenging. Most of the crash estimation techniques (in general and for work zones) are developed for long-term expected crash frequency. The major challenge while applying the conventional crash estimation techniques for estimating the crashes of marking operations is the short-term nature of marking operations. NCHRP research results digest 339 (Improving the safety of mobile lane closures), indicates some of the problems associated with safety analysis (*36*). The report indicates that several DOTs do not conduct safety analysis of mobile operations since it is difficult to isolate relevant crashes from the crash database. The report also presents the results of some previous studies. According to a study conducted in New York (*37*), intrusion accidents accounted for 10% of all work zone accidents and 8% of serious injury crashes.

A FHWA report on road user costs (*33*) describes a methodology to evaluate the crash costs associated with work zones (of longer duration). The following are some potential difficulties associated with the methodology described in report for application to the current study.

1. Obtaining the crash data: It is sometimes difficult to obtain the crash data for multiple years on a particular stretch of road. The crash data reduction to usable form needs skilled personnel to develop the database. This is also a significant effort to undertake for each pavement marking project.
2. The crash modification factors (CMFs) developed in the studies are for long-term work zones. It is not appropriate to use these CMFs when analyzing the work zones of short duration as in the case of marking operations.

Methodology Adopted for This Study

The researchers chose to use a methodology used by Young (*11*) to evaluate the crash costs. The study calculates the cost of the crash per vehicle mile travelled (VMT) based on the available data. The study reports that there are 20%–30%

more accidents in a work zone than on a general road. This study assumes a similar methodology to estimate the costs associated with the accidents. The 2010 NHTSA report (*38*) indicates a value of $242 billion for the economic costs of all motor vehicle crashes. The total value of the societal harm from the motor vehicle crashes has been reported as $836 billion. As per the bureau of transportation statistics report in 2010, the total VMT travelled in year 2010 was 2,999,506 million vehicle miles. By using these values, the crash cost per VMT is estimated at 28.18 cents. This cost has been evaluated using the cost associated with total societal harm. The duration of the work zone, the traffic volume at the time of the work zone, the estimated percentage increase in crashes, and the average crash cost of 28.18 cents per vehicle mile traveled are used to estimate the impact on crash costs that are the result of the pavement marking work zone operations.

Administrative Costs. Administrative costs cover the agency costs associated with contracting, installing, monitoring, and maintaining the marking asset. More frequent and higher demand activities required by an agency to deal with an asset will result in higher administrative costs associated with that asset. Ted Miller (*26*) gave the following equation to calculate the annualized application cost of the markings as

$$C = M + P + E + \text{ADMIN}$$

where:

M= annualized materials costs, including binder, beads, and fuel;

P=annualized personnel costs, including wages, fringe benefits, and per diem when striping
crews are away from home overnight;

E=annualized costs of equipment and storage facilities; and

ADMIN = annualized contract letting, monitoring, and other administrative costs.

The study found that TxDOT estimates the cost of contract letting, monitoring, and administrative costs as 5% of total contract cost while North Carolina DOT estimates this as 7% of total contract cost (*26*). A Texas A&M Transportation Institute (TTI) report (*39*) gave an approximate breakdown of the administrative costs. The TTI report discussed in-depth the need for including the administrative and supervision costs in the total cost of the project. The report gives an administrative cost range of 1.6% to 6.7%. The contracting process has several steps including, project identification, prioritization of projects, revision of contracts/bids, and contract management. Most of these costs are hidden costs and are not directly accounted for in any types of costs.

Another TTI report that focused on performance-based pavement marking contracting (*40*) surveyed the districts in Texas on the burden with regard to managing marking/marker assets. The report states that an inspector was present

Table 3. Summary of Pavement Marking Service Life Information from the Literature

	Range	Typical values
Paint	3 months–3 years	1 year
Thermoplastic	2 years–10 years	3–4 years
Epoxy	1–5 years	3–4 years
Tape	2–10 years	6 years
Polyurea	2–5 years	3–4 years
Methyl methacrylate	2–8 years	–
High build paint	3 years	–

Note: Cells with only a – indicate that typical value was not established.

for approximately 70% of the time when new markings were applied.

In a recent email exchange with a State DOT official it was indicated that the administrative and supervision costs are close to 35% of the total project cost for pavement marking projects.

It is suggested that an administrative cost value of between 5% and 35% of the total project cost be used when looking at the total costs of pavement marking materials. The user of the tool has to enter the appropriate percentage for each marking type based on judgment (the same percentage of admin costs should not be used for all marking types).

Service Life Considerations in the Tool

Pavement Marking Service Life. The service life of the markings found in literature (*2–25*) is summarized in Table 3. Typical ranges for the main types of marking are indicated as well as a typical expected service life for the marking types. Like the cost information the service life information is quite variable and is impacted by many factors.

Additional Considerations to the Pavement Marking Service Life. Table 3 summarizes the expected service life and provided the ranges for the various types of markings. However, the following factors can potentially affect the service life of the markings and should be considered when selecting an estimated service life:

1. Road surface type
2. Color of the markings
3. Thickness of the markings
4. Traffic volume (AADT)
5. Installation quality
6. Climatic conditions (winter maintenance activities)
7. Replacement criteria (retroreflectivity level, fixed cycle, etc.)
8. Remaining road surface life

Remaining Service Life of the Road Surface. The remaining service life of the road surface is a critical aspect to consider for restriping activities as the road surface ages. The usage of expensive markings with longer service lives is feasible in the

PAVEMENT MARKING COST CALCULATOR				Texas A&M Transportation Institute Developed by the Texas A&M Transportation Institute
Date:	8/1/2017			
Analyst:	TTI			
Project Description:	4 lane divided highway			
Comments:	example			
SITE FACTORS				
Traffic Volume (AADT)	55000		Road Surface	Asphalt
Number of Lanes (each direction)	2		Winter Maintenance	No
Remaining Road Surface Life / Analysis Time Frame (in years)	14		Marking Material Width	4 inch
Marking Quantity Estimate (ft)	300000		Marking Color	White
Grooving (Yes/No)	No		Current Markings Applied	None
Special Requirements (Yes/No)	No		Crash Cost $ per VMT	0.281813015
Marking Removal (Yes/No)	No		Annuity Factor, A(t,r)	14
K-Factor (average for the work zone duration)	5.0%		Hourly Volume (vph)	2750
Time of Day for the Work Zone (Day/Night)	Day			
Type of Work Zone (Mobile/Lane Closure)	Lane Closure			
Work Zone % Increase in Crashes	35%			
FHWA Cost Contributions	80/20			
Discount Rate (%)	3.00%			
Interest Rate (%)	0.00%			
Traffic Delay Cost ($) [From Installation Delay Cost Sheet]	0.347903358			

Figure 1. Screenshot of the pavement marking lifecycle cost tool user inputs.

long-term only if the remaining service life of the road surface is long. If the expected service life of the road surface is short, it may be advisable to use a less-durable marking. The user of the tool has to enter the remaining road service life or the analysis period in the tool.

Life Cycle Cost Analysis

Life cycle cost analysis is a method for evaluating a pavement marking by determining the annual cost of owning and maintaining the pavement marking asset. The cost considers the initial cost of the marking, costs associated with maintaining the asset until it needs to be replaced, interest on capital expenses, other costs incurred such as traffic delay or crash costs during installation, and the expected service life of the asset. Annualizing the costs and considering all cost sources allows markings with different expected service lives to be compared for cost effectiveness.

Discount Rate. The discount rate is needed to evaluate the present value of the investment for each type of marking over the service life of the road.

Interest Rate. The interest rate is needed to evaluate the annuity factor which is needed to evaluate the equivalent annual cost (EAC) for each type of markings.

FHWA Cost Contributions. Construction projects are often funded by both the state and the FHWA, where the FHWA generally contributes either 80% (80/20) or 90% (90/10) of the construction costs. The maintenance costs in the subsequent years are generally paid entirely by the state agencies.

Tool Inputs and Results

The screenshot of the input and output of the tool can be found in Figures 1 and 2, respectively. The user has to input the values for certain parameters in the cells while values for certain parameters have to be selected from dropdown lists (green shaded cells).

Site Factors

The site factors have two separate input parameters on the left- and right-hand sides. The parameters on the left-hand side directly affect the costs in the tool while the parameters on the right do not directly affect the costs in the tool. However, the user has to use engineering judgment to input values of certain parameters in the tool based on the values of parameters in the right-hand side.

Parameters Affecting Cost Directly in the Tool (Left-Hand Side)

- Traffic Volume (AADT)
- Number of Lanes (each direction)
- Remaining Road Surface Life/Analysis Time Frame (in years)
- Marking Quantity Estimate (ft)
- Grooving (Yes/No)
- Special Requirements (Yes/No)
- Marking Removal (Yes/No)
- *k*-Factor (average for the work zone direction)
- Time of Day for the Work Zone (Mobile/Lane Closure)
- Type of Work Zone (Mobile/Lane Closure)
- Work Zone % Increase in Crashes
- FHWA Finance Scheme

	COST FACTORS	Marking Material 1	Marking Material 2	Marking Material 3	Marking Material 4
Primary Costs	Marking Material	Paint	Thermoplastic	Epoxy	Tape
	Duration of 1-mile of work zone (hours)	0.15	0.2	0.2	1
	Estimated Material Unit Costs ($/ft)	$0.10	$0.30	$0.50	$2.50
	No Special Requirements. No Additional Costs				
	No Marking Removal. No Additional Costs				
	Total Primary Costs	$0.10	$0.30	$0.50	$2.50
Secondary Costs	Administrative Costs & Agency Costs (% of total project costs)	40%	25%	20%	10%
Societal Costs	Work Zone Safety Impact Costs ($/ft)	$0.01	$0.01	$0.01	$0.05
	Traffic Delay Costs ($/ft) - During maintenance period only	$0.05	$0.07	$0.07	$0.35
	Total Installation Cost (excluding Grooving)	$0.20	$0.45	$0.68	$3.15
	No Grooving	Paint	Thermoplastic	Epoxy	Tape
	No Grooving. Leave Blank				
	No Grooving.Leave Blank				
	PV of Grooving - State Costs	$0.00	$0.00	$0.00	$0.00
	LIFE CYCLE COST ANALYSIS				
	Marking Material	Paint	Thermoplastic	Epoxy	Tape
	Marking Material Service Life (years)	1	3	4	7
	PV - State Costs - Including all Costs	$2.25	$1.68	$1.90	$3.71
	PV - State Costs - Updated Costs (to be used for analysis)	$2.19	$1.61	$1.83	$3.36
	EAC- State Costs-including all costs($/ft/year)	$0.16	$0.12	$0.13	$0.24
	Federal Costs(Only Primary Costs for Initial Installation)	$0.08	$0.24	$0.40	$2.00
	PV- State Costs-Only Primary Costs	$1.08	$1.03	$1.29	$2.53
	EAC- State Costs-Only Primary Costs($/ft/year)	$0.08	$0.07	$0.09	$0.18
	Project Cost-State Costs- including all costs($)	$658,062	$483,030	$548,351	$1,008,606
	Project Cost-Federal Costs- including all costs($)	$24,000	$72,000	$120,000	$600,000
	Total Project Cost($)	$682,062	$555,030	$668,351	$1,608,606
	Primary Cost-State Costs - only Primary costs($)	$325,049	$307,838	$386,891	$759,819
	% of Total Costs Spent by Federal Government	3.52%	12.97%	17.95%	37.30%

Figure 2. Screenshot of the pavement marking lifecycle cost tool output.

- Discount Rate (%)
- Interest Rate (%)
- Traffic Delay Cost ($/ft From Installation Delay Cost Sheet)

Parameters Not Affecting Cost Directly (Right-Hand Side)

- Road Surface
- Winter Maintenance
- Marking Material Width
- Marking Color
- Current Marking Applied

Parameters Directly Output by the Tool

- Crash Cost $ per VMT
- Annuity Factor, $A(t,r)$
- Hourly Volume (vph)

Most of the rows are titled to ensure they are self-explanatory. The user can also hover over the individual row to obtain additional details on the input parameters in the form of comments. The cells in gray are directly evaluated by the tool and the user should not enter any values for cells in gray. The value of delay has to be input by user by looking at values from another sheet in the tool and thus is marked in blue.

Cost Factors

The user has to select the marking materials to be compared in the cost factors table. The user has to input the values for the duration of 1 mile of work zones for each marking type using engineering judgment. The material cost depends on several factors as discussed earlier and the user has to input the estimated value. The user can also choose a value that is provided in Table 1. The special requirements and marking removal costs need to be input only if they are used in that project and these cells remain gray when no input is needed from the user. The percentage values for the administrative and agency costs should be entered by user based on judgment or standard agency practices. The percentage is expected to be different for the different material types. The values in the other cells are automatically populated based on the other inputs.

The user has to input the values in the grooving table only when grooving is used in the project. The present value calculations are done automatically by the tool. The table turns gray when no grooving is selected and the user should not input any values in the cells in gray.

Parameters to Be Input the User in Cost Factors

- Marking Material
- Duration of 1 mile of Work Zone (hours)
- Estimated Material Unit Cost ($/ft)
- Additional Installation Costs: Special Requirements ($/ft) (only when case arise)
- Additional Installation Costs: Marking Removal ($/ft) (only when case arise)
- Administrative Costs and Agency Costs (% of total project costs)
- Marking Material in Groove ($/ft) – only when grooving exists
- Service Life of Groove (in Years) – only when grooving exists
- Cost of Grooving ($/ft) – only when grooving exists

Parameters Directly Output by the Tool

- Total Primary Costs
- Total Installation Cost (excluding Grooving)
- Present Value (PV) of the Grooving – only when grooving exists

Life Cycle Cost Factors and Analysis

Service Life of the Markings. The user only has to input the service life of the markings to factor it into the analysis. The service life of a marking depends on several factors as discussed earlier. The user has to exercise engineering judgment while entering a value. The user can also use a value provided in the Table 3.

Analysis. Based on the input values the tool outputs the EAC for the different considerations as shown in Figure 2. The tool also outputs the total cost of the project and the costs to the state and federal agencies. The EAC ($/ft/year) enables the user to compare the markings to see the cost per year for different markings and rank them accordingly. The tool only compares the life cycle costs of the different kinds of markings based on input conditions. The user has to exercise engineering judgment and also take it into account the practices of the local jurisdictions before making a choice of marking for a particular project.

Parameters Output by the Tool

- PV and EAC – State Costs: The present value and EAC of the marking that has to be paid by the state agencies
- Federal Costs: Cost paid by the federal government (only applies to construction projects)
- Project Costs: State and Federal
- Total project cost and % of total costs spent by the federal government
- Primary Cost: State Costs

Example Results

Figure 2 shows an example of results obtained for one set of input conditions. The current project is a construction project and thus there is 80% funding from the federal government (FHWA cost contribution = 80/20). The analysis period is 14 years. The type of the work zone is a lane closure and it is during the day. The appropriate value of the delay can be looked up in the tool and entered into the correct input location. The duration of 1-mile of work zone are entered based on judgment (application speeds will be different for different markings). The service life of the markings is also different for different markings and entered based on judgment.

The results are directly output by the tool. The equivalent annual cost (EAC) can be used to compare the different markings based on monetary value. In the current example the total installation cost (primary cost) in $/ft for paint, thermoplastic, epoxy, and tape are 0.20 (0.10), 0.45 (0.30), 0.68 (0.50), 3.15 (2.50), respectively. The EAC for the state total cost (primary cost) for the analysis period in $/ft is 0.16 (0.08) for paint, 0.12 (0.07) for thermoplastic, 0.13 (0.09) for epoxy, and 0.24 (0.18) for tape. The results of the analysis show that though the initial cost (considering all costs) of the durable markings may appear higher, the life cycle EAC become comparable (in the case of tape) and lower (in the case of thermoplastic and epoxy). In the case of construction projects, the more expensive markings result in the overall percentage of costs paid by the federal government being higher. In this example, the tape is the most expensive marking, but the percentage of cost paid by the federal government is the highest at 37.3%.

Summary

The cost tool incorporates many different cost factors to obtain a total cost of installing and maintaining a pavement marking asset. When comparing alternative treatments, it is not as simple as just comparing their cost per foot and expected service life. Other factors, such as traffic delay and administrative costs, can play a significant role in the overall cost of an individual type of marking.

Durable markings have several advantages over nondurable markings. These advantages include less frequent contracting, less frequent delays to traffic, and less frequent reductions in safety due to striping activities. The cost tool allows a user to determine the impact of these advantages. These advantages do not consider the visibility performance of the durable markings, which is typically superior to that of nondurable markings.

Not only does the frequency of the striping operations affect the overall costs, but the speed of application also has a large impact when considering crash costs and traffic delay costs. Markings that are applied more quickly will result in less traffic delay and less work zone exposure, resulting in reduced crash costs. An ideal marking would be applied quickly and balance cost and durability.

The tool does have a few shortcomings as pointed out within the tool. With additional research these shortcomings could be addressed to improve the overall effectiveness of the tool. Even with the shortcomings the tool serves as a means of providing a better understanding of the costs that practitioners may want to consider when determine the most cost-effective pavement marking material to apply to their roadways.

References

1. Carlson, P., E. Park, and C. Andersen. Benefits of Pavement Markings. *Transportation Research Record: Journal of the Transportation Research Board*, 2009. 2107: 59–68.
2. Martin, P., J. Perrin, S. Jitprasithsiri, and B. Hansen. A comparative analysis of the alternative pavement marking materials for the state of Utah. Department of Civil and Environmental Engineering, University of Utah, 1996.
3. Cottrell, B. H., and R. A. Hanson. Determining the effectiveness of pavement marking materials. Virginia Transportation Research Council, Charlottesville, VA, 2001.
4. New York State Dept. of Transportation. Pavement marking material guidelines. New York State Dept. of Transportation, Materials Bureau, Albany, NY, 2002.
5. Garvey, P. M., S. J. Damin, and M. T. Pietrucha. Pavement Markings State-of-the-Practice Study (No. FHWA-PA-2008-005-PSU 008), 2008.
6. Cuelho, E., J. Stephens, and C. McDonald. A Review of the Performance and Costs of Contemporary Pavement Marking Systems (No. FHWA/MT-03-001/8117-17), 2003.
7. Thomas, G. B., and C. Schloz. Durable, cost-effective pavement markings. Phase I: Synthesis of current research (Iowa DOT Project TR-454), 2001.
8. Bischoff, A. L., and H. G. Hawkins. Pavement Marking Retroreflectivity and Service Life Issues (No. 400931). Texas Transportation Institute, Texas A&M University, 2004.
9. Montebello, D., and J. Schroeder. Cost of pavement marking materials (No. MN/RC-2000-11), 2000.
10. Migletz, J., J. Graham, D. Harwood, and K. Bauer. Service life of durable pavement markings. *Transportation Research Record: Journal of the Transportation Research Board*, 2001. 1749: 13–21.
11. Lee, Y. J. Life Cycle and Economic Efficiency Analysis Phase II: Durable Pavement Markings (No. MD-11-SP808B4P), 2011.
12. Jiang, Y. Durability and retro-reflectivity of pavement markings (synthesis study). *Joint Transportation Research Program* 235, 2008.
13. Donnell, E. T., V. Karwa, and L. Sasidharan. Retroreflectivity Database Study (No. PA-2009-007-PSU 004). Pennsylvania Department of Transportation, 2009.
14. Fu, H., and C. G. Wilmot. Assessing performance of alternative pavement marking materials (No. TAR 08-4TA). Louisiana Department of Transportation and Development, Louisiana Transportation Research Center, 2008.
15. Carlson, P. J., J. Miles, and S. Chalmers. *Strategy for Evaluating Durable Pavement Marking Materials for High Traffic Areas and Mountain Passes*. Transportation Research Board, Washington, D.C., 2006.
16. Fitch, J. Pavement Marking Durability Statewide (No. 2007-3), 2007.
17. Migletz, J., and J. L. Graham. *NCHRP Synthesis 306: Long-Term Pavement Marking Practices.* Transportation Research Board of the National Academies, Wahington, D.C., 2002.
18. Dwyer, C., W. Vavrik, and R. Becker. Evaluating Pavement Markings on Portland Cement Concrete (PCC) and Various Asphalt Surfaces. Report ICT-13-033. *Applied Research Associates*, 2013.
19. Carlson, P. J., J. D. Miles, A. M. Pike, and E. S. Park. Evaluation of Wet-Weather and Contrast Pavement Marking Applications: Final Report (No. FHWA/TX-07/0-5008-2), 2007.
20. Sarasua, W., L. Bell, and W. J. Davis. Estimating the Lifecycle of Pavement Markings on Primary and Secondary Roads in South Carolina (No. FHWA-SC-12-01). South Carolina Department of Transportation, Columbia, SC, 2012.
21. Lu, J. J. Performance of Traffic Markings in Cold Regions. Final Report (No. INE/TRC 95.03), 1995.
22. KDOT, Kansas Department of Transportation. Pavement Marking Policy, 2002. https://www.ksdot.org/bureaus/burtrafficsaf/traffic/pmp02.pdf (accessed March 14, 2017).
23. TxDOT, Texas Department of Transportation. Pavement Marking Handbook, 2004. http://onlinemanuals.txdot.gov/txdotmanuals/pmh/pmh.pdf (accessed March 14, 2017).
24. Migletz, J., J. L. Graham, D. Harwood, K. M. Bauer, and P. L. Sterner. Evaluation of All-Weather Pavement Markings. Contract FHWA C, 1038, 3524-04, 2000.
25. Migletz, J., J. K. Fish, and J. L. Graham. Roadway delineation practices handbook (No. FHWA-SA-93-001), 1994.
26. Miller, T. R. Benefit–cost analysis of lane marking. *Transportation Research Record: Journal of the Transportation Research Board,* 1992. 1334: 38–45.
27. Pike, A. M., and J. D. Miles. *NCHRP Synthesis 759: Effective Removal of Pavement Markings.* Transportation Research Board of the National Academies, Washington, D.C., 2013.
28. Carlson, P., E. S. Park, C. Andersen, B. Kuhn, A. Pike, Miles, J. Pike, and R. J. Porter. Pavement Marking Demonstration Project: State of Alaska and State of Tennessee—Report to Congress (No. FHWA-HRT-09-039), 2010.
29. Lagergren, E., J. Bertsch, and D. Fernald. I-90 Pavement Marking Material Test. Presented at 85th Annual Meeting of the Transportation Research Board, Washington, D.C., 2006.
30. Hawkins, N., O. Smadi, Z. Hans, and J. Resler. To Groove or Not to Groove, That is the Question. Presented at 86th Annual Meeting of the Transportation Research Board, Washington, D.C., 2007.
31. Weng, J., and Q. Meng Estimating Capacity and Traffic Delay in Work Zones: An Overview. *Transportation Research Part C: Emerging Technologies,* Vol. 35, 2013, pp. 34–45.
32. Weng, J., and Q. Meng. Ensemble Tree Approach to Estimating Work Zone Capacity. *Transportation Research Record: Journal of the Transportation Research Board,* 2012. 2286: pp. 56–67.
33. Mallela, J., and S. Sadavisam. Work Zone Road User Costs: Concepts and Applications. U.S. Department of Transportation, Federal Highway Administration, 2011.

34. TxDOT road user costs for the year 2017. http://ftp.dot.state.tx.us/pub/txdot-info/cst/ruc2017.pdf (accessed on April 12, 2017).
35. NHTS 2009 Avg. vehicle occupancy http://nhts.ornl.gov/tables09/fatcat/2009/avo_TRPTRANS_WHYTRP1S.html (accessed April 12, 2017).
36. Hadayeghi, A., and B. Malone. *NCHRP Research Results Digest 339: Improving the Safety of Mobile Lane Closures.* Transportation Research Board of the National Academies, Washington, D.C., 2007.
37. Bryden, J. E., L. B. Andrew, and J. S. Fortunewicz. Intrusion Accidents on Highway Construction Project. *Transportation Research Record: Journal of the Transportation Research Board*, 2000. 1715: 30–35.
38. National Highway Traffic Safety Administration. The Economic and Societal Impact of Motor Vehicle Crashes, 2010 (Revised). *Annals of Emergency Medicine*, Vol. 66, No. 2, 2015, pp. 194–196.
39. Garcia-Diaz, A., and F. Cediel-Franco. Evaluation of in house maintenance contract costs for guardrail, rest areas, pavement marker, stripping and seal coats. Texas Transportation Institute, FHWA/TX-87/380-1F, Vol. I, 1988.
40. Pike, A., P. Songchitruksa, S. Geedipally, D. Kang, and I. Damnjanovic. Evaluating the Effectiveness of Performance Based Pavement Marking Maintenance Contracts in Texas. Texas A&M Transportation Institute, FHWA/TX-14/0-6705, 1, 2014.

The Standing Committee on Signing and Marking Materials (AHD55) peer-reviewed this paper (18-01533).

All research conducted for this paper was conducted by the Texas A&M Transportation Institute of the Texas A&M University System. The contents of this paper reflect the views of the authors, who are responsible for the facts and accuracy of the data presented herein. The contents of this paper do not necessarily reflect the official views or policies of the Texas A&M Transportation Institute, or any State or Federal agency. This report does not constitute a standard, specification, or regulation.

TRR
JOURNAL OF THE TRANSPORTATION RESEARCH BOARD

Article

Transportation Research Record
2018, Vol. 2672(12) 158–166

DOI: 10.1177/0361198118767412
journals.sagepub.com/home/trr
SAGE

The Effect of Additives on the Low Temperature Ice-Melting Capacity of NaCl

Kine Nilssen[1], Alex Klein-Paste[1], and Johan Wåhlin[1]

Abstract

Winter maintenance of high-traffic volume roads requires chemicals for the strategy of obtaining sufficient friction. Challenges often arise in very cold temperatures because the effect is reduced. To improve the low temperature ice-melting capacity of road salt, additives are sometimes used. There is a lack of knowledge as to how these additives affect low temperature ice-melting capacity. The tests often use SHPR H-205.1 and H-205.2; however, recent research has shown that the SHRP tests produce inaccurate results. This paper uses calorimetry to study the effect of additives on the low temperature ice-melting capacity of sodium chloride (NaCl). Magnesium chloride ($MgCl_{2)}$, calcium chloride ($CaCl_{2)}$, potassium formate (KFo), calcium magnesium acetate (CMA), and sugar (sucrose) were added to NaCl at −18°C. The chemicals were also tested individually. The measurements were done in a recently improved custom-made calorimeter, described in the paper. The results showed that solid chemicals had a far higher melting capacity than solutions. In a solid state NaCl had the highest ice-melting capacity, however, the results differed for solutions, where brine had the lowest. $CaCl_2$ had the highest ice-melting capacity of the solutions, melting 285% more than NaCl at −18.6°C. This was followed by $MgCl_2$, KFo, mix NaCl/$CaCl_2$, mix NaCl/CMA, CMA, and mix NaCl/$MgCl_2$. The improvements using chlorides ($MgCl_2$ and $CaCl_2$) depended on the amount of additive. KFo had a destructive influence on melting capacity when mixed with NaCl, whereas mixing CMA with NaCl produced a higher melting capacity than the two components individually. Sugar froze at −18°C.

Winter road maintenance is challenging in cold temperatures. Sodium chloride (NaCl) is the most frequently used chemical for deicing and anti-icing operations. Although throughout most of the winter NaCl is able to maintain ice-free roads, problems arise in areas where temperatures get very cold. The eutectic point of NaCl is −21.2°C (−6.2°F) with a eutectic concentration of 23.2% (*1*); however, winter maintenance operators' many years of field experience has made them aware of the fact that NaCl progressively ceases to work as temperatures drop, and they experience "very little effect" at temperatures as high as −10°C to −15°C. NaCl's ability to melt ice, called its ice-melting capacity (g_{ice}/g_{deicer}), is drastically reduced with decreasing temperatures. To counteract this reduction, additives, e.g., other chemicals, have been used for many years to improve the performance of deicing and anti-icing chemicals, among other things to improve low-temperature performance (*2*). However, the degree to which additives can improve ice-melting capacity is not clear.

Research results are ambiguous when it comes to the melting capacity of deicing/anti-icing chemicals and additives in cold temperatures. Several research results have shown that magnesium chloride ($MgCl_2$) solution (*3–6, 8*) and calcium chloride ($CaCl_2$) solution (*4, 6, 8*) have a higher ice-melting capacity than NaCl at −18°C. Others have shown that $CaCl_2$ has a very low melting capacity at −18°C *(7, 20)*. Using $MgCl_2$ to pre-wet, NaCl has been by Koefod et al. shown to increase ice-melting capacity (*21*). The improvement from using $MgCl_2$ as an additive has been shown to be supplemental to the total ice-melting capacity, meaning that the new level of ice-melting capacity is simply the sum of the ice-melting capacity of the individual components (*9*). Nonetheless, as regards other additives, it remains inconclusive whether a synergistic boost (the result exceeds the sum of two components) is possible. For instance, agro-based products, and complex chloride mineral (CCM) products are increasingly being employed in snow and ice operations (*2*), and manufacturers have claimed benefits for low-temperature performance. Adding corn-based derivatives to $MgCl_2$ has been shown to improve and exceed the ice-melting capacity of $MgCl_2$ at −18°C (*7*). Others report decreasing ice-melting capacity when glycerol was mixed in with $MgCl_2$ (*10*), and no increase in melting capacity for agro-based products compared with NaCl (*2, 4*). Interestingly, CCM

[1]Norwegian University of Science and Technology, Trondheim, Norway

Corresponding Author:
Address correspondence to Kine Nilssen: kine.nilssen@ntnu.no

products have not shown a higher ice-melting capacity in cold temperatures (*2*).

The reason for ambiguous results is the test methods used for measuring ice-melting capacity. These tests are usually performed using SHRP H-205.1 and H-205.2, which are still the standard method for measuring ice-melting capacity (*2, 11*). However, recent research has shown these test methods' low level of accuracy (*12*), and the fact that the results found in literature are so inconsistent is just another indicator of this point. Ice-melting capacity is a physical property, and should provide consistent results independent of the test methods used.

There are many deicing/anti-icing chemicals with a lower eutectic point than NaCl, but this is not synonymous with having a higher melting capacity in cold temperatures (*13*). Danilov et al. found low eutectic temperatures (as low as −30°C), by mixing different ratios of magnesium and calcium nitrates, without having a higher melting capacity than NaCl before exceeding the eutectic point of NaCl (*14*). Later on, they tested sodium acetate, potassium acetate, magnesium acetate, calcium acetate, ammonium acetate, sodium formate, potassium formate (KFo), NaCl, and $CaCl_2$ (*15*). Some of these tested chemcials have a very low eutectic point; but yet the result showed that NaCl had the highest melting capacity (down to −20°C). However, potassium acetate, ammonium acetate, and KFo were able to melt ice down to −50°C, and $CaCl_2$ to −40°C (*15*). The tests were performed using visual polythermal analysis.

Therefore, the deicer's melting capacity at a given temperature seems to be the result of complex compositions of several qualities. As more and more commercial products with mixed compositions are manufactured, the need to be able to measure their performance is increasing. For example, a new test method for measuring ice-melting capacity was published in 2016 (*17*), using calorimetry. An improved version of this calorimeter was built during this study and used to determine the ice-melting capacity of NaCl with additives at −18°C. Common deicers were used individually and as additives: $MgCl_2$, $CaCl_2$, KFo, calcium magnesium acetate (CMA) and sucrose ($C_{12}H_{22}O_{11}$). The goal of the study was to find out whether these chemicals had a higher ice-melting capacity than NaCl at low temperatures, both individually and as additives to NaCl.

Method

Design of the Calorimeter

While the new calorimeter follows the same principal as an earlier version presented in Nilssen et al. (*16*), it has two new features to improve its accuracy. The mixing of ice and solution was performed vertically instead of horizontally because this produced a better blend of the ice and solution. The horizontal mixing process led to difficulties because the ice started floating to the top. Moreover, the heater was placed inside the deicing/ice mixture instead of underneath the reactor with the result that when the heater was placed under the reactor, it had to heat up the insulation, air, and reactor bottom before penetrating the solution, which caused lost energy (heat).

The calorimeter is illustrated in Figure 1 and consisted of the following:

(a) A cylindrical, insulated, hollow container separated into one bottom part and one lid. The insulation was a 15-cm thick extruded polystyrene layer covered with glass fiber to prevent water penetrating the insulation.
(b) A 0.5-L volume plastic reactor inside the hollow container, which is where the mixing of ice and solution took place.
(c) One heater and two temperature sensors drawn through the insulated lid of the container and placed inside the plastic reactor. One temperature sensor monitored the heating element. It was attached inside a steel block along with the heating element and programmed to cut the power if the heater became overheated to prevent fire. The other temperature sensor measured the temperature inside the ice-solution mixture.
(d) A mixing table was fitted to the container and constructed to roll the entire container 180° back and forth. This movement ensured the proper mixing of the ice and solution blend.
(e) A 35-mm hole in the insulation was made for inserting a deicing solution during testing; otherwise, it remained closed.
(f) Wires connected the electrical elements of the heater, temperature sensors, and mixing table to a power supplier and data program.

The calorimeter was tilted to one side by a few degrees. All the equipment was placed in a large walk-in cold storage room, temperature controlled (±1°C) to minimize heat flow through the insulation.

Ice was inserted in the plastic reactor (b) and the solution was inserted through a hole in the lid (e). When the deicer came in contact with the ice, the melting process began. This melting requires energy, which is normally taken from the surrounding heat, causing the temperature to drop. Usually, heat quickly flows in, bringing the temperature back to its initial level: the system/concentration has then reached equilibrium. However, in this case, since the melting process was taking place inside an insulated calorimeter, no heat was flowing in after the temperature dropped. Therefore, to simulate this heat flow, a heater was inserted into the calorimeter, which supplied energy (heat) to bring the temperature back to the initial temperature. The calorimeter constantly monitored and logged the temperature and energy flow. As the latent heat of fusion for ice is 334 kJ/kg (*1*), the amount of melted ice could be calculated from the amount of energy added by the heater. Figure 2 illustrates the

Figure 1. Illustration of the calorimeter.

calorimeter, power supply, control application from a computer (LabVIEW), and the data logger. The container was then removed between the addition of each chemical, to be cleaned and prepared for a new chemical or mix. Prior to testing, a known quantity of ice was placed in the container and brought to the desired test temperature. The temperature sensors inside the container and insulation were used to determine if a constant temperature was reached. At the start of a test, the software stored the temperature of the ice as the initial/setpoint temperature. The solution's mass and temperature were then entered manually to correct for differences between the temperature of the solution and the setpoint temperature. The control software kept the heating block at 0.5°C above the setpoint, until the melting process was almost finished. At the end of the procedure this reduced to 0.3°C.

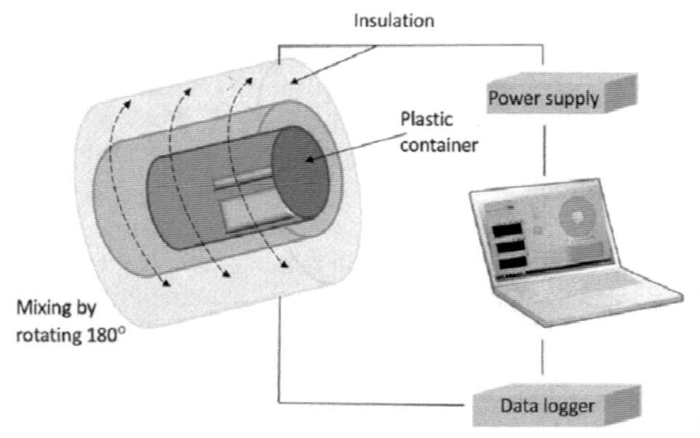

Figure 2. Illustration of the calorimeter, power supply, data logger, and PC control software.

Test Procedure

A total of 63 tests were performed at approximately −18°C. Mixes of NaCl and $MgCl_2$, $CaCl_2$, KFo, CMA and $C_{12}H_{22}O_{11}$ were tested in the calorimeter. Each of the additives/deicing chemicals was also tested individually. Ice was cooled down inside the calorimeter and the deicing chemicals inserted through an opening at the start of each test. The ice comprised crushed ice cubes approximately 2–6 mm in size.

Preparing Additives/Deicing Chemicals

Different solutions were prepared by using the required amount of solutes dissolved in distilled water. A concentration representing a freezing point of −27°C was chosen for $MgCl_2$, $CaCl_2$, KFo, and CMA. This was under the eutectic concentration to prevent precipitation, and was the lowest before reaching CMA's eutectic concentration. According to the literature, this was comprised of 20w% for $MgCl_2$ (*17*), 25w% for $CaCl_2$ (*17*), 35w% for KFo (*17*) and 35w% for CMA (*18*). With regard to sugar, a freezing point of −19°C was chosen, as this was close to the test temperature and would not unnecessarily increase the concentration as it got very thick; this was to correspond to a 70w% solution (*19*). The eutectic point for NaCl (−21°C) was used with the corresponding, eutectic concentration of 23w%. Data for the tested deicers are shown in Table 1. NaCl, KFo, CMA,

Table 1. Data Eutectic Point and Freezing Point Concentration: NaCl (*17*), $MgCl_2$ (*17*), $CaCl_2$ (*17*) KFo (*17*), CMA (*18*), and $C_{12}H_{22}O_{11}$ (sucrose) (*19*)

Deicer	Chemical symbol	Eutectic temperature	Eutectic concentration	Freezing point concentration (−18°C)
Sodium chloride	NaCl	−21°C	23.4%	21.5%
Magnesium chloride	$MgCl_2$	−33°C	21.8%	16.5%
Calcium chloride	$CaCl_2$	−50°C	30.5%	20.3%
Potassium formate (KFo)	KCO_2H	−51°C	48%	27.5%
Calcium magnesium acetate	CMA	−27.5°C	32.5%	27.5%
Sucrose	$C_{12}H_{22}O_{11}$	−30°C	80%	70%

Table 2. Performed Ice-Melting Capacity Tests

Deicer	Initial concentration	Temperature	Freezing point	Number of repetitions
NaCl	23%	−17.3°C (0.9°F)	−21°C	14
$MgCl_2$	20%	−18.9°C (−2°F)	−27°C	6
Mix 80/20 NaCl/$MgCl_2$	23% NaCl + 20% $MgCl_2$ (18.4w% NaCl + 4w% $MgCl_2$)	−18.8°C (−1.8F)		6
$CaCl_2$	25%	−18.6°C (−1.5°F)	−27°C	6
Mix 80/20 NaCl/$CaCl_2$	23% NaCl + 25% $CaCl_2$ (18.4w% NaCl + 5w% $CaCl_2$)	−19.4°C (−2.9°F)		6
KFo	35%	−18°C (−0.4°F)	−27°C	6
Mix 80/20 NaCl/KFo	23% NaCl + 35% KFo (18.4w% NaCl + 7w% KFo)	−18.9°C (−2°F)		6
CMA	35%	−18.9°C (−2°F)	−27°C	6
Mix 80/20 NaCl/CMA	23% NaCl + 35% CMA (18.4w% NaCl + 7w% CMA)	−18.3°C (−0.9°F)		6
Sucrose	70%	−18°C (−0.4°F)	−19°C	1

and sugar had unknown water content and were therefore placed in a dryer at 70°C for 5 days to extract all water. No water content in the delivered NaCl and sugar was observed after drying. The 23w% NaCl solution was prepared by starting with 230 g of solid NaCl and then adding distilled water until a total weight of 1,000 g was reached. The same procedure was followed for the dried CMA, KFo, and sugar. Solid $MgCl_2$ and $CaCl_2$ was (and is normally) delivered as hexahydrates ($MgCl_2$-$6H_2O$) and dihydrates ($CaCl_2$-$2H_2O$). This water content needs to be included. An $MgCl_2$ hexahydrate ($MgCl_2$-$6H_2O$) consists of 46.8% $MgCl_2$ and 53.2% H_20. A $CaCl_2$ dihydrate consists of 75.5% $CaCl_2$ and 24.5% H_20. To obtain the correct amount of solutes to make a 20w% $MgCl_2$ solution, 300 g of $MgCl_2$-$6H_2O$ was weighted, this equals 140.58 g of $MgCl_2$, and 402.87 g of distilled water was added. To make the 25w% $CaCl_2$ solution, 300 g of $CaCl_2$-$2H_2O$, which is equal to 226.53 g of $CaCl_2$, was added to 606.12 g of distilled water.

Six repetitions were performed for each chemical or mix, 14 for NaCl and one for sugar, resulting in a total of 63 tests. Approximately 90 g of ice were used in each test. The first test using a new chemical was performed with approximately 40 g of solution. The consecutive five tests using the same chemical were performed with the same ice and solution from previous tests by adding approximately 20 g of solution. This was done out of practical consideration for the purpose of achieving satisfactory mixing in the beginning and without surpassing the container's volume capacity at the test's conclusion. An overview of the tested solutions is presented in Table 2.

Converting Results from Liquid to Solid Ice-Melting Capacity

Although testing in the calorimeter was performed with deicing solutions, the result can be converted to represent solid chemicals by using the equilibrium concentration (the freezing point concentration). Equation 1 describes how the equilibrium concentration is calculated using the results from the calorimeter, and Equation 2 how ice-melting capacity of solid chemicals is calculated:

$$c_{eq} = \frac{m_{solid}}{m_{solid} + m_{melted\ ice} + m_{water\ in\ product}} \tag{1}$$

where

c_{eq} = freezing point concentration (%)
m_{solid} = mass of solid deicer (g)
$m_{melted\ ice}$ = mass of melted ice and water in solution (g)

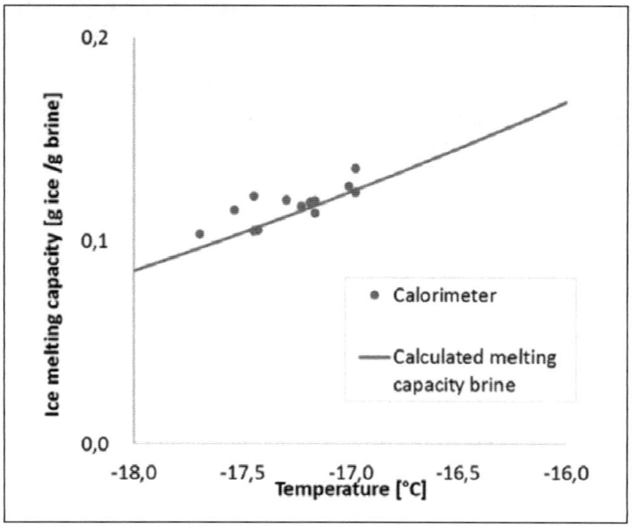

Figure 3. Melting capacity of theoretical NaCl and measured with the calorimeter.

$m_{water\ in\ product}$ = mass of water in solution or in product (g) = unrealized melting capacity = the water used to dissolve the deicer (g) (*16*)

$$m_{melted\ ice} = \frac{m_{solid}}{c_{eq}} - m_{solid} - m_{water\ in\ product} \tag{2}$$

where

$m_{melted\ ice}$ = mass of melted ice (g). If m_{solid} +$m_{water\ in\ product}$ is 1, then $m_{melted\ ice}$ can be defined as the ice-melting capacity for solid chemical (IMC_{solid}) (g/g), because ice-melting capacity is defined as melted ice per gram of melted deicer

$m_{water\ in\ product}$ = mass of water in solution = unrealized melting capacity = the water used to dissolve the deicer (g) (*16*)

m_{solid} = mass of solid deicer (g)

Results

Accuracy of the Calorimeter

Accuracy was found by comparing measured data with calculated melting capacity. NaCl has a known phase diagram, which makes it possible to calculate the melting capacity (*12*). Fourteen tests were performed using 23% NaCl (brine) for the accuracy test. The scope of the testing involved cold temperatures, and the temperature ranged between −17.0°C and −17.7°C. The amount of ice used was approximately 100 g per test, and the amount of solution approximately 20–40 g (to test the practical challenges when varying the solution content).

Figure 3 shows a graphical illustration of the accuracy, measuring melted ice in grams. Red dots indicate the measured ice melt in the calorimeter, and the blue line represents the calculated melting capacity. The calorimeter is able to measure from 97% to 115% of the actual melted water; the average error was 104%. The absolute error ranges between −0.001 g to 0.016 g per gram of brine. When converted into solids by using Equations 1 and 2, this gives an accuracy of between 99.6% and 101.8%.

Ice-Melting Capacity of Deicing Solutions and Additives

Figure 4 shows the ice-melting capacity in gram-per-gram solution. Figure 4*a* shows all data, and Figure 4*b* shows average data. Both are included because there are some differences in setpoint temperature as well as melting capacity. The blue line represents the calculated melting capacity of NaCl solution (brine) (*12*). $MgCl_2$ (20%), $CaCl_2$ (25%), and KFo (35%) had a higher melting capacity than brine at −18°C. Sugar (70%) froze and had zero ice-melting capacity. Using $MgCl_2$ (20%), $CaCl_2$ (25%), and CMA (35%) as additives to brine in an 80/20 brine/additive-mix produced higher ice-melting capacity. KFo (35%) as an additive decreased the melting capacity of the brine.

A significance analysis was performed with a 95% confidence interval on the difference between measured ice melt for NaCl in the calorimeter and the calculated ice melt for NaCl, and the difference between measured ice melt with each chemical/additive and calculated ice melt for NaCl. See Table 3. Table 3 also shows the improvement in ice-melting capacity compared with brine.

Ice-Melting Capacity of Solid Chemicals and Additives

Regarding a winter maintenance perspective, it is important to note that the melting capacity of solutions (the numbers on the left-hand side of Figure 4) are low. The ice-melting capacity of solutions is lower compared with solid chemicals because of the water content. In Figure 5, the ice-melting capacity has been converted to represent solid chemicals (the water content is subtracted), using equations 1 and 2. Note that the results for $MgCl_2$ are given in hexahydrates and the results for $CaCl_2$ in dihydrates. The blue line represents calculated ice-melting capacity for solid NaCl (*12*).

All of the tested chemicals in a solid state had less melting capacity than NaCl. The melting capacity of solid NaCl was also reduced if additives were used. A *t*-test was performed with a 95% confidence interval, and the observed lower melting capacity was significant. See Table 4.

Discussion

The Calorimeter

The new calorimeter showed more accurate results than the earlier published version. The method accuracy for solutions ranged from 97% to 115% and for solid chemicals from

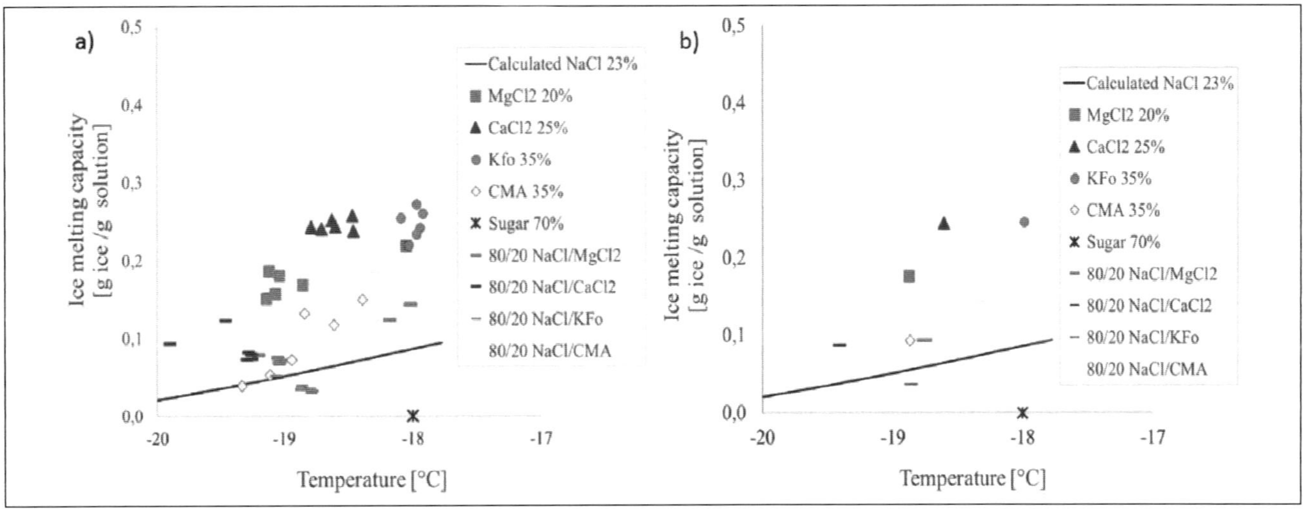

Figure 4. Ice-melting capacity of deicing solutions: (*a*) all data and (*b*) average data.

Table 3. Difference in Ice-Melting Capacity between Different Deicers and Brine

Deicer	Temperature (°C)	IMC (g/g)	More (+)/less (−) melted ice than brine g	%	Statistically significant
$MgCl_2$	−18.9	0.17	+0.12	+228	Yes
80/20 NaCl/$MgCl_2$	−18.8	0.09	+0.03	+56	Yes
$CaCl_2$	−18.6	0.24	+0.18	+285	Yes
80/20 NaCl/$CaCl_2$	−19.4	0.09	+0.05	+144	Yes
KFo	−18.0	0.25	+0.16	+187	Yes
80/20 NaCl/KFo	−18.9	0.04	−0.02	−33	Yes
CMA	−18.9	0.09	+0.04	+61	No
80/20 NaCl/CMA	−18.3	0.14	+0.07	+97	Yes
Sugar	−18.0	0	–	–	–

IMC = ice-melting capacity.

99.6% to 101.8%. There is a small tendency to overestimate the heat. This might come from heat being lost in the wires to heat them up, furthermore, the calculated 20J to heat up the element might be a little high as it contains holes with air. One time-saving improvement was the ability to run a new test after the ice/deicer solution had achieved equilibrium without taking the calorimeter out of the freezer. One negative point was the lack of the cryostat contained in the previous version that facilitated more accurate control of the setpoint temperature. It was difficult achieving the same setpoint temperature for each new test because the temperature in the walk-in cold storage room varied from day to day, which made it difficult to compare the deicers with each other. In this paper, the results were compared with calculated NaCl. In further research, the calorimeter could be used to draw complete phase diagrams, which will make comparisons easier.

Improving the Ice-Melting Capacity of NaCl by Using Additives

The improved version of the calorimeter was used to research the effect of additives on ice-melting capacity of NaCl in −18°C. The data showed that solid NaCl had a high ice-melting capacity relative to other solid deicing chemicals at low temperatures, but a lower ice melting capacity than other solution when used as brine (23% NaCl solution). Of the individual chemicals tested, the ice-melting capacity expressed on a solid chemical basis ($MgCl_2$-$6H_2O$, $CaCl_2$-$2H_2O$, KFo, CMA, and sugar), NaCl had the highest ice-melting capacity. Using them as additives to solid NaCl will also decrease the melting capacity. $MgCl_2$-$6H_2O$ and $CaCl_2$-$2H_2O$ contain high amounts of water in a solid state, which contributes to their lower ice-melting capacity in a solid state. The data also confirm that there is no linear correlation between freezing point and ice-melting capacity (*13*). The

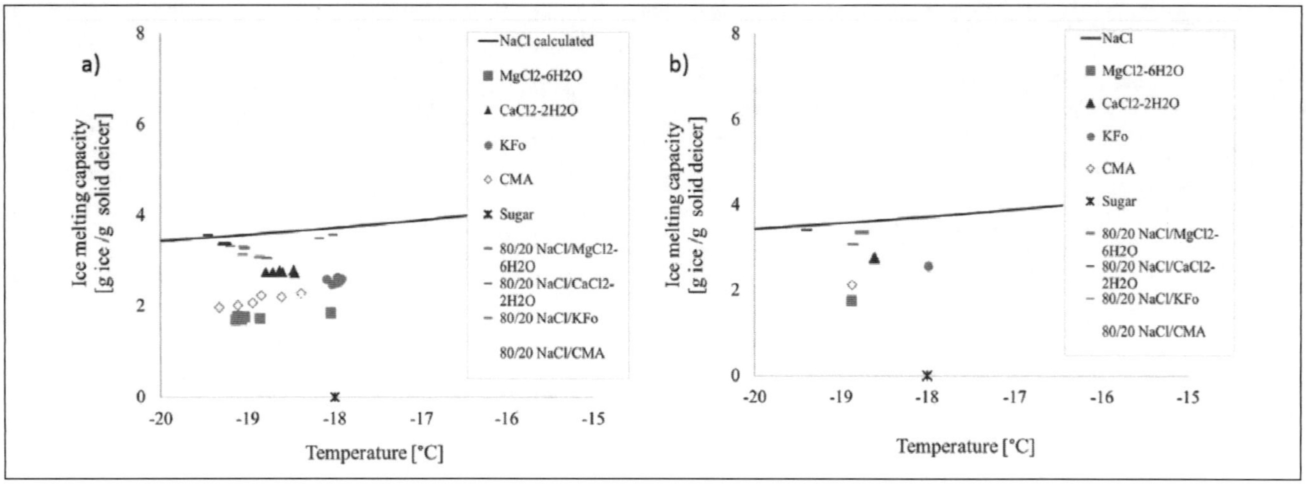

Figure 5. Ice-melting capacity solid chemicals: (*a*) all data, and (*b*) average data.

Table 4. Melted Ice Compared with NaCl

Deicer	Temperature (°C)	IMC (g/g)	More (+)/Less (−) melted ice than NaCl g	%	Statistically significant
$MgCl_2$-6H_2O	−18.9	1.75	−1.83	−51	Yes
80/20 NaCl/ $MgCl_2$-6H_2O	−18.8	3.36	−0.24	−7	Yes
$CaCl_2$-2H_2O	−18.6	2.76	−0.87	−24	Yes
80/20 NaCl/ $CaCl_2$-2H_2O	−19.4	3.42	−0.1	−3	Yes
KFo	−18.0	2.56	−1.16	−31	Yes
80/20 NaCl/ KFo	−18.9	3.08	−0.51	−14	Yes
CMA	−18.9	2.12	−1.46	−41	Yes
80/20 NaCl/ CMA	−18.3	2.27	−1.14	−38	Yes
Sugar	−18.0	0	na	na	na

IMC = ice-melting capacity.

reason why NaCl has a high melting capacity at −18°C is that it has a low equilibrium concentration. This can be seen in Equation 3 (*12, 14*). A high initial concentration and low freezing concentration result in a high melting capacity.

$$IMC = \frac{Ci - Cfreeze}{Cfreeze} \quad (3)$$

where

IMC = the ice-melting capacity (g/g)

C_i = the initial concentration (chemical without water content). NaCl;100. $MgCl_2$; 46.2

C_{freeze} = concentration where a solution freezes (%)

In solutions the results are reversed. $CaCl_2$, $MgCl_2$, KFO, Mix NaCl/$CaCl_2$, Mix NaCl/CMA, CMA, and mix NaCl/$MgCl_2$ had a higher ice-melting capacity than NaCl (the CMA results were not significant). Using additives such as brine increases the melting capacity, except in the case of sugar and KFo. The sugar froze, and the mix of NaCl with KFo had a significantly lower melting capacity than NaCl. Regarding additives, mixing $CaCl_2$ solution with brine gave the highest ice-melting capacity, increasing it by 144%. In solutions, the water content in solid $MgCl_2$ and $CaCl_2$ becomes part of the solution, and they benefit from their combination of a low freezing concentration and high solubility at low temperatures and therefore achieve a much higher ice-melting capacity relative to an NaCl solution.

The test visualized the drastic decrease in ice-melting capacity when solutions were used instead of solid chemicals: brine only melts 2% of what solid NaCl at −18°C does. The other tested chemicals melted 4 to 12% in solution compared with solid form. The reason is naturally that solutions contain water, and water does not contribute to melting in temperatures below 0°C. This is also the reason why pre-wetting results in lower melting capacity.

Previous studies (*16, 21*) have implied that the benefits of using $MgCl_2$ as an additive to NaCl are supplemental to the two compounds. That was not confirmed in this research; however, the results did show that there is an increase in melting capacity when $MgCl_2$ and $CaCl_2$ are added to brine, and that the melting capacity of the mix is situated somewhere between the melting capacity of the two compounds. Contrastingly, with respect to KFo and CMA, the picture was different; KFo had a very high melting capacity at −18°C, and this might be because of a low eutectic point and high solubility in low temperatures. However, when KFo was added to NaCl, the results were detrimental. The ice-melting capacity was reduced to a lower level, not only for KFo but also for NaCl. Although the reason for this is unknown, it might be that the quality of these two chemicals together makes them detrimental to melting capacity. Mixing CMA and NaCl also produced an unexpected result: the mix increased the melting capacity to a higher level than the two components individually. Danilov et al. found that mixing potassium acetate with KFo produced a doubling in ice-melting capacity than the two components on their own (*15*). More research conducted on acetates would therefore be very interesting.

The solubility of $MgCl_2$ and $CaCl_2$ at −18°C is higher than used in this research. Solubility according to the extended UNIQUAC modelis 23.9% for NaCl at −17.3°C, 27.3% for $MgCl_2$ at −18.9°C and 34.5% for $CaCl_2$ at −18.6°C (*22*). Calculations were carried out, according to Equation 2 to determine what the ice-melting capacity would be if maximum solubility was used. C_{eq} was used from the calorimeter results.

$$IMC_{NaCl\,(23.9\%)}: 0.237/0.206 - 1 = \underline{0.15\ g/g\ solution}$$
$$IMC_{MgCl2\,(27.3\%)}: 0.273/0.170 - 1 = \underline{0.61\ g/g\ solution}$$
$$IMC_{CaCl2\,(34.5\%)}: 0.345/0.201 - 1 = \underline{0.72\ g/g\ solution}$$

NaCl used as 23.3% instead of 23% at −17.3°C increases the ice-melting capacity from 0.11 g/g to 0.15 g/g (36%). The ice-melting capacity of $MgCl_2$ will increase 258%, from 0.17 g/g used as a 20% solution at −18.9°C to 0.61 g used as a 27.6% solution; and the ice-melting capacity for $CaCl_2$ increases 200%, from 0.24 g/g used as a 25% solution at −18.6°C to 0.72 used as a 34.5% solution. Using $MgCl_2$ and $CaCl_2$ as additives to a 23% NaCl solution probably follows the same trend as with lower solubility, increasing the melting capacity of the brine. This calculation was not done for CMA and KFo as we did not have data for maximum solubility at −18°C. However, the melting capacity for solutions is still low compared with solid chemicals, even with maximum solubility. A comparison between the published data (*18*) and the calorimeter was made. For $MgCl_2$, $CaCl_2$, and KFo, the data were equal to one another. However, for CMA and sugar the difference was far greater than the uncertainty of the calorimeter. This indicates that the previously published phase diagrams for CMA and sugar are not correct.

Former results and experience with NaCl suggests that it is "ineffective" in cold temperatures; nevertheless, as these data point out, this is not due to the melting capacity, which is actually high, even when close to its eutectic point compared with other deicing chemicals. This confirms the findings from Koefod et al. (*20*). The "ineffectiveness" experienced is probably due to other mechanisms, such as its low melting rate (NaCl works very slowly) in cold temperatures (*9*). This means that the overall performance might still improve by using additives. More research is needed on, for example, ice-melting rate to give an answer on the most "effective" chemical in cold temperatures.

Conclusion

In this study, improvements have been made to the previously described calorimeter (*16*), and it can now measure ice-melting capacity in cold temperatures with an accuracy between −0.001 g to 0.016 g, 97–115%. It was used in this paper to measure the ice-melting capacity of deicing chemicals and additives to NaCl in cold temperatures. Additives were mixed with NaCl at −18°C in an 80/20 mix (80% NaCl with 20% additive). The additives were also tested individually. The calorimeter was built for testing liquid chemicals, but the results were also converted to solid chemicals.

Of all tests performed, solid NaCl had the highest ice-melting capacity, with a 31% higher ice-melting capacity than the second best solid deicer, $CaCl_2$-$2H_2O$. Additives to solid NaCl resulted in a lower ice-melting capacity. The tests showed that ice-melting capacity is drastically reduced when solutions are used instead of solid chemicals. The solutions' ice-melting capacity was only 3 to 10% of the solids' ice-melting capacity. But if solutions do need to be used, brine is the solution with the poorest ice-melting capacity of all the tested chemicals at −18°C. Additives to brine increased the brine's ice melting capacity from 56 to 144%, except when adding KFo or sugar. KFo and sugar decreased the ice-melting capacity of brine, with sugar the solution actually froze. Using maximum solubility for $MgCl_2$ and $CaCl_2$ results in a more than 200% higher ice-melting capacity for the solutions than the calorimeter showed; however, the ice-melting capacity is still low compared with solid chemicals.

Sugar froze and had zero ice-melting capacity at −18°C.

Acknowledgments

This work was funded by the Norwegian Public Roads Administration. The authors thank Bent Lervik, Jan Erik Molde, Frank Stæhli, and Tage Wessum for their technical support during the design, construction, and modification of the calorimeter, as well as their assistance during our experiments.

References

1. Haynes, W. M., and T. Bruno. *CRC Handbook of Chemistry and Physics*, 95th ed. Online. Taylor and Francis Group, LLC., Calif., USA, 2014.

2. Muthumani, A., L. Fay, D. Bergner, and S. Xiaming. *Understanding the Effectiveness of Non-Chloride Liquid Agricultural By-Products and Solid Complex Chloride/Mineral Products*. Clear Roads Technical Report CR13-12. Western Transportation Institute, Montana, USA, 2015.
3. Gerbino-Bevins, B. M., C. Y. Tuan, and M. Mattison. Evaluation of Ice-Melting Capacities of Deicing Chemicals. *Journal of Testing and Evaluation*, Vol. 40, No. 6, 2012, pp. 1–9.
4. Fay, L., and X. Shi. Laboratory Investigation of Performance and Impacts of Snow and Ice Control Chemicals for Winter Road Service. *Journal of Cold Regions Engineering*, Vol. 25, No. 3, 2011, pp. 89–114.
5. Nixon, W. A., G. Kochumman, J. Qiu, L. Qiu, and J. Xiong. *Evaluation of Deicing Materials and Corrosion Reducing Treatments for Deicing Salts*. IIHR Technical Report Number 43. Iowa Highway Research Board, Iowa, USA, 2007.
6. Druschel, S. J. *Salt Brine Blending to Optimize Deicing and Anti-Icing Performance*. DOT Technical Report MN/RC 2012-20. Minnesota State University, Mankato, Minn., USA. 2012.
7. Bytnar, S. C. De-icing Composition and Method, U.S. Patent No. 7,563,386, 2009.
8. Alger, R., and J. Haase. *Analysis of the Benefits of Bulk Pre-wetting Solid NaCl with Several Different Liquids*. DOT Technical Report RC-1473. Michigan Tech University, Houghton, Mich., 2006
9. Koefod, S., R. Mackenzie, and J. Adkins. Effect of Prewetting Brines on the Ice-Melting Rate of Salt at Very Cold Temperatures. *Transportation Research Record: Journal of the Transportation Research Board*, 2015. 2482: 67–73.
10. Talor, P., G. Kasthurirangan, J. G. Verkade, W. Kuldeep, and K. Sunghwan. *Development of an Improved Agricultural-Based Deicing Product*. InTrans Project Report 09-310. Iowa State University, Institute of Transportation. Iowa, USA, 2010.
11.Chappelow, C. C., D. A. McElroy, R. R. Blackburn, D. Darwin, F. G. Noyelles, and E. Locke Carl. *Handbook of Test Methods for Evaluating Chemical Deicers*, SHRP-H-332nd ed. Strategic Highway Research Program, National Research Council, Washington, D.C., 1992.
12. Nilssen, K., A. Klein-Paste, and J. Wåhlin. Accuracy of Ice Melting Capacity Tests: Review of Melting Data for Sodium Chloride. *Transportation Research Record: Journal of the Transportation Research Board*, 2016. 2551: 1–19.
13. Koefod, S. *Transportation Research Circular EC126/ SNOW08-008: Eutectic Depressants. Relationship of Eutectic, Freezing Point, and Ice Melting Capacity in Liquid Deicers*. Transportation Research Board of the National Academies, Washington, D.C., 2008.
14Danilov, V. P., E. A. Frolova, D. F. Kondakov, L. I. Avdyushkina, and V. T. Orlava. Low-Temperature Anti-icing Reagents in the Ca(NO3)2-Mg(NO3)2-CO(NH2)2-H2O System and their Properties. *Theoretical Foundations of Chemical Engineering*, Vol. 45, No. 5, 2011, pp. 697–701.
15Danilov, V. P., E. A. Frolova, D. F. Kondakov, L. I. Avdyushkina, A. V. Bykov, and V. T. Orlava. Low-temperature Anti-icing Reagents in Aqueous Salt Systems Containing Acetates and Formiates. *Theoretical Foundations of Chemical Engineering*, Vol. 46, No 5, 2012, pp. 528–535.
16. Nilssen, K., A. Klein-Paste, J. Wåhlin, and M. Delapaz. Use of Calorimetry to Measure Ice-Melting Capacity. *Transportation Research Record: Journal of the Transportation Research Board*, 2017. 2613: 1–7.
17. Melinder, Å. *Thermophysical Properties of Aqueous Solutions Used as Secondary Working Fluids*. PhD thesis. Royal Institute of Technology, KTH Department of Energy Technology, Stockholm, Sweden, 2007.
18. Ketcham, S., L. Minsk, R. Blackburn, and E. Fleege. *Manual of Practice an Effective Anti-Icing Program - A Guide for Highway Winter Maintenance Personnel*. Publication FHWA-RD-96-202. U.S Department of Transportation, USA, New Hampshire, 1996.
19. Young, F. E., and F. T. Jones, Sucrose Hydrates. The Sucrose–Water Phase Diagram. *The Journal of Physical Chemistry*, Vol. 53, No. 9, 1949, pp. 1334–1350.
20. Koefod, S., R. Mackenize, and J. Adkins. Effect of Prewetting Brines on the Ice Melting Rate of Salt at Very Cold Temperatures. *Transportation Research Record: Journal of the Transportation Research Board,* 2015. 2482: 67–73.
21. Koefod, S., J. Adkins, and M. R. Akin. Alternative Approaches to Measuring Deicer Ice-Melting Capacity: Optimization of Test Accuracy and Precision. In *Transportation Research Circular E-C162: Winter Maintenance and Surface Transportation Weather*. Transportation Research Board of the National Academies, Washington, D.C., 2012, pp. 432 -442.
22. Thomsen, K. *Aqueous Electrolytes Model Parameters and Process Simulation*. PhD thesis. Technical University of Denmark, Center for Energy Resources Engineering. Lyngby, Denmark, 1997.

The Standing Committee on Winter Maintenance (AHD65) peer-reviewed this paper (18-00107).

Transportation Research Record
2018, Vol. 2672(12) 167–178

DOI: 10.1177/0361198118797804
journals.sagepub.com/home/trr
SAGE

Bundling Bridge and Other Highway Projects: Patterns and Policies

Yu Julie Qiao[1], Jon D. Fricker[1], Samuel Labi[1], and Trevor Mills[2]

Abstract

The concept of project bundling—combining multiple projects into a single contract—continues to gain momentum. Unanswered questions from past studies include how to quantify project bundling benefits. This paper addresses the issue using a data-driven approach that involves initial comparisons of the unit average project costs of single and bundled projects, preliminary investigations of the potential influential variables of project cost, the development of project cost statistical models, and an analysis of past and possible future bundling strategies. The study found that the primary drivers of project cost include the project size (economies of scale), bundling strategy (economies of bundling), and bidding market conditions (economies of competition). It is also shown that the similarity between different project types in terms of their constituent pay items is an influential factor on project cost. The study confirmed that larger contracts yield economies of scale but may also lead to less competition (discouraging all but the largest firms from bidding). This finding suggests the existence of a contract-size threshold beyond which the unit project cost increases with increasing project cost. By analyzing past bundling strategies, the study identified the most frequent combinations of work categories in the past bundled contracts. The study also carried out scenario analyses using hypothetical bundling strategies applied to actual past projects and determined that bundling projects using a carefully designed strategy can yield significantly reduced overall contract costs.

An earlier study (*1*) indicated that the Indiana Department of Transportation (INDOT) practice of grouping (or bundling) projects into multiple-project contracts (MPCs) usually resulted in lower unit costs. Because the practice had not yet become common within INDOT, the data on such contracts were limited. Nevertheless, it was possible to confirm that economies of scale exist, i.e., unit costs decline as project size increases. However, the declining trend in unit costs appeared to level off when bridge projects reached 20,000 sq ft of deck area and when road projects exceeded 10–20 lane miles. Other possible influences on unit costs, e.g., number of bidders and time of year for bid letting, were examined, without conclusive results. Because of limited data, only a few work types within the bridge and road work categories yielded statistically significant results in the earlier study. This paper reports on a follow-up study that took advantage of INDOT's increased use of project bundling and the data that have been assembled since the conduct of the first study.

Background

In the survey by McCarthy et al. (*2*), nearly half of the surveyed states reported that they had adopted the strategy of project bundling in their construction projects. The FAST Act added a Bundling of Bridge Projects provision as 23 U.S.C. 144(j), which promoted bundling practices in order to "save costs and time by encouraging states to bundle multiple bridge projects as 1 project" (*3*). Highway construction costs can vary greatly, depending on project types and site conditions. According to past studies, the main factors determining the cost of a construction project include the size and duration of project, costs of materials, competition and market conditions, project-specific factors, and macroeconomic factors (*4*). While the agency has control over some of these factors, it has no control over others. For example, factors such as design, specifications, project scope, and size of the contract, are controllable, whereas it would be impossible to control global demand for materials or macroeconomic factors. The number of bidders cannot be controlled or known in advance but can be predicted with some uncertainty.

[1]Lyles School of Civil Engineering, Purdue University, West Lafayette, IN
[2]Indiana Department of Transportation, Indianapolis, IN

Corresponding Author:
Address correspondence to Jon D. Fricker: fricker@purdue.edu

This study examined INDOT data for single-project contracts and multiple-project (bundled) contracts, looking for patterns that could justify project bundling and guide future use of the strategy.

Factors Affecting Project Cost

Data used in this study were provided by INDOT in the form of the INDOT Official Bid Tabulation database, the INDOT Site Manager database, and the INDOT Scheduling Project Management System (SPMS) database. All the above datasets were merged to yield a large dataset, containing comprehensive contract information for projects awarded from 2008 to early 2017 in the State of Indiana. There are 4611 projects and 2528 contracts in the merged dataset. There were 784 bundled contracts in the dataset, containing 2865 projects. The projects were categorized into six project types (i.e., bridge, road, traffic, small structure, miscellaneous, and utility), and were further classified into 36 different work types. The project types in the Bridge and Road work categories are presented below:

Bridge work category:

- B1: New Bridge
- B2: Bridge Replacement
- B3: Superstructure Replacement
- B4: Deck Replacement
- B5: Bridge Widening
- B6: Bridge Deck Overlay
- B7: Thin Deck Overlay
- B8: Misc. Bridge Rehab & Repair

Road work category:

- R1: New Road Construction
- R2: Added Travel Lanes
- R3: Patch & Rehab Pavement
- R4: Partial 3R
- R5: Road Rehabilitation (3R/4R)
- R6: Wedge & Level Only
- R7: Sight Distance Correction
- R8: Shoulder Rehab & Repair
- R9: Pavement, Other
- R10: Pavement Replacement
- R11: Intersection Improvement
- R12: Interchange Work

In this paper, the project cost is measured at three levels:

- At the contract level, the cost is the award amount of a contract (contract award). It is also the sum of the project awards for all projects in that contract.
- At the project level, the cost is the project award for each project in a single project contract or in a bundled contract.
- At the pay item level, the cost is measured by the cost of each pay item for a project. The sum of the cost of all the pay items is equal to the project award for that project.

Design-build projects were excluded from this study because the payment for such projects is in one lump sum, and the item cost is the project award.

Unit costs for all projects in the dataset were computed, based on work type and whether the projects were part of a bundle. Figure 1 suggests that, for all but one Bridge work type (B3: Superstructure Replacement), bundling leads to a lower unit cost. However, six of the Road work types shown have higher unit costs when bundled. To understand these findings, the remainder of this section describes preliminary data analyses for factors that may affect (or be affected by) project costs when bundled.

Project Size (Economies of Scale)

In microeconomics, the term *economies of scale* refers to economic efficiencies that result from carrying out a process on a larger scale (*5*). In the construction industry, it has been recognized that economies of scale play an important role in determining highway construction project costs (*6*). Of the eight Bridge work types, we show in Figure 2*a* the plot for the three work types with the most observations in our dataset. The dark (blue) stars are for single-project contracts; the light (green) plus signs are for projects in bundled contracts. Some scatter of the data points can be observed. Nevertheless, the trend of decreasing project unit costs with increasing contract size (bridge deck area) is clear. Although bundling does not have the same positive (beneficial) effect on Road projects as it seems to have on Bridge projects (see Figure 1), economies of scale are still present. Figure 2*b* presents plots for the three most common Road work types, but the other work types show similar patterns.

Project Bundling (Economies of Bundling)

The benefits of project bundling have been recognized by numerous public agencies. Washington DOT found that bundling local federal aid projects at the planning phase can enhance efficiency in pricing, contract administration and construction efforts (*2*). It was also determined by some DOTs that combining multiple projects together can lead to reduced paperwork, staff time, resources, and

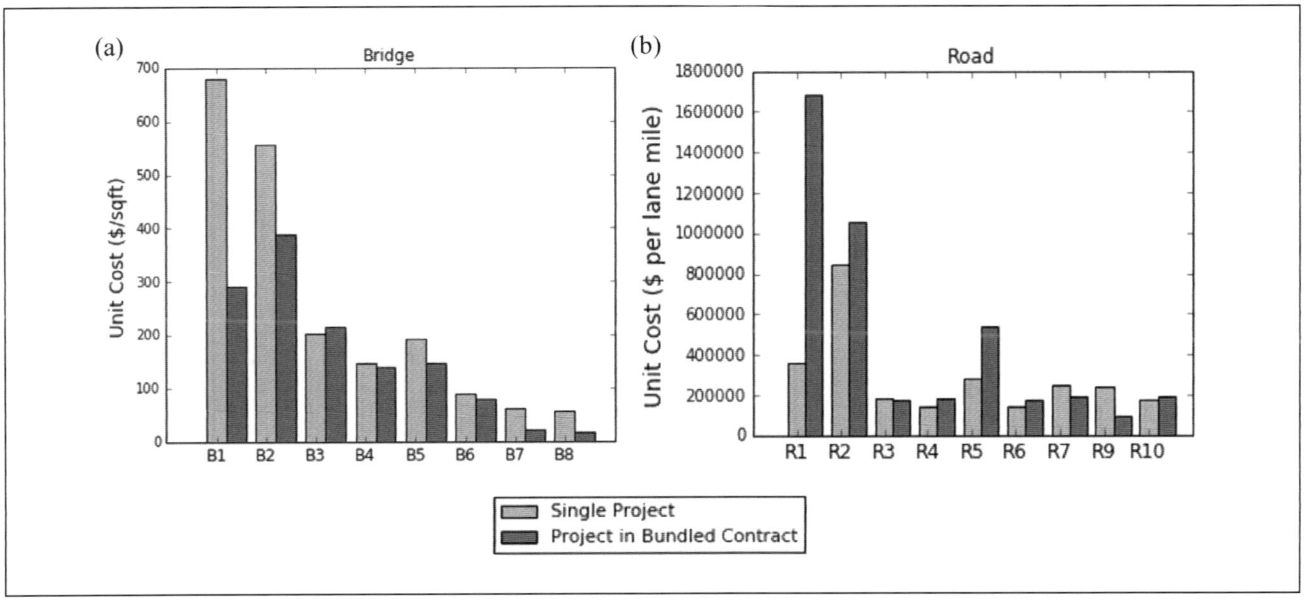

Figure 1. Comparison of average unit cost for projects in a single-project contract and a bundled project contract for (*a*) bridge work and (*b*) road work.

overall project cost. In addition, project bundling can enhance the cost-effectiveness for the maintenance of traffic activities during the construction period. For individual projects located along a corridor, a coordinated bundled contract consisting of these projects may reduce the duration of workzones, compared with individual contracts for each project that would lead to separate road use restrictions at different times for the different projects (*1*). The reduction in workzone duration would not only bring benefits to the agency by reducing bid amount and risk of injury/fatality associated with construction, but it can also benefit the road users and neighborhoods in terms of reduced traffic interruption. This, in turn, reduces travel time and environmental degradation caused by construction activities, such as air and noise pollution.

Figure 2*a* shows a trend in which the light green plus signs (bundled projects) have a lower unit cost than the unbundled projects (dark blue stars). Figure 3*a* shows the corresponding relationship between unit cost and number of projects for the three most common Bridge work types. Again, there is a downward trend in unit cost as the number of projects increase. However, the considerable range (vertical) of unit cost at each number of projects may cause problems when building a unit cost model. For Road work types, only R3, R4, and R7 show any tendency toward economies of bundling. See Figure 3*b*. It appears that, for certain road projects, the economies of scale dominate the economies of bundling. This means that, for these road projects, the cost savings associated with one large project in a single-project contract outweigh the cost increase associated with unbundling or no bundling.

Market Competition (Economies of Competition)

Competition for a contract is desirable for an agency, because it typically leads to higher quality and reduced cost of the delivered product (*7*). It is, however, difficult to ascertain the number of bidders for a contract in advance, because market competition is affected by several factors outside the agency's control. Typically, the number of bidders interested in a particular project is determined by the project type, location and size of the project, and the timing of the project letting. Two relationships were studied:

- Number of bidders vs. contract award: The plot for all contracts (i.e., all six project categories) in Figure 4*a* shows that the four contracts over $100 million attracted three, seven, three, and four bidders. The figure shows that the maximum number of bidders generally decreases as the contract cost increases. This indicates that larger contracts generally lead to lower market competition.
- Number of bidders vs. number of projects: According to Figure 4*b*, when the number of projects gets very large, the maximum number of bidders is low. As in Figure 4*a*, there is a general decreasing trend and a higher variation when the number of projects (or contract award) is small. (Each dot in Figure 4*b* can represent numerous

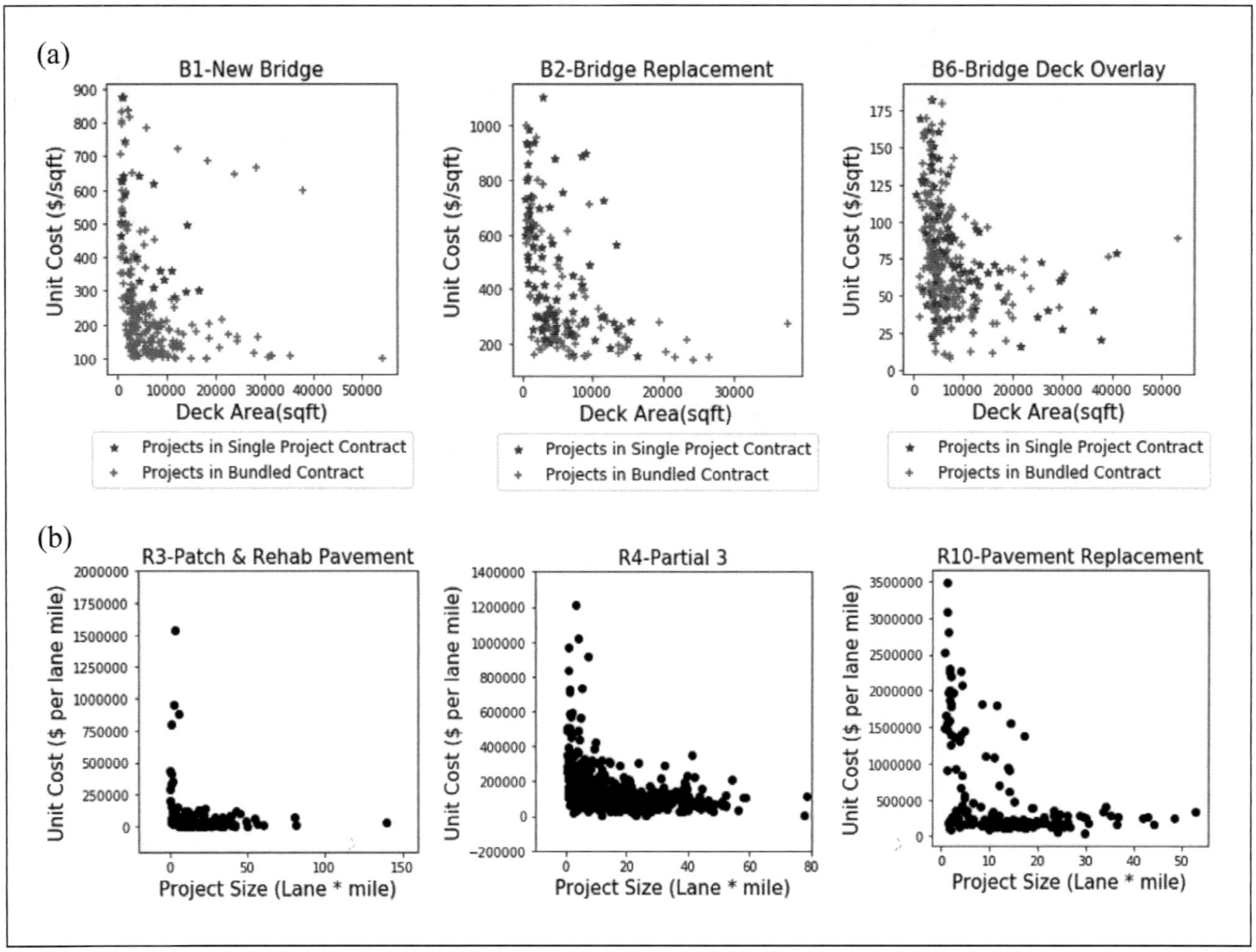

Figure 2. Economies of scale, unit cost vs. project size for (*a*) the most common bridge work types in single and bundled contracts and (*b*) the three most common road work types.

contracts that have the same number of bids and same number of projects.)

Although contract award would be a more accurate measurement of project size (because the number of projects does not take project type and cost into account), the number of projects measure is more useful for certain applications. The contract award is what we seek to predict, but it is not known until the contract is awarded. The number of projects is the variable that can be controlled during the planning phase, and it can be used to predict the number of bidders and the project cost before a contract is awarded.

An attempt was made to develop a regression model that can predict the average number of bidders in a contract. However, a satisfactory model for the average number of bidders could not be found after trying several different combinations of variables and various functional forms. Some dots in Figure 4*b* represent a greater number of projects than other dots do.

However, useful results were achieved with a deterministic upper-bound model. To create the upper-bound model, the highest number of bidders for each number of projects is plotted in Figure 4*c*. This was done for each project type. After trying to fit various functional forms to the data points, the models listed below as Equations 1–6 were chosen. These upper-bound models can be used to estimate the maximum number of bidders for a contract with a given number of projects and can be therefore used to predict the lower bound on project unit cost. In this paper, the upper-bound models were used only in the bundling scenarios analysis for the purpose of comparison, and will not affect the quality of the project cost models (where the actual—not the estimated—number of bidders was used as the independent variable). When estimating the actual project cost developed in this paper

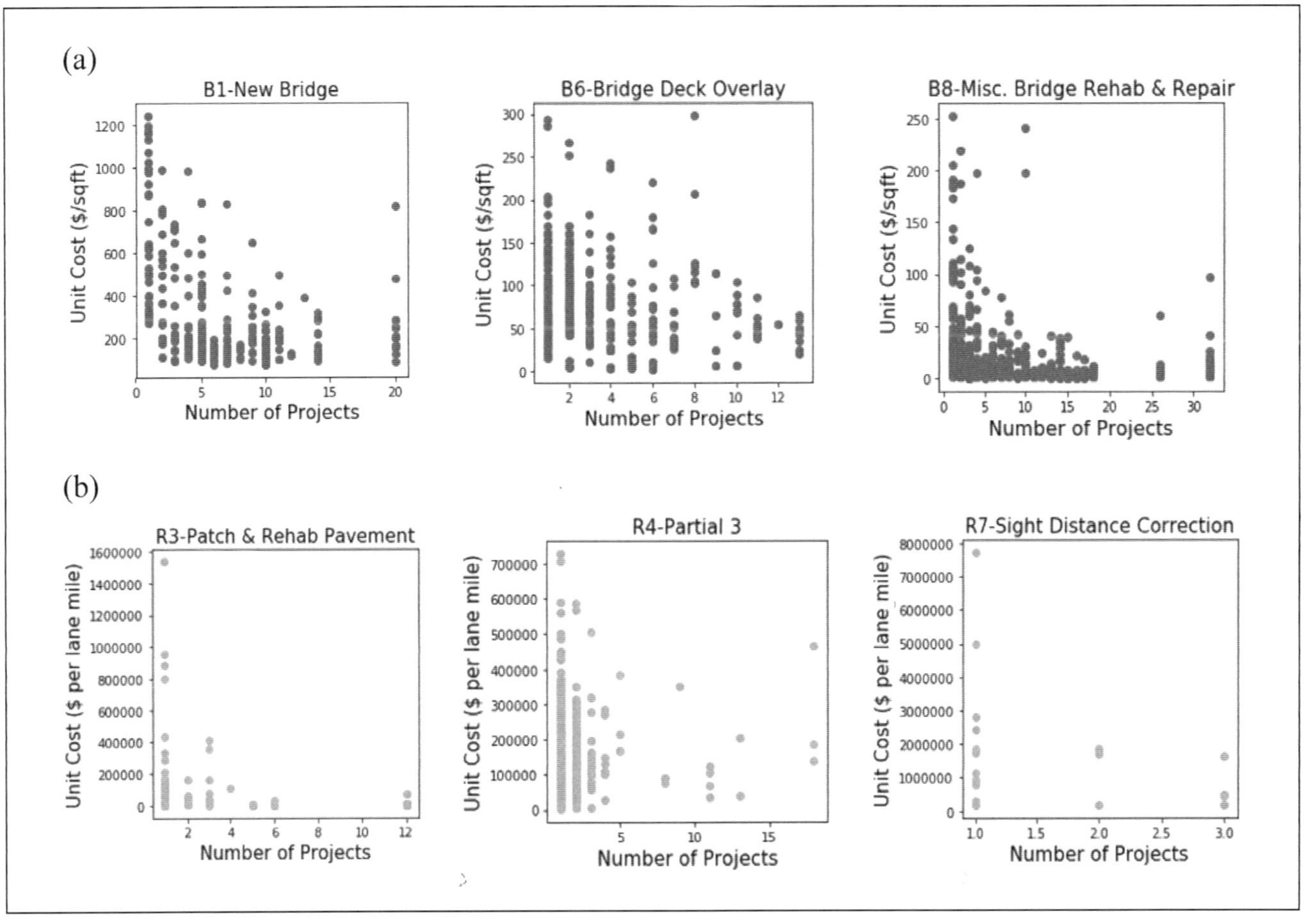

Figure 3. Economies of bundling, unit cost vs. number of projects for (*a*) bridge work types and (*b*) road work types.

(instead of the optimal cost) based on the project cost models, one can use a better estimate of the number of bidders, instead of using the upper bound value:

Bridge projects: number of bids = 13.038 – [3.631*ln(number of projects)](R2 = 0.869) (1)

Road projects: number of bids = 11.167– [2.038*ln(number of projects)](R2 = 0.479) (2)

Traffic projects: number of bids = 9.7143–[1.667*(number of projects)2]–[0.0952*number of projects] (R2 = 0.875) (3)

Small structure projects: number of bids = 15.538–[1.517*ln(number of projects)](R2 = 0.664) (4)

Miscellaneous projects: number of bids = 11.575–[4.778*ln(number of projects)](R2 = 0.922) (5)

All contracts: number of bids = 15.710–[4.037*ln(number of projects)](R2 = 0.737) (6)

Project Similarity (Compatibility)

Projects that involve similar materials and resources can be considered compatible and thus are prime candidates for simultaneous delivery through project bundling. In other words, projects that share a greater number of common pay items might earn more cost savings if they are bundled together, compared with projects with

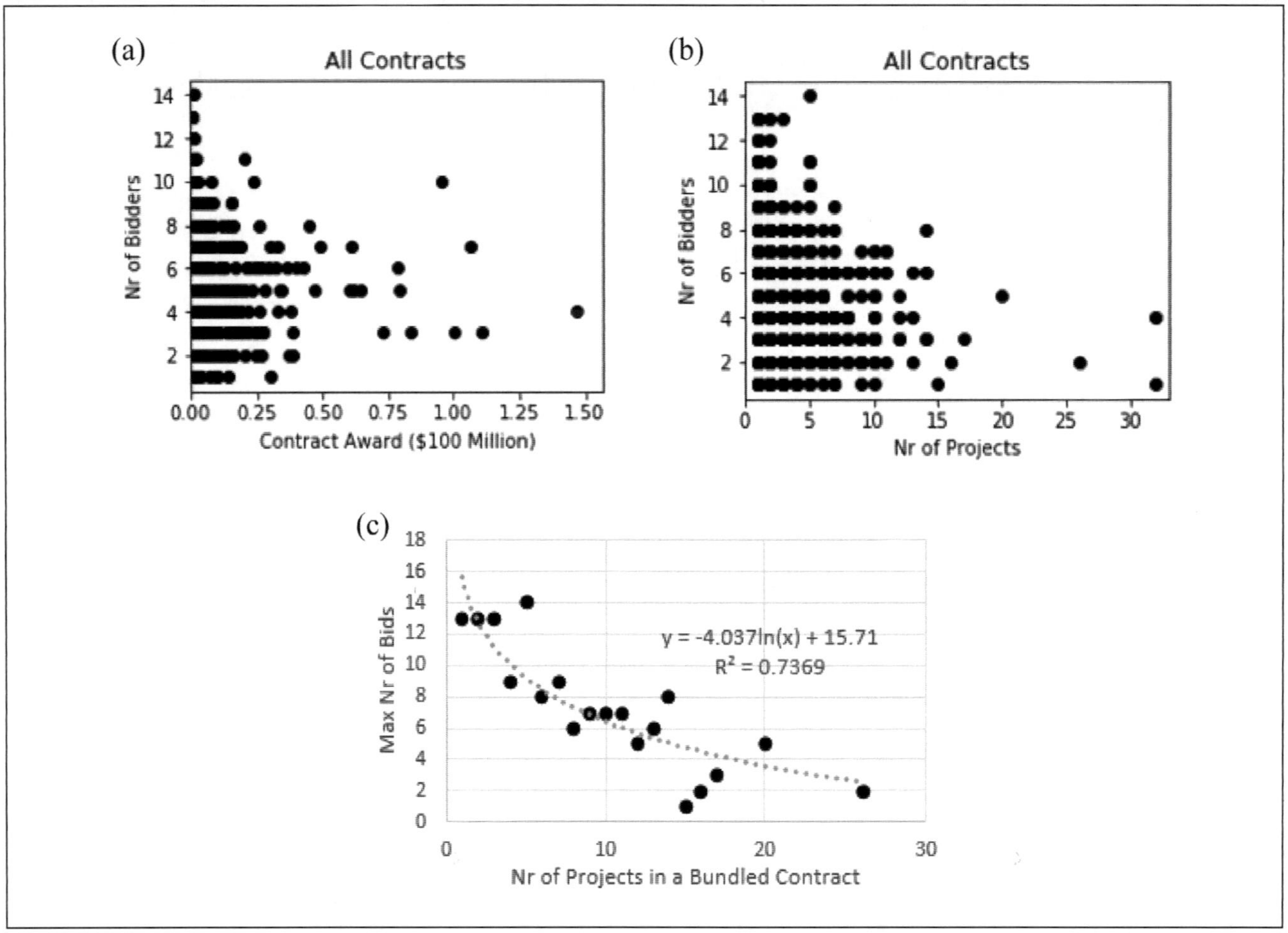

Figure 4. Trend observed for number of bidders and the upper-bound model to predict the number of bidders: (*a*) number of bidders vs. contract award (1.0 = $100 million); (*b*) number of bidders vs. number of projects in a bundle; (*c*) fitting a curve through maximum number of bids points.

different pay items. For example, bridge deck overlay and thin deck overlay might be more similar to each other as work types in a multiple-project contract than to other project types. Having similar work types in a contract is more likely to lead to cost savings. To provide a quantitative way to supplement the judgment of an agency's project schedulers, it was hypothesized that projects that have more pay items in common would be more compatible. We developed the average similarity distance (ASD) variable to capture such similarity across projects. The ASD is calculated as the ratio of the sum of the similarity distances between each pair of project types in the contract, to the total number of pairs.

To measure the potential compatibility of projects being bundled, the similarity of the distribution of pay items within project awards was quantified using ASD; a higher ASD means lower similarity or higher incompatibility. See the report (*8*) on which this paper is based for details. Table 1 shows only that portion of the ASD matrix that concerns Bridge work types. The shorter the similarity distance, the greater the similarity between two different project types. Work types B4: Deck Replacement and B5: Bridge Widening have the smallest similarity distance (0.14) in Table 1. This suggests that they are the most compatible Bridge work types. Work types B1: New Bridge and B7: Thin Deck Overlay could be combined under one contract, but the activities that are indicated by their pay items (ASD = 0.534) suggest that B1 and B7 are not good candidates for bundling. Although similar pay items do not necessarily equate to compatible projects, the ASD results are reasonable and correspond well to the bundling strategies that INDOT has been employing. More importantly, it can identify possibilities for project combinations that had not been considered compatible in the past.

Other Factors That May Affect Unit Cost

Bridge work types had the most observations and the most reliable project size data (deck area). Therefore, in

Table 1. ASD Matrix among Bridge Work Types

Project type	B1	B2	B3	B4	B5	B6	B7	B8
B1	0	0.140	0.240	0.267	0.242	0.394	0.534	0.398
B2	0.140	0	0.174	0.199	0.189	0.315	0.492	0.342
B3	0.240	0.174	0	0.074	0.098	0.294	0.507	0.361
B4	0.267	0.199	0.074	0	0.112	0.298	0.508	0.358
B5	0.242	0.189	0.098	0.112	0	0.305	0.517	0.362
B6	0.394	0.315	0.294	0.298	0.305	0	0.261	0.351
B7	0.534	0.492	0.507	0.508	0.517	0.261	0	0.476
B8	0.398	0.342	0.361	0.358	0.362	0.351	0.476	0

our bid to show how unit costs were affected by factors such as spatial proximity, road functional classes, letting season, and project location, we focused on these work types.

Spatial Proximity. Multiple projects in the same corridor are generally expected to benefit from project bundling, because it offers contractors more flexibility to move and coordinate resources. Various measures of geographic proximity were tried, but no meaningful relationship could be developed.

Letting Season. Does time of year affect the strategy of bidders, which would affect project unit costs? Using project data for the Bridge work types with the largest sample size, plots were made of project unit cost vs. deck area relationships (unit cost pattern) in terms of four seasons in a calendar year. The projects were assigned to seasons based on the letting date information given in the SPMS dataset. The scatter plots give no evidence that the unit cost in a certain season is different from that of the others.

Project Location. INDOT has six highway districts, with some differences in topography, traffic characteristics, and climate. For New Bridge, Bridge Replacement and Deck Replacement projects, no clear difference was found between the unit cost patterns of projects located in different districts. However, for Deck Replacement, Bridge Deck Overlay, Thin Deck Overlay, the projects in northern or central districts have somewhat higher unit cost than projects in southern districts when the deck area is relatively small.

Cost Models

Bridge Work Cost Model

In order to test various bundling strategies, it is necessary to develop models that estimate unit costs in bundled and unbundled circumstances. For Bridge work types, it was found that the total cost (project award) model generally yields results that are statistically superior to the unit cost model. Therefore, the regression model was built first for a project's total cost (Equation 7 below), which can be transformed into Equation 8 by taking the exponential form on both sides of the equation. The total cost model in Equation 8 can then be further transformed into a unit cost model by dividing the estimated project award amount by the project's bridge deck area (Equation 8):

$$\ln(\text{project award}) = \beta_0 + \sum_{i=1}^{N} \beta_i * ln(V_i) \tag{7}$$

$$\text{Project award} = e^{\beta_0} * \prod_{i=1}^{N} V_i^{\beta_i} \tag{8}$$

$$\text{Unit cost} = \frac{e^{\beta_0} * \prod_{i=1}^{N} V_i^{\beta_i}}{\text{Deck area}} = e^{\beta_0} * V_1^{\beta_1 - 1} * \prod_{i \neq 1}^{N} V_i^{\beta_i} \tag{9}$$

where β_i is a parameter representing the ith estimated coefficient and V_i is the ith influential variable. Table 2 shows the parameter values for work type B1: New Bridge. Qiao et al. (*8*) contains the parameter values for all bridge, road, traffic, and miscellaneous work types for which adequate data were available. Notice that the variables corresponding to parameters β_1, β_2, and β_3 relate to economies of scale, economies of bundling, and economies of competition, respectively. The positive sign for parameter β_4 indicates that project cost increases as the incompatibility of projects increases.

Project Cost Estimation

Sensitivity analysis was carried out to show how the unit cost changes with changes in one variable, when all other variables are kept unchanged. Figure 5 presents sample sensitivity analyses on the three major factors—economies of scale (deck area), economies of bundling (number of projects) and economies of competition (number of bidders)—for eight bridge project types.

Table 2. Total Cost Models for Bridge Project Types (2015$)

B1: New Bridge (209 observations)					
Variable	Parameter	Coefficients	*T*-values	*p*-values	*r*-squared
Constant	β_0	8.677	13.638	4.20E-27	0.589
Deck area (V_1)	β_1	0.72	11.732	2.64E-22	
Number of projects (V_2)	β_2	−0.389	−6.071	1.23E-08	
Number of bids (V_3)	β_3	−0.062	−1.424	2.73E-01	
ASD (V_4)	β_4	0.067	1.61	1.10E-01	

However, in reality, some of these variables are not independent of each other. As discussed previously, the number of bidders is affected by the project size, i.e., a larger project can lead to fewer bidders, which means weaker market competition. To account for this relationship, the upper-bound model (Equations 1–6) was used to predict the number of bidders for a given number of projects in the sensitivity analyses.

Bundling Strategies

An Evaluation of Past Bundling by INDOT

An evaluation of project bundling by INDOT in past projects was carried out to assess the impact of different bundling strategies on the project cost, using the developed project cost models. (The rest of these are presented in the study report (*8*).) Due to its superior data availability and cost models, Bridge work was selected to demonstrate the analysis. The analysis was conducted in two ways: (i) assuming past single projects are bundled and (ii) assuming that past bundled projects are either unbundled or bundled with new bundling strategies.

Assuming Past Single Projects Had Been Bundled

All the stand-alone bridge projects in two selected corridors were used for this analysis. This was done to compare the original project cost of these single projects with the new project cost estimated by using the cost models, assuming that projects are bundled with a specified bundling strategy.

Interstate 69 Corridor. From 2008 to 2017, there were 10 stand-alone bridge projects completed along the I-69 corridor. The total cost of all these projects is the sum of all project awards ($10,883,024). Several bundling strategies were tried, each time applying the appropriate models to the case at hand. The first three scenarios were:

1. *All Projects Bundled into a Single Contract.* Equation 1 was used to compute the estimated number of bids:

   Number of bids = 13.038 − [3.631 * ln(10)] = 13.038 − [3.631*2.3026] = 4.67.

   This number is left in non-integer form because it is an estimate and rounding to integers will distort the impact of this factor when compared across the other contracts. Equation 8 was applied to each of the 10 projects using number of projects = 10, the coefficients for the project work type (see Table 2), and the project deck area. The 10 resulting costs were summed to reach a total estimated project award of $8,921,390 for the bundle (see Table 3). Bundling lowers the cost but having 10 projects in a contract instead of one also reduces the estimated number of bidders.
2. *Projects Bundled by Letting Year.* Bundle the four projects that were let before 2012 as one contract and the remaining six as another contract. This reduces the adverse effect of scheduling some projects sooner than necessary. Here, there is less flexibility in bundling, but there are more bidders, so the estimated des award is only slightly less than for Scenario 1.
3. *Projects Bundled on the Basis of Their Similarity.* According to Table 2, project work types B2, B3, and B4 have pairwise ASD values of 0.199 or lower, so the five B2: Bridge Replacement, B3: Superstructure Replacement, and B4: Deck Replacement projects are bundled into one contract. That leaves five B6: Bridge Deck Overlay projects. The sum of the estimated project awards (from the model) with this bundling strategy indicates that taking project compatibility into account can lead to lower costs.

US Highway 31 Corridor. A similar analysis was conducted for the seven stand-alone bridge projects along the US-31 corridor in 2008–2017. The original total project cost (project award) of the 10 single projects was $12,080,933. Scenarios 4 and 5 in Table 3 follow the same strategies as Scenarios 1 and 3 for I-69. To investigate further the effect of bundling by project similarity, Scenario 6 was devised to randomly assigned projects to two bundles such that the number of projects (and therefore number of bids) would be the same as in Scenario 5. Again,

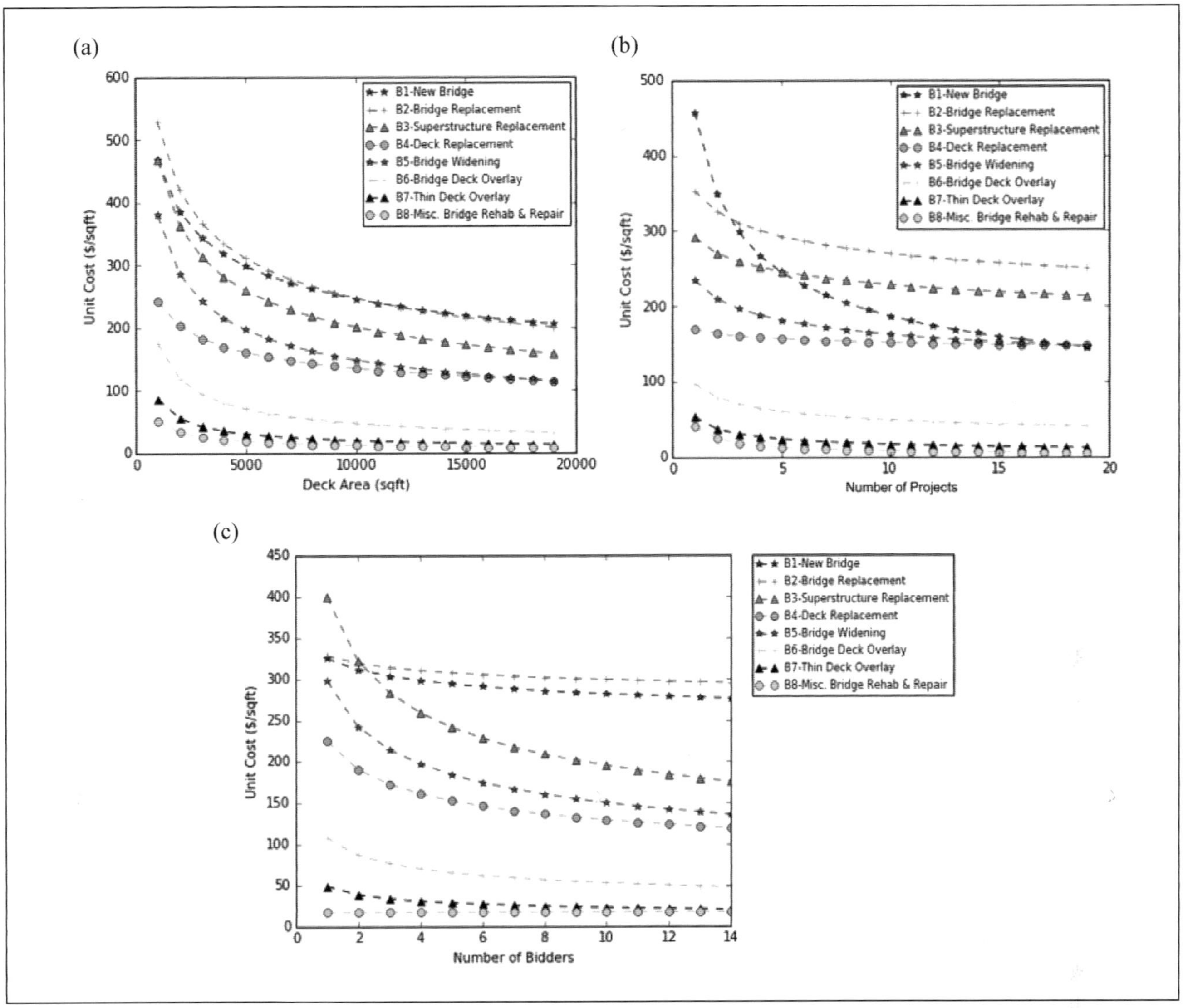

Figure 5. Sensitivity analysis results for the three major factors showing effect of (*a*) deck area, (*b*) number of projects, and (*c*) number of bidders.

Table 3. Comparing Costs for Various Bundling Strategies

Scenario strategies and details	Number of projects	Estimated number of bids	Estimate of project award
Scenarios for I-69 corridor ($10,883,024 unbundled)			
1. All projects bundled as a single contract	10	4.67	$9,190,408
2. Two contracts, bundled by letting year	4 + 6	8.0 + 6.5	$9,056,951
3. Two contracts, bundled on the basis of their similarity	5 + 5	7.2 + 7.2	$8,834,707
Scenarios for US31 corridor ($12,080,933 unbundled)			
4. All projects bundled as a single contract	7	5.97	$7,067,895
5. Two contracts, bundled on the basis of their similarity	3 + 4	9.1 + 8.0	$7,095,977
6. Two random bundles, three and four projects each	3 + 4	9.1 + 8.0	$7,271,474

bundling projects with higher similarity into one contract is the more cost-effective strategy.

Assuming That Past Bundled Projects Are Unbundled or Rebundled

In this section, scenario analyses are carried out on three contracts randomly selected from those that were actually bundled in the past. The contracts had eight, four, and seven bridge projects, respectively. The actual total project cost (project award) of all bridge projects of the three randomly selected bundled contracts is \$19,082,148. After applying the cost estimation models to each project, the estimated (or predicted) total project award is \$16,327,768. There are several possible reasons for this difference:

a. The cost models use an upper bound on the expected number of bidders, which usually underestimates the project cost.
b. The projects actually bundled may have been bundled for reasons other than cost only.
c. The cost models are statistically acceptable, but the scatter in the data and the presence of other factors must be taken into account.

For these reasons, the cost models should not be relied upon for accurate estimates of the magnitude of cost savings. However, the relative values of the results seem to be not only reasonable, but a good basis for comparing bundling strategies. Because the scenarios to follow will be analyzed using cost model estimates, it would be fairer to compare scenario results with the \$16,327,768 estimated original value.

The scenario analysis here seeks to ascertain how the total contract cost will change if these projects had been either (a) unbundled or (b) bundled using a different bundling strategy. The scenarios summarized in Table 4 are:

1. *Unbundling of Bundled Projects*. As standalone projects, the 19 projects would each expect to receive 13 bids. After applying the appropriate cost model to each project, the sum of the 19 estimated costs is \$20,117,761.
2. *Four Contracts, Bundled by Project Similarity*. Project compatibility proves to be a valuable characteristic. More than \$1 million is saved with respect to the \$16,327,768 estimated original value.
3. *Four Contracts, Bundled Randomly*. This random bundling case is more costly than the three bundles used by the agency and the similarity-based Scenario 8.
4. *Three Contracts, Bundled by Project Similarity*. Three compatible bundles had lower cost than four compatible bundles formed from the same 19 projects.
5. *Three Contracts, Bundled Randomly*. Although this is just one random example of three bundles, the cost decrease in forming three larger random bundles rather than four smaller bundles (Scenario 9) is significant.
6. *Two Contracts, Bundled by Project Similarity*. The bundles are getting fewer and larger. Using the ASD, 13 of the 19 projects were placed in one of the two bundles. The positive impact of project compatibility outweighs the costs of having a larger number of projects and a lower number of bids in one bundle.
7. *Two Contracts, Bundled Randomly (Unbalanced)*. The size of the two random bundles in this scenario were intended to match the bundle sizes in the similarity-based Scenario 12. As expected, it is verified that random bundles are more expensive than bundles of compatible projects.
8. *Two Contracts, Bundled Randomly (Balanced)*. To investigate further the impact of bundle size, the same projects in Scenario 13 were reallocated,

Table 4. Comparing Costs for Various Unbundling and Rebundling Strategies

Scenarios (\$19,082,148 actual original cost, \$16,327,768 by model)	Number of projects	Est. number of bids	Estimate of project award
1. All projects unbundled as separate contracts	19	13 each	\$20,117,716
2. Four contracts, bundled by project similarity	3 + 3 + 7 + 6	9 + 9 + 6 + 7	\$15,061,743
3. Four contracts, bundled randomly	3 + 3 + 7 + 6	9 + 9 + 6 + 7	\$17,257,785
4. Three contracts, bundled by project similarity	6 + 7 + 6	7 + 6 + 7	\$14,481,562
5. Three contracts, bundled randomly	6 + 7 + 6	7 + 6 + 7	\$15,193,676
6. Two contracts, bundled by project similarity	6 + 13	7 + 4	\$13,041,827
7. Two contracts, bundled randomly (unbalanced)	6 + 13	7 + 4	\$14,985,185
8. Two contracts, bundled randomly (balanced)	10 + 9	5 + 5	\$14,195,652
9. Bundle all projects into one contract	19	2	\$13,677,940

so that the bundle sizes were as equal as possible. The small resulting shift in number of bidders helped lower costs somewhat.

9. *Bundle All Projects into One Contract.* As the number of bundles has been decreasing, so has the estimate of project award. One bundle is the logical next step, but the total cost increases. A maximum of two bidders are expected.

Lessons from Analysis of Bundling Scenarios

The exact values from the cost models may not be precise, but the results from a variety of analyses are consistent:

- Fewer, bigger bundles are more cost-effective, subject to some constraints.
- If a bundle gets too big, fewer bids (and a higher cost) can be expected.
- Project compatibility in a bundle, as measured by the ASD, leads to lower contract costs.

Discussion and Conclusions

To address the issues associated with cost savings from project bundling, this paper describes a data-driven approach that involves:

- comparisons of the unit project cost of single and bundled projects using an average cost approach
- preliminary investigations on the potential influential variables for overall project cost
- the development of project cost statistical models
- an analysis of past and possible future bundling strategies

The study found that the primary drivers of project cost are the project size (economies of scale), bundling strategy (economies of bundling), and bidding market conditions (economies of competition). It is also shown that the similarity between different project types, based on the cost compositions of project pay items, is an influential factor on project cost.

The study also confirmed that larger contracts yield economies of scale, but they may also lead to less competition (by discouraging all but the largest firms from bidding on the work). This finding suggests that there appears to be a threshold of contract sizes beyond which the unit project cost might start increasing. It was also determined that, due to the uncertainty of market competition, the project cost estimated using the overall project cost models can vary to a certain degree, depending on the marginal effect of the number of bidders on project cost. The study's results verified that project types in the same work category have greater compatibility (similarity) in terms of their constituent pay items, compared with those in different work categories. The cost models confirmed that higher similarity between projects bundled in a contract can lead to lower project cost.

The study also carried out scenario analyses using hypothetical bundling strategies applied to actual past projects, and determined that a carefully designed bundling strategy is critical in achieving significantly reduced overall contract costs. As more data become available, future research in this important area of project management can attempt to build models that: (a) more thoroughly consider the geographical distance variable, (b) predict more reliably the number of bidders for a contract, and (c) consider all plausible bundling combinations. Furthermore, to identify the optimal strategy, instead of exhaustive enumeration, a method that produces an "optimal" bundling strategy from a reduced set of combinations would present the decision makers with choices that might not otherwise be considered.

Often, in order to bundle projects (and in anticipation of achieving the resulting savings), an agency may carry out certain projects a few years earlier or later than their programmed dates. However, deferring projects or carrying them out before their due dates is not cost-effective. Therefore, there exists a tradeoff between optimal timing and bundling, and in any given situation, the net effect of the bundled projects (which includes those deferred or done earlier) could be positive or negative. The models in this paper can help estimate this net effect and thus could influence agency decisions as to whether, how, and when to bundle their projects. In cases where projects are rescheduled so they can be bundled with others in a different year, future research could investigate the tradeoff between the cost reduced by project bundling and the cost (agency and user) increase due to project delay.

Acknowledgments

This work was supported by the Joint Transportation Research Program administered by the Indiana Department of Transportation and Purdue University. The Study Advisory Committee members were Tim Wells (Project Administrator), co-author Trevor Mills (Business Owner) and Scott Adams, all of INDOT. The authors gratefully acknowledge the extraordinary contributions of INDOT's Mann Omran, who provided the data needed to complete such a study, and Louis Feagans, who served on the Study Advisory Committee.

Author Contributions

The authors confirm contribution to the paper as follows: study conception and design: TM, JDF, SL; data collection: TM; analysis and interpretation of results: YJQ, SL, JDF, TM; draft manuscript preparation: JDF, SL, YJQ. All authors reviewed the results and approved the final version of the manuscript.

References

1. Xiong, Y., J. D. Fricker, and S. Labi. Bundling or Grouping Pavement and Bridge Projects: Analysis and Strategies. *Transportation Research Record: Journal of the Transportation Research Board*, 2017. 2613: 37–44.
2. McCarthy, L. A., Mensching, and A. Horgan, A. *NCHRP Synthesis 414: Effective Delivery of Small-Scale Federal-Aid Projects*. Transportation Research Board of the National Academies, Washington, D.C., 2011.
3. U.S. House of Representatives 2015. SEC. 1111. Bundling of Bridge Projects, Section 1111 of Subtitle A of Title I (Federal Aid Highways), FAST Act Amendments to Section 144 of title 23, United States Code. https://transportation.house.gov/uploadedfiles/crpt-114hrpt-hr22.pdf. Accessed November 14, 2017.
4. Damnjanovic, I., S. Anderson, A. Wimsatt, K. Reinschmidt, and D. Pandit. *Evaluation of Ways and Procedures to Reduce Construction Cost and Increase Competition*. Report 0-6011-1. Texas Transportation Institute, College Station, TX, 2009.
5. Pearson, J. N., and J. D. Wisner. Using Volume, Economics of Scale Benefits Long-Term Productivity. *Industrial Management*, Vol. 35, No. 6, 1993, pp. 13–13.
6. Bhargava, A., S. Labi, and K. C. Sinha. *Development of a Framework for Ex Post Facto Evaluation of Highway Project Costs in Indiana*. FHWA/IN/JTRP-2009/33. Purdue University, West Lafayette, IN, 2010.
7. Lamb, D., and A. Merna. Development and Maintenance of a Robust Public Sector Comparator. *The Journal of Structured Finance*, Vol. 10, No. 1, 2004, pp. 86–95.
8. Qiao, Y. J., J. D. Fricker, and S. Labi. *Capital Program Cost Optimization Through Contract Aggregation Process*. Draft Final Report. Joint Transportation Research Program, Purdue University, West Lafayette, IN, February 19, 2018.

The Standing Committee on Maintenance and Operations Management (ADH10) peer-reviewed this paper (18-05299).

Article

TRR
JOURNAL OF THE TRANSPORTATION RESEARCH BOARD

Transportation Research Record
2018, Vol. 2672(12) 179–186

DOI: 10.1177/0361198118781136
journals.sagepub.com/home/trr
SAGE

Highway Asset Deterioration Rates

William Rasdorf[1] **and Ali Almalki**[1]

Abstract

This paper is about highway asset deterioration rates. Typically, these are difficult to determine. However, using a data collection method adopted in North Carolina (and the resulting data obtained over a 5-year timespan), an approach was developed to determine accurate deterioration rates for several highway assets. This was not previously done, and thus, deterioration was not included as a factor in estimating future maintenance needs for these assets despite the ready availability of the necessary data. This paper focuses on unpaved shoulders and outlines how their deterioration rates were obtained and how such rates can also be obtained for other assets. The resulting deterioration rates are presented and compared over various road systems and geographical regions. With this new knowledge, deterioration rates can now be included in life cycle asset analysis, resulting in more accurate condition predictions and maintenance budgeting. The paper focuses on highway assets and does not address bridge or pavement maintenance.

Jha & Maji defined deterioration as "the amount of degradation of the condition from its original condition" (*1*). Infrastructure in general, and roadways in particular (ex. shoulders, ditches, cross line pipes, etc.), deteriorate over time due to weather and other factors. To plan maintenance and rehabilitation activities for roadway assets, a deterioration model is needed (*2*). Such a model can aid transportation agencies in establishing a maintenance plan and prioritizing maintenance activities when funds are limited.

The North Carolina Department of Transportation (NCDOT) has been collecting condition data on 11 highway assets beginning in 2012 and continuing through the present. The 11 highway assets are:

1. Unpaved shoulders
2. Ditches
3. Cross line pipes (blocked)
4. Cross line pipes (damaged)
5. Curbs and gutters (blocked)
6. Drainage boxes (blocked or damaged)
7. Brush & trees
8. Turfs
9. Pavement markings
10. Words & symbols
11. Raised reflective pavement markers

This paper focuses on unpaved shoulders and demonstrates how a deterioration rate can be determined based on actual field condition evaluations. This approach is possible only because of the availability of this data. In contrast, our findings indicate that such condition data availability is rare for highway assets such as these. Thus, most of the work by others is based on models that approximate deterioration or do not even take it into account. However, to create realistic maintenance plans and budgets, one must accurately incorporate accurate deterioration rates. The unique aspect of the work presented here is that it is based on actual field-collected condition data for a relatively comprehensive set of highway assets.

Background

According to Alyami & Tighe, an infrastructure asset's deterioration is affected by many factors, including the environment, periodic maintenance, traffic volume,geographic location, and material properties (*3*). To establish an accurate deterioration model for assets, all affecting factors must be considered. In addition, all current data and all historical data need to be accurate and precise. One key problem is that deterioration and prediction models have uncertainties (*1*, *4*). Durango and Madanat addressed deterioration uncertainty. That is, they developed maintenance and repair policies that do take into account deterioration even when deterioration rates are uncertain (*4*). In another paper Madanat et al. also addressed deterioration uncertainty, this time utilizing probabilistic models (*5*).

A few transportation agencies have developed deterioration models to predict changes in infrastructure conditions (*6*). Some agencies use mathematical models to

[1]North Carolina State University, Department of Civil, Construction, and Environmental Engineering, Raleigh, NC

Corresponding Author:
Address correspondence to William Rasdorf: rasdorf@ncsu.edu

Table 1. Division 3 Secondary Route Condition Data (Original)

(1) Year	(2) Secondary route mileage	(3) Beginning year condition rating	(4) Beginning year good mileage	(5) End year condition rating	(6) End year good mileage	(7) Improved mileage	(8) Deteriorated mileage	(9) Percent deterioration
2012	4,301.96	88	3,785.73	90	3,871.77	1,585.33	−1499.29	−34.85
2013	4,310.23	90	3,879.21	98	4,224.03	201.38	143.44	3.33
2014	4,330.50	98	4,243.89	97	4,200.59	267.07	−310.37	−7.17
2015	4,344.78	97	4,214.43	93	4,040.64	135.43	−309.22	−7.12
2016	4,359.05	93	4,053.91	88	3,835.96	247.20	−465.15	−10.67
								Avg = −11.30

predict the future condition of pavements according to past deterioration rates. A variety of modeling techniques have been introduced in pavement management systems to analyze pavement performance and predict pavement deterioration over time. The following models have been adopted and implemented by several transportation agencies (*7*):

- General linear model (simple regression method)
- Mixed linear model
- Markovian chain-based model
- Bayesian regression model

Other techniques such as regression, mathematical models, or Markovian chains have been used by several researchers to model deterioration of bridge assets (*1*). Madanat et al. implemented two econometric methods to develop models to predict the beginning and progression of highway pavement cracking over time (*8*). However, deterioration models have not been developed for unpaved shoulders. In fact, they have not been developed for any of the 11 highway assets mentioned above.

An increase in the deterioration rate of an asset will increase the total cost of maintenance work. Additionally, as the deterioration rate increases, maintenance work will need to be initiated sooner. However, a decrease in the deterioration rate results in a decrease in total maintenance cost. With no change to the deterioration rate, the total cost of maintenance will decrease only when a lower future condition rating is acceptable (*3*).

The deterioration rate of an asset depends on the maintenance activities applied over its life span (*9*). For infrastructure assets, most deterioration rate models are calculated using one of only two different scenarios: "do nothing" and "do maintenance work." In the first scenario, a future condition rating is predicted assuming that no work is done to maintain or improve the asset. In the second scenario it is assumed that at least some work is performed (*10*, *2*). The researchers also stated that a deterioration model can be either a straight-line or a declining-balance model. A straight-line model assumes that an asset's condition deteriorates over its useful life at a steady rate per year. In a declining-balance model, asset condition deteriorates more at the beginning of an asset's life and less at the end. The main criticism of the straight-line model is that all assets are assumed to have the same rate of deterioration and the model does not consider whether or not the assets are being maintained. Consequently, the straight-line model is not precise because it ties the asset's deterioration to only its age (*10*).

Garza and Krueger proposed simulation equations that can be used to simulate the dynamics between pavement deterioration and allocated maintenance budget (*11*). The proposed equations can be implemented by transportation agencies to plan for short- and long-term maintenance strategies subject to a specific desired condition rating along with a specific allocated budget. Thompson et al. provided a guide to help transportation agencies estimate life expectancies of transportation assets such as pavement, bridges, traffic signs, and culverts (*2*).

Condition Data and Highway Mileage

The data used for this study was obtained from NCDOT's maintenance management system (MMS) which is a part of NCDOT's overall asset management system (AMS) that tracks maintenance activities, expenditures, and resources (*12*). NCDOT collects roadway condition data annually to evaluate the overall condition of its assets. Some of this data is presented in Table 1.

For our study, unpaved shoulder condition data from six different divisions of North Carolina (NC) were obtained and analyzed. Condition data were derived from an annual field sample of between 21,000 and 22,000 tenth mile highway segments. For each sample, condition data were determined for a total of 11 highway assets. The divisions whose data were used for the study were: Divisions 3 and 4 in the coastal region, Divisions 5 and 10 in the piedmont, and Divisions 13 and 14 in the mountains.

NCDOT's data collection process measures the amount of mileage, within each surveyed tenth mile highway segment, for which the condition of each asset is acceptable (and for which the condition is unacceptable) according to performance criteria established by NCDOT's maintenance condition assessment procedure (MCAP). The mileages for which the criteria are met and the mileages for which the criteria are not met are summed for all the sampled sections and divided by the total number of sample miles, resulting in a numerical condition rating. Thus, the condition rating is a measure of the highway mileage that satisfies a given criteria. The actual field data directly results in a condition rating (columns 3 and 5) which, in turn, results in a total number of "good" (meets the performance criteria) miles (columns 4 and 6).

Table 1 shows a sample of condition data for unpaved shoulders in Division 3 for secondary routes. The first column marks the year. The second column shows the number of existing miles for each route in each year. However, for secondary roads in Division 3 we were only able to obtain measured mileage data for three years (2012, 2013, and 2014). Thus, we estimated 2015 and 2016 mileages by assuming an increase in miles for 2015 and 2016 that was proportional to the change in miles over the years 2012, 2013, and 2014. These mileages (4,344.78 and 4,359.05) are displayed in the 2015 and 2016 rows respectively, in the second column of Table 1. In the other sampled divisions, no new roads were constructed during these three years. For those divisions, the number of miles in 2015 and 2016 was equal to the mileage recorded for 2014.

Methodology

This section outlines the approach we used to determine the amount and rate of deterioration of unpaved shoulders for each road system. The following paragraphs describe the steps taken to obtain the deterioration rates.

Columns 1 and 2 of Table 1 show the year and the number of existing roadway (route) miles in that division during that year, respectively. In the third column of Table 1, is the condition rating of the unpaved shoulders at the beginning of each year (column 3). The beginning of year condition rating is the same as the condition rating from the end of the previous year (column 5). The number of miles in good condition at the beginning of the year is shown in column 4. These values were obtained by multiplying existing roadway mileage by beginning year condition rating and dividing by 100.

The fifth column indicates the condition rating at the end of that year. The number of miles in good condition at the end of each year is displayed in column 6. End of year mileage is the result of multiplying the existing mileage for the year by the end of year condition rating.

Every year NCDOT improves its roads by performing maintenance work that affects the condition of each of its roadway assets. Column 7 indicates the number of miles that were improved or repaired by the NCDOT during the fiscal year. This data is based on actual maintenance expenditures during the year and is obtained from the MMS. Unpaved shoulders (as well as other roads assets) deteriorate over time due to weather and other factors.

Column 8 of Table 1 shows the roadway miles that deteriorated during the current (between beginning and end) year. Column 9 reports the deterioration rate for that year as a percentage. All data pertaining to good miles at the beginning of the year, at the end of the year, and any improvements made, are known to the NCDOT, have been obtained from the MMS, and are known inputs to our model. The unknown is deterioration mileage, from which one can calculate the percentage of miles that have deteriorated. The following deterioration relationship was used to determine and report the deteriorated mileage in column 8 of Table 1.

$$\text{End of Year Good Miles} - \text{Beginning of Year Good Miles} = \text{Improoement} + \text{Deterioration}$$

The only unknown in this relationship is deterioration (miles). The number of deteriorated miles was divided by the route mileage (column 2) to obtain the percentage of deterioration shown in column 9. Finally, the average percentage of deterioration in each road system was calculated and displayed at the bottom of column 9. This same procedure was executed for each of the other six selected divisions.

Below is an example of a set of calculations that show how the data was used to complete Table 1. In this example, we use the condition data collected in 2014, corresponding to the middle row in the table. Note that although we are illustrating only NC Division 3 secondary roads we have actually assessed interstate and primary roads and we have done so for 6 of the 14 total divisions, including two divisions each from NC's coast, piedmont, and mountain regions. Summary results for all these analyses are included below.

Deterioration Calculations

The total number of Division 3 secondary route miles (synonymous with Shoulder miles) during 2014 was 4,330.50 miles. The Beginning of Year Condition Rating was 98%. The Beginning of Year Good Mileage was obtained by multiplying the secondary route mileage by the Beginning of Year Condition Rating.

$$\text{Beginning of Year Good Mileage} = 4,339.50 \times 98\% = 4,243.89 \text{ miles}$$

The End of 2014 Condition Rating in this division for secondary routes was 97%. The End of Year Good Mileage was also calculated by multiplying the secondary route mileage by the End of Year Condition Rating.

$$\text{End of Year Good Mileage} = 4,330.50 \times 97\% = 4,200.89 \text{ miles}$$

The number of miles which were improved during 2014 in the Division 3 secondary route was 267.07 miles based on MMS maintenance records. The Deteriorated Mileage was then obtained by subtracting the Beginning of Year Good Mileage and the Improved Mileage from the End Year Good Mileage.

$$\text{Deteriorated Mileage} = 4{,}200.89 - 4{,}243.89 - 267.07 = -310.37 \text{ miles}$$

The results show that 310.37 highway miles deteriorated during 2014 in Division 3 on secondary routes. The last step was to calculate the Percent Deterioration, which is the Deteriorated Mileage divided by the secondary route mileage.

$$\text{Percent Deterioration} = -310.37 \div 4{,}330.50 = -7.17\%$$

In this example we can see that Division 3 secondary routes deteriorated by around 7% over the 2014 fiscal year. This value is displayed in the 2014 row, in the last column of Table 1.

The average percent deterioration for all years can be obtained in one of two ways. The values in column 9 may simply be averaged (−34.85 + 3.33 − 7.17 − 7.12 − 10.67)/5 = 11.30% to obtain this value or the values in column 8 may be summed and divided by the sum of the values in column 2 (−1,499.29 + 143.44 − 310.37 − 309.22 − 465.15)/(4,301.96 + 4,310.23 + 4,330.50 + 4,344.78 + 4,359.05) = 11.30%.

It is important to note that although these calculations appear to be correct there are field considerations that introduce small errors. One such consideration is the quantity of good miles. The secondary route mileage can increase in any given year as shown in Table 1 column 1 for Division 3. This increase is a known value and is always small. Any new mileage created during a year is included in this total for the year. But both the beginning and end of year good mileage are determined by a sampling of field measured miles. This can result in a slight discrepancy in good miles. If the field survey is completed after the new mileage is in place the results are fully representative. But if the field survey is completed prior to the new mileage being in place then that new mileage is not included in the year's good miles inventory resulting in a small error in the results.

Data Quality

Table 1 represents the original uncorrected deterioration data. Columns 2, 3, 5, and 7 contain the field measured data. The shaded columns (4, 6, 8, and 9) contain the calculated values whose derivations were illustrated in the last section.

Data Error Identification

Two types of errors were found in the data: unrealistic values and out-of-range values. Unrealistic values are those having a positive deterioration. A percent deterioration cannot be a positive number because infrastructure assets do not repair themselves. Instead, in a natural state they deteriorate over time and, therefore, a deterioration rate is expected to be a negative value.

The second type of error is one that is an out-of-range value, defined as an atypically large percent deterioration when compared with others for the same division and route system, and it indicates, in fact, an error in the data.

There are three major causes of data errors such as these: errors that occur when entering the data, failure to report the accomplished work, and inconsistent measurement units. The first possible cause of error is keyboard error which can arise out of pressing the wrong key when entering data into the computer or attaching an extra digit to a value (1,585 instead of 585). Numerous other keyboard errors are also possible. Another cause of error is a failure to report some or all of the accomplished work. The maintenance crew may have reported the amount of materials and time spent to accomplish the work but did not report its quantity. If a field is left blank, the data is recorded with a value of zero, which is clearly an error.

The third cause of data errors is inconsistent measurement units. The amount of unpaved shoulders that was improved is usually measured and recorded in linear shoulder miles (LSM). In some cases, the amount of improvement is measured and recorded in linear shoulder feet (LSF) and then converted to LSM. This can be done incorrectly, or it may be overloaded entirely. This inconsistency in units of measurement leads to data error because the person who is responsible for entering data or for the conversion approving data values may not be aware of the unit of measurement error.

Condition data and the amount of work accomplished are usually collected and recorded daily and are approved weekly. The data operator entering the data may not have the expertise or background to understand all the data and identify the nature of errors that are related to it.

The next section and Figure 1 illustrate the procedures implemented to correct these types of data errors.

Correcting Data Errors

In any given data set (a total of five years of values), unrealistic and out-of-range data can occur individually, in multiples, or in combination. The combination of both an unrealistic value and an out-of-range value is illustrated in Table 1 and is, in large part, why we chose this particular table for illustration.

In Table 1, we see that the majority of the percentages for deterioration range from −7.12 to −10.67, except for the value −34.85 in the first row in the year 2012 and the value 3.33 in row two for the 2013 year. The 34.85 percentage is atypically large when compared to others for the same division and route system and is, in fact, an error. It is known through detailed independent checking and verification that that amount of work shown in column 7 (Improved Mileage) was not actually done. The improved mileage associated with this error (obtained by solving for improvement given this deterioration rate) is shown as 1,585.33 miles. The other improved mileages ranged from 201.38 to 267.07. Because

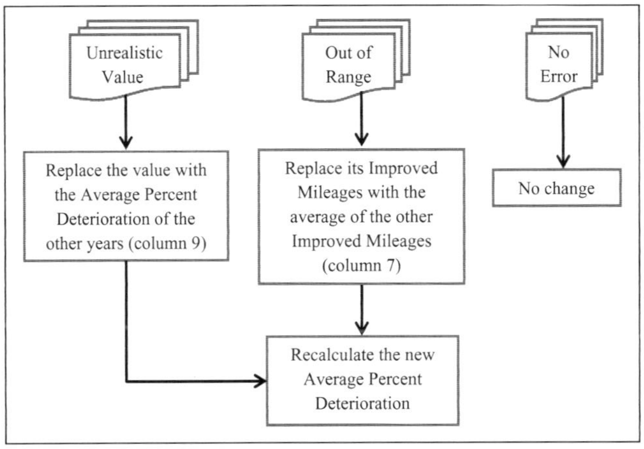

Figure 1. Data error correction process.

the improved mileages are assumed to be somewhat consistent over the years for the same division and route system, the value of 34.85 was identified as an outlier and we verified this with NCDOT. Of course, the origin of the error was considered, but is not further addressed in this paper.

There are two common methods to correct for outliers (out-of-range errors): the inter-quartile range (1.5 × IQR) method and the mean plus/minus three standard deviations method. In the former, IQR = Q3 − Q1, where Q1 (lower quartile) means larger than one-quarter of the data and Q3 (upper quartile) means larger than three-quarters of the data. In this method, anything below (Q1 – 1.5 IQR) or above (Q3 + 1.5 IQR) is defined as an outlier (*13*). The second method consists of identifying outliers by determining an interval spanning over the mean, plus or minus three standard deviations. In this method, anything below $(\bar{x}-3\sigma)$ or above $(\bar{x}+3\sigma)$ is defined as an outlier (*14*).

Practically, most statistical problems solved to correct for outliers using the two methods illustrated above must contain a data set whose population is 20 or above. In our case, given the limited amount of data available (we have only five years of condition data), these methods could not be used. Therefore, we used the methods discussed below for identifying out-of-range values.

To correct the out-of-range value (−34.85 in the first row for the year 2012) we replaced the 1,585.33 miles with the average of the other four improved mileages (201.38 miles, 267.07 miles, 135.43 miles, and 247.20 miles) which resulted in an average improved mileage of 212.77 miles. Using this value in the relationships presented in the previous sections, the new deteriorated mileage (−2.95 miles) was obtained and is shown in Table 2.

Table 1 also displays an example of an unrealistic percent deterioration. In this table we see a positive 3.33% deterioration for primary routes in division 3 in the second row for the year 2013. Statistically, the proper way to correct this unrealistic value is to average the other four deterioration rates (–2.95, –7.17, –7.12, –10.67), thus obtaining a new value of –6.98. The new values are displayed in Table 2.

Table 2. Division 3 Secondary Route Condition Data (Corrected)

Year	Secondary route mileage	Improved mileage	Deteriorated mileage	Percent deterioration
2012	4,301.96	212.77	−126.73	−2.95
2013	4,310.23	201.38	143.44	−6.98
2014	4,330.50	267.07	−310.37	−7.17
2015	4,344.78	135.43	−309.22	−7.12
2016	4,359.05	247.20	−465.15	−10.67
				Avg = −6.98

Results

Table 3 shows the deterioration rates for unpaved shoulders on NC's routes for each division and cumulatively for the six divisions included in the study. These rates represent the average percent deteriorations (last column of Table 2) for the corrected condition data. The results show that secondary roads in Divisions 4 and 13 had the highest deterioration rate at more than 8% whereas interstate roads in Division 10 had the lowest deterioration rate at less than 2%.

To obtain the overall annual average percent deterioration, the values in the deteriorated mileages column of Table 2 were summed and divided by the sum of route mileages for each road system in all six divisions. The resulting average percent deterioration for all six divisions is –2.04%, –5.08%, and –6.05% for the interstate, primary, and secondary systems, respectively, as displayed in the last row of Table 3. Interestingly, it is apparent from Table 3 that each of the primary and secondary roads deteriorate at an average rate that ranges between two and a half to three times that of interstate roads for all divisions for all five years.

Table 3 also shows the average percent deterioration for each road system in coastal, piedmont, and mountain regions. (Note that Divisions 3 and 4 are coastal divisions, Divisions 5 and 10 are piedmont divisions, and Divisions 13 and 14 are mountain divisions.) Findings show that secondary roads in the mountains clearly had the highest average percent deterioration (about 8%), although those at the coast were similarly high at about 7%, while the interstate roads in the piedmont and the mountains had the lowest average percent deterioration (less than 2%).

In the last column of Table 3 the combined average percent deterioration for each division and region along all road systems is shown. To obtain this percentage, the values in the deteriorated mileages column of Table 2 were also summed and divided by the sum of route mileages for each individual division and region, and then for all six divisions combined along the three road systems. Findings from the table show that Divisions 4 and 13 had the highest combined average percent deterioration (about 8%) whereas Division 5 had the lowest combined average percent deterioration (about 3%).

The last column of Table 3 also verifies that regionally the unpaved shoulders in the coastal and mountain regions had the highest average percent deterioration by approximately

Table 3. Annual Deterioration Rates for Unpaved Shoulders in NC by Division (2012–2016)

Division	Average percent deterioration			Combined average percent deterioration
	Interstate	Primary	Secondary	
3	–2.54	–5.67	–6.98	–4.57
4	–5.38	–7.44	–8.51	–8.17
5	–3.09	–3.10	–3.09	–2.97
10	–1.45	–3.10	–5.13	–4.72
13	–2.23	–6.06	–8.67	–8.09
14	–4.01	–7.35	–6.53	–6.59
Coast (3 & 4)	–2.90	–5.54	–6.84	–6.49
Piedmont (5 & 10)	–1.79	–2.68	–3.97	–3.73
Mountains (13 & 14)	–1.29	–6.71	–7.63	–7.35
All divisions	–2.04	–5.08	–6.05	–5.79

Note: There are 14 total highway divisions.

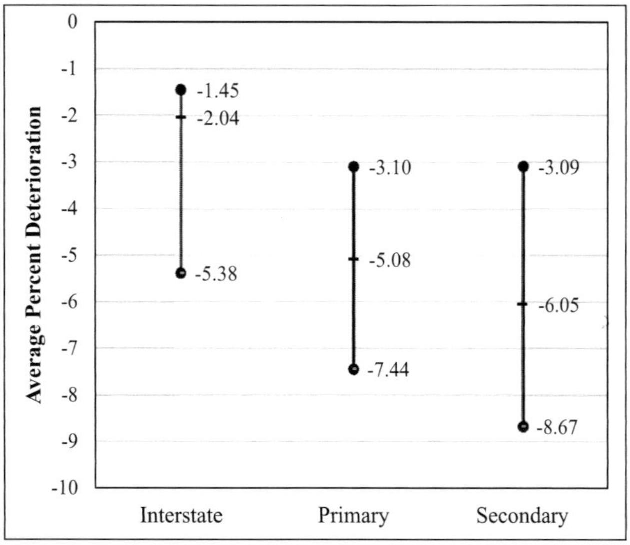

Figure 2. Corrected deterioration rates for unpaved shoulders in NC routes.

6.5% and 7%, respectively, along all three road systems: interstate, primary, and secondary. It is also regionally apparent that the unpaved shoulders in the piedmont region deteriorate at a lower rate than in the coastal and mountain regions by a little over 50%. Lastly, the combined average percent deterioration for all six divisions along the three road systems together is about 6%.

Figure 2 shows the range of corrected deterioration rates for unpaved shoulders in NC for the three road systems. Minimum, average, and maximum values are shown for each road system. The negative sign denotes deterioration (a positive number would indicate a condition improvement). The larger the number the greater the deterioration rate.

Validation of Condition Data

To validate the results shown in Table 3, condition data from three other divisions were assessed. The divisions include Division 1 in the coastal region, Division 7 in the piedmont, and Division 11 in the mountains. Table 4 shows the deterioration rates for unpaved shoulders in these three divisions derived in the same manner as those in Table 3.

To allow the reader to more easily compare Table 4 with Table 3, Table 5 is taken from the bottom half of Table 3 and is shown below.

The average percent deterioration for interstate roads in all of Divisions 1, 7, and 11 (–4.05) (in the bottom row of Table 4) compares reasonably with the previously studied divisions (–2.04) in Table 5 for the interstate route system. Similarly, the primary and secondary route systems compare favorably (–6.37 to –5.08 and –6.93 to –6.05).

Furthermore, the combined average percent deterioration for Division 1 in the coastal region (–8.67) (as shown in Table 4) compares reasonably well with the previous two divisions 3 and 4 (–6.49) (in Table 5). Also, the piedmont and mountain divisions compare favorably (–4.36 to –3.73 and –7.4 to –7.35).

In addition, the combined average percent deterioration for all three divisions along the three road systems together (–6.80) (in Table 4) compares reasonably well with that in Table 5 for all six divisions (–5.79). Thus, in all cases, these results are an indicator that the original sampling of divisions (6 out of 14) produces results that mimic those for the remaining eight divisions. No exceptionally surprisingly different results were found.

Summary and Conclusions

The objective of this paper was to provide an overview of the methodology used to determine the amount of deterioration of unpaved shoulders for NC road systems and to present the deterioration results. Deterioration models for unpaved shoulders had not previously been developed for NC roads and are not otherwise available for the types of day-to-day roadway assets identified in this paper. The research was motivated by the need to have a better understanding of roadway asset deterioration rates so that NCDOT can better

Table 4. Deterioration Rates for Unpaved Shoulders in Three of NC's Divisions (2012–2016)

Division	Average percent deterioration			Combined average percent deterioration
	Interstate	Primary	Secondary	
Coast (1)	−2.09	−8.95	−8.58	−8.67
Piedmont (7)	−4.69	−2.15	−4.85	−4.36
Mountains (11)	−3.30	−6.68	−7.56	−7.40
All divisions (1, 7, 11)	−4.05	−6.37	−6.93	−6.80

Table 5. Deterioration Rates for Unpaved Shoulders in NC's Regions (2012–2016)

Division	Average percent deterioration			Combined average percent deterioration
	Interstate	Primary	Secondary	
Coast (3 & 4)	−2.90	−5.54	−6.84	−6.49
Piedmont (5 & 10)	−1.79	−2.68	−3.97	−3.73
Mountains (13 & 14)	−1.29	−6.71	−7.63	−7.35
All divisions	−2.04	−5.08	−6.05	−5.79

prioritize its maintenance work. It was made possible by the availability of a unique and valuable set of data.

Field-collected condition data for unpaved shoulders from six different NCDOT divisions (from 2012–2016) was used to develop a deterioration model. Field data quality problems were identified. Two particular types of data quality errors (unrealistic and out-of-range values) were found, discussed, and addressed. The procedures implemented to correct these errors were described. The proper statistical data adjustments were determined and applied and a representative final data set to be used for future analysis was presented.

Finally, condition data from three additional divisions (one each in coastal, piedmont, and mountain regions) were assessed to validate the results obtained from the six original divisions. The findings show that the average percent deteriorations for the three new divisions compare reasonably well with those for the six divisions for both road systems and regions.

The results showed that unpaved shoulders on primary and secondary roads deteriorate two and a half to three times more, respectively, than on interstate roads. The results also showed that secondary roads had the largest average annual percent deterioration (by approximately 6%) over all divisions. In addition, it was shown that unpaved shoulders in the coastal and mountain regions deteriorate at about the same rate (about 7%) while unpaved shoulders in the piedmont region deteriorate at a much lower rate of about 4%. It was also shown that unpaved shoulders in the six NCDOT divisions deteriorated by 5.79% over the period 2012–2016 along the three road systems.

It should be pointed out that asset age was not used as a parameter in this work. Instead, we had other data that led us to a deterioration rate based on actual field measurement and assessment. Thus, we determined an annual average deterioration rate and a 5-year average. It seems plausible that deterioration should increase with age. However, our results did not verify this postulation. Further study of the impact of age is warranted and recommended.

These results are important and significantly affect NCDOT's maintenance process. A shortcoming in previous NCDOT models was the assumption of a 100% deterioration rate if no improvements were made in any year. This work, based on an exceptionally robust data set, provides a realistic deterioration rate based on five years of asset performance. These rates can be used by other state DOTs to reasonably predict future conditions of unpaved shoulders as well as to develop condition-budget models. The methodology can be applied to each state's unique data to evaluate local deterioration rates. The results obtained can be used as very good guidelines for deterioration rates, especially given the MMS basis for the data.

Acknowledgments

The authors would like to thank Lonnie Watkins and Matthew Whitley of the NCDOT for their help regarding the data collection and analysis for this study as well as for their insight and guidance.

References

1. Maji, A., and M. Jha. Modeling Highway Infrastructure Maintenance Schedules with Budget Constraints. *Transportation Research Record: Journal of the Transportation Research Board*, 2007. 1991: 19–26.
2. Thompson, P. D., K. M. Ford, A. Mohammad, S. Labi, A. Shirolé, and K. Sinha. *Guide for Estimating Life Expectancies*

of Highway Assets. NCHRP Project 08-7. Transportation Research Board of the National Academies, Washington, D.C., 2011.
3. Alyami, Z., and S. Tighe. Development of Maintenance and Rehabilitation Program: Pavement Assets under Performance-Based Contracts. *Transportation Research Record: Journal of the Transportation Research Board*, 2013. 2361: 1–10.
4. Durango, P. L., and S. M. Madanat. Optimal Maintenance and Repair Policies in Infrastructure Management Under Uncertain Facility Deterioration Rates: An Adaptive Control Approach. *Transportation Research Part A*, Vol. 36, 2002, pp. 763–778.
5. Madanat, S., M. Karlaftis, and P. McCarthy. Development of Probabilistic Infrastructure Deterioration Models with Panel Data. *Journal of Infrastructure Systems*, Vol. 3, No. 1, 1997, pp. 4–9.
6. Thompson, P. Estimating Asset Deterioration and Life Expectancy by Using Levels of Service. *Transportation Research Record: Journal of the Transportation Research Board*, 2012. 2285: 19–26.
7. Bardaka, E., J. E. Haddock, and S. Labi. Using Enhanced Econometric Techniques to Verify the Service Life of Asset Interventions. *Transportation Research Record: Journal of the Transportation Research Board*, 2014. 2431: 16–22.
8. Madanat, S., S. Bulusu, and A. Mahmoud. Estimation of Infrastructure Distress Initiation and Progression Models. *Journal of Infrastructure Systems*, Vol. 1, No. 3, 1995, pp. 146–150.
9. Richard, W. P. Durability, Maintenance and the Price of Used Assets, *Economic Inquiry*, Vol. 17, No. 2, 1979, pp. 197–217.
10. Falls, L. C., R. Haas, and S. Tighe. Asset Service Index as Integration Mechanism for Civil Infrastructure. *Transportation Research Record: Journal of the Transportation Research Board*, 2006. 1957: 1–7.
11. de la Garza, J. M., and D. A. Krueger. Simulation of Highway Renewal Asset Management Strategies. In *Proceedings of the International Workshop on Computing in Civil Engineering* (L. Soibelman, B. Akinci, and H. Shon, eds.), *American Society of Civil Engineers Pittsburgh, PA*, 2007.
12. Howell, R., and M. Whitley. *Maintenance Management System User Manual*. North Carolina Department of Transportation, Raleigh, NC, 2012.
13. Ott, R. L., and M. T. Longnecker. *An Introduction to Statistical Methods and Data Analysis*. Nelson Education, Toronto, Ontario, Canada, 2015.
14. Bernard, P., C. Ley, C. Leys, L. Licata, and O. Klein. Detecting Outliers: Do Not Use Standard Deviation Around the Mean, Use Absolute Deviation Around the Median. *Journal of Experimental Social Psychology*, Vol. 49, No. 4, 2013, pp. 764–766.

The Standing Committee on Maintenance and Operations Management (ADH10) peer-reviewed this paper (18-00004).

The opinions expressed in this paper are those of the authors and do not necessarily reflect the views of the NCDOT.

Article

TRR
JOURNAL OF THE TRANSPORTATION RESEARCH BOARD

Transportation Research Record
2018, Vol. 2672(12) 187–197

DOI: 10.1177/0361198118792327
journals.sagepub.com/home/trr
SAGE

Developing a Pre-Task Safety Briefing Tool for Kentucky Maintenance Personnel

Zamaan Al-Shabbani[1], Roy Sturgill[2], and Gabriel B. Dadi[1]

Abstract

The dangerous work environment in the construction industry and the inherent high risks associated with its work make it the focus of safety training and regulations. Highway construction and maintenance has unique hazards but seemingly less directly applicable safety standards, regulations, and programs. Department of Transportation (DOT) employees working in highway maintenance are exposed to a variety of unique hazards specifically associated with their work and not relating to the adjacent traffic. Highway site safety does not receive sufficient attention in terms of safety research and programs. As part of the efforts of the Kentucky Transportation Cabinet (KYTC) to improve safety for their employees, this paper describes the data-driven design of a pre-task safety briefing tool. By analyzing previous incident data of KYTC maintenance workers and identifying frequent hazards present within their typical work operations and the reasons behind such hazards, the final product of this study is a tool that is relatable and relevant to KYTC maintenance workers. The tool presents these hazards along with the associated reasons and the appropriate safety practices to avoid or mitigate the associated risk. The data analysis of this study highlighted that human factors and ergonomics play a significant role in the injuries related to the subject population, yet there is little guidance or standards for addressing these factors. The goal of this safety tool is to improve the safety performance of KYTC maintenance crews by increasing workers' safety awareness.

The construction industry is well known for its dangerous nature. Within the larger construction industry, highway construction and maintenance has its own unique hazards and risks. Work adjacent to high-speed passing traffic, large construction and maintenance equipment, massive amounts of material movement, and extreme environmental conditions are some of the different types of hazards found at highway work sites.

Safety is the responsibility of all project stakeholders from top management to the labor workforce. In highway construction and maintenance work, this responsibility extends from the field staff to the executives of the state departments of transportation (DOTs). Workers, supervisors, and everyone on the jobsite should be well trained and knowledgeable to recognize potential hazards when present. At the management level, leadership should express a commitment to provide a safe work environment for everyone with the ultimate goal of providing a safe environment with zero fatalities and injuries. However, this goal is not easy to achieve or maintain and today's statistics express a different reality. Although state DOTs have taken significant steps to improve safety performance and provide a safe working environment for their employees, the statistics show that highway construction and maintenance is still characterized by high rates of injuries and fatalities. According to the Bureau of Labor Statistics, there were 844 worker fatalities in road work zones between 1995 and 2002 (*1*), 962 fatalities between 2003 and 2010 (*2*), and 609 fatalities between 2011 and 2015 (*3*). Road-worker fatalities consistently accounted for 2% of all work-related fatalities in the nation, with no descending trend (*4*). These numbers do not capture those workers who survived or close calls. Therefore, the industry is still far from achieving its safety goal.

Kentucky is one such state that manages the safety of their unique transportation system and has relatively high work zone incident rates. The Kentucky Transportation Cabinet (KYTC), an executive branch agency, manages over 27,000 miles of this transportation system. KYTC is responsible for supervising the development and maintenance of the transportation system across the commonwealth. With more than 2000 maintenance employees, KYTC regularly performs a variety of

[1]University of Kentucky, Lexington, KY
[2]Kentucky Transportation Center, Lexington, KY

Corresponding Author:
Address correspondence to Roy Sturgill: roy.sturgill@uky.edu

maintenance operations across the state. This exposes their maintenance workers to different types of hazards that could easily lead to incidents. KYTC continues to work on improving their safety performance to provide a safe working environment and to protect their employees and the public. In fact, KYTC has achieved significant improvement in their safety performance. In a July 2016 safety report, KYTC reported that their total recordable incidents rate (TRIR) declined from 5.5% to 4.2% since 2010, total days away from work declined 20%, worker compensation claims declined 30%, and the total number of claims decreased by 43%. In addition to the safety performance improvement, a significant reduction of \$2.5 million per year was achieved by reducing spending on worker compensation claims from \$4 million to \$1.5 million per year (*5*). However, KYTC realized that their safety performance does not meet their desired goals. While KYTC states a goal of a TRIR of 5% or less, their latest report revealed that their TRIR is higher than 15% when considering the appropriate population and focusing more appropriately on field staff and removing office work hours.

The safety culture within highway construction, especially in the region, leaves much room for improvement. Complacency abounds in an environment full of uncontrollable hazards. Other states and regions have lower incident rates. For example, due to the adoption of new safety culture and setting a goal of zero incidents rate ten years ago, the DOT in the state of Texas, which has approximately 6500 maintenance employees, achieved significant improvements in safety performance reaching 1.35 Occupational Safety and Health Administration (OSHA) Incident Rate. In 2015, the total number of injuries in maintenance crews of Texas DOT was 168 compared with 316 injuries in KYTC maintenance crews inferring there is a learning opportunity for Kentucky's highway industry.

Previous studies have been conducted to address safety issues for KYTC employees. In a study conducted at the Kentucky Transportation Center (KTC), Hopwood and Palle (*6*) provided a review of KYTC safety issues related to construction activities for both KYTC and contractor personnel. After interviewing KYTC resident engineers, as well as surveying and interviewing district safety coordinators, the authors recommended creating new safety programs and training, partnering with contractors and the Kentucky Department of Labor, promoting changes in KYTC policies, and increasing the role and support of safety coordinators. Another study conducted by Hancher et al. (*7*) sought to address research questions with a focus on the hazards associated with work zones and vehicular traffic. In this study, using surveys and focus group discussions, the authors addressed safety concerns and identified best safety practices for highway maintenance workers. The survey sought ideas and feedback from KYTC and private highway construction and maintenance workers. Based on feedback, recommendations were provided, such as closed cab tractors for mowers, LED stop signs in work sites, and additional lighting for nighttime work. These projects highlighted the need for safety education and training and helped to inform and refine the current study.

As part of the efforts of the KYTC to improve safety for their employees, this paper describes the data-driven design of a pre-task safety-talk tool. This tool is based on previous incident data of KYTC maintenance workers making it relatable and true to form for KYTC. The tool is intended to be used prior to any workday task among ten of the typical work operations of KYTC maintenance employees found in their Field Operations Guide (*8*). The goal of this safety tool is to improve the safety performance of KYTC maintenance crews by increasing workers' awareness of potential hazards to expect at the work site and introducing safety controls to be practiced to prevent or minimize the possibility of encountering such hazards. There are three root causes behind work incidents including failure to identify unsafe condition, failure to accurately perceive the risk, and/or deciding to act unsafe regardless of the identified hazards and perceived risk (*9*). Using this tool will help KYTC maintenance employees to identify usafe conditions present within the context of their work operations and weight the associted risk with such conditions. This in turn could lead to or promote safe behavior among workers. In addition, the tool can be used as a part of the larger systematic approach to address safety culture by educating workers to reduce the unsafe behavior. Unlike previous safety programs that provide general safety education, this tool provides safety education specific to the task to increase workers' awareness of potential frequent hazards in their daily work. As this study highlights the lack of safety standards, policies, and regulations targeting highway maintenance workers, it does not represent an additional study in work zone traffic control safety, as attention has been focused on this area. The interest of this study is the safety practices of those that perform highway maintenance work with a focus on OSHA and Kentucky Occupational Safety and Health Office (KYOSH) standards. This rationale comes from four of the five worker fatalities occurring in KYTC work zones in the past several years being work related and not traffic related. In addition to the development of the tool, this study sheds light on two main issues: the lack of safety policies, practices, and regulations targeting highway maintenance workers; and the involvement of human factors and ergonomics as significant contributors to the high rate of incidents in this unique workforce.

Objectives

The purpose of this study is to improve safety performance of KYTC maintenance workers by increasing workers' awareness of the potential hazards present within their work operations and presenting best practices to prevent or minimize the possibility of incidents associated with such hazards.

The following tasks were undertaken to achieve the objective:

1. Review available literature and other DOT safety programs to identify similar effective safety measures that could be of use in designing a safety tool;
2. Analyze historical health and safety data (data provided was from 2005 to 2015) of KYTC maintenance workers to understand the nature of hazards, incidents, their consequences, and potentially the incident causes;
3. Identify available safety best practices to address the hazards identified from the data analysis; and
4. Design a pre-task safety tool that can be used prior to the start of any workday operation to increase workers' awareness of the potential hazards and preventative safety controls.

The previous tasks led to the development and delivery of the pre-task safety tool described in following sections of this paper. The implementation of this tool will potentially be monitored through future research efforts.

Federal Safety Standards and Regulations

At the federal level, nationwide policies and standards are imposed to minimize risks for the public and workers in roadway construction and maintenance work zones. The Manual on Uniform Traffic Control Devices (MUTCD) presents the mandatory work zone practices (*10*). These standards include guidance on areas such as the setup of temporary traffic controls in work zones. The focus of federal standards associated with highway construction and maintenance is toward controlling traffic in work zones due to the high number of fatalities associated with vehicular accidents. This focus on vehicle worker interaction to reduce workers' and drivers' risk in highway work zones is a trend in highway construction safety research in general. The OSHA presents standards and regulations for the overall construction industry with little specifics on highway operations. These standards also apply to state DOTs but do not always cover the entirety of the work involved in highway construction and maintenance. Twenty-one states have developed individual state specific safety and health plans that cover local and state government workers, including DOT employees (*11*). In Kentucky, this body is the KYOSH which is still generalized when considering highway construction and maintenance. It is apparent that there are minimal federal safety standards and practices directly addressing the safety of highway construction and maintenance workers. As a result, and due to the uniqueness of each state transportation system, state DOTs often develop their own safety programs and practices.

Effective Safety Practices in Other States

In an effort to improve safety performance, state DOTs have dedicated numerous resources and implemented a variety of safety controls to create a safe work environment for their employees. Based on studying and analyzing the history of worker injuries and fatalities, state DOTs seek to understand the nature of safety hazards, the root causes that led to incidents, the types of incidents, the associated results, and what could be done to prevent such incidents. Therefore, reporting and archiving becomes a core pillar in most safety improvement programs. In any safety initiative of this type, data on injuries and fatalities becomes the main source to understanding safety issues and creating or improving safety controls. Data-driven safety controls have proven to be effective in different states across the country.

A recent National Cooperative Highway Research Program (NCHRP) synthesis study reviewed existing state DOT safety programs with interviews of six states to identify the effective safety controls and programs across the nation. The six states interviewed were California, Maine, North Dakota, Oregon, South Carolina, and Washington. The states' safety programs were explored and discussed to find examples of safety improvements and highlight elements of these programs viewed as effective (*11*). The safety programs varied because each state DOT had developed their safety program to address their specific conditions.

For example, California has its "Design for Safety Initiative," which is a data-driven program. This program focuses on using data to identify safety aspects requiring improvement and to inform landscape architects and engineers of such issues, so they might address them in the design phase. This is a proactive measure that helps to minimize the risk of potential hazards in work zones. This program also seeks ideas and feedback from maintenance workers and communicates these ideas to designers to help them better understand which actions need to be taken to improve safety performance (*12*). This program resulted in high continuous support of leadership, which increases the overall program fund from \$1.9 million in 2010 to \$90 million in 2017 (*11*). This increase in funding and support indicates that leadership realized the benefits of such initiatives. Another advantage of this program is the change in design and the modification in the construction plans not only to improve safety but also to save construction and maintenance expenses.

Maine initiated the "Safety Idea Incentive Program" in 2012 in which DOT employees are encouraged to participate as crews in the development of the safety program through safety discussions. Every month, the safety ideas are collected and evaluated to determine which idea is the most valuable and applicable. Every member of the winning crew receives 50 points, which is equivalent to $25. This program continued from 2012 to 2014 and resulted in many safety best practices which have been implemented with less intervention from management since the ideas come with buy-in from the work crews. Although this program lasted for three years only, it led to many safety ideas that were implemented, such as creating the poisonous plant pocket guide, providing the tick removal kits with the first-aid kits, and marking the sidewalks at snow and ice drop zones (*11*).

North Dakota created the "Leading Indicators Initiative," in which the state DOT recorded and analyzed different leading indicators, such as employee participation in self-inspections, first-aid training, or employee suggestion programs. Other leading indicators, such as employees' participation in safety audits were evaluated through employee activity in safety programs. The main objective of this program is to adopt a new safety culture that looks for proactive controls other than counting the consequences and evaluating lagging indicators. One advantage of this program was a 50% reduction in insurance premium over the last six years (*11*).

One common aspect between these practices is a data-driven decision approach. Most existing state programs are based on some combination of past health and safety data and feedback or ideas from employees. This highlights the importance of data accuracy in future decision making. In fact, data-driven decision approaches have been emphasized at the federal level. FWHA has made data-driven decision approaches a policy priority (*11*). This policy is evidenced by FHWA developments in the Highway Safety Improvement Program (HSIP), which as part of the Fixing America's Surface Transportation (FAST) Act became a much more data-driven decision program. Regardless of the effectiveness of the safety programs and practices, it can be concluded that data-driven decision making is an effective approach to designing safety controls to improve safety performance, as in the case of this study.

Methodology

This study seeks to develop a tool with the intent of improving safety in KYTC maintenance operations. The expectation was that analysis of the recordable incident data would present areas of safety need. The literature review would provide best practices to mitigate those safety needs. The tool would present this information in a clear and concise form such that employees would begin the workday knowledgeable of the typical hazards present for the operations to be performed, and they would be equipped with knowledge of effective practices to mitigate those hazards. To achieve this, the study methodology included collection and in-depth analysis of KYTC incident data relative to maintenance operations, organizing this according to typical maintenance operations, and then organizing best practices according to the hazards present for typical operations. These steps are further defined in the following sections and were amalgamated to form a comprehensive tool for implementation with KYTC maintenance employees.

Data Collection

KYTC works with the Risk Management Services Company to collect and track recordable incidents and their associated severity, lost time, and costs. KYTC recordable incident data was collected for a ten-year period (2005–2015). This data, while including incidents for all KYTC employees, was pared down to only KYTC maintenance employee incidents. The data was categorized according to the code of National Council of Compensation Insurance (NCCI) and included a total of 3876 claims. The available information included claimant information, such as age and work title, and incident information, such as the location of incident, the date and time, causes of incident, and an incident description. Some of the data entries were missing information. Where possible, the data was reviewed such that it could be included in the following analysis, but a few records had to be removed due to lack of detail. The remaining data, provided in Microsoft Excel, still included over 3000 records. The data detail was a leading factor in recommending an improved process and requirements for incident data collection and reporting.

Data Analysis

The data was analyzed through three phases in working toward the development of a safety tool. The data was first analyzed for trends regarding the maintenance operations during which the incident occurred. These incidents were further analyzed to draw out the main causes of the incidents within the separate operations. Finally, effective safety practices were aligned with these causes and incidents to present methods to mitigate the hazards present.

Phase I: Incident Categorization by Maintenance Operation. According to the KYTC Field Operations Guide for Maintenance, KYTC has 17 maintenance work operations. The purpose of this phase was to categorize the incident data according to these 17 work operations. Doing so would work toward the design of a tool that would provide best practices to KYTC maintenance workers according to their maintenance work operations. Unfortunately, there was no systematic reference within the incident data that drew a link between the claim and the work operation it belonged to. Therefore, the research team manually categorized the injuries according to the work operations through a careful examination of the information provided within the incident data, such as the incident description and causes, incident location, and claimant work title. The incident description and causes were especially helpful in this intensive data analysis.

Due to the missing information in some of the claims and the lack of accuracy regarding the collection and recording of the data, there appeared to be data available only aligning with ten operations of the 17 KYTC typical maintenance work operations shown in Table 1. The seven remaining operations include herbicide and pesticide application, landslide repairs, rocks fall and sinkhole work, structural repairs, roadside landscaping, ditching and ditch cleaning, and traffic control maintenance. The lack of information in the data for the remaining seven operations resulted in the tool not currently being applicable to them. Again, this highlights the need to further standardize incident recording and reporting practices and potentially establishing "operation" as a field for incident data.

Through categorization of the data, the research team was able to convert some of the qualitative aspects of the incident data into quantitative data and enable the determination of the frequency of incidents within each of the ten selected operations. In addition, categorizing the data paved the way to identify the top frequent causes behind incidents within each work operation.

Table 1. Selected KYTC Maintenance Work Operations

Selected KYTC typical maintenance work operations
1. Concrete repairs and Bridge maintenance
2. Equipment maintenance
3. Guardrail maintenance
4. Litter and debris removal
5. Mowing
6. Pipe/drain clearing and replacement
7. Road and Shoulder repairs
8. Sign inventory and replacement
9. Snow and ice removal
10. Tree and brush trimming

Phase II: Identification of Leading Incident Causes. After the data categorization and organization according to the ten applicable maintenance operations, the research team worked to identify the most frequent causes leading to the incidents within each of the selected operations. These causes were classified according to NCCI codes. Some similar causes, such as lifting, holding or carrying, object being lifted, etc. were combined since they represent similar activities and are related to similar injuries. Microsoft Excel was used to identify frequent causes according to the NCCI code followed by manual identification of the causes to confirm the results. Within each of the selected operations, five to six of the most frequently cited causes were identified. Collectively, 12 causes were identified as the most frequent causes for all ten of the operations and data available. These causes and their frequency over the data period (2005–2015) are shown in Table 2.

Some of these causes, such as lifting, appear in all the operations; lifting ranks at the top of this list with the highest frequency (64.1 incidents per year). Many of the incidents appear to be caused by human factors and ergonomics. For example, most of the "fall from a different level" incidents were caused by either exiting a vehicle or getting in or out of a truck bed. In these incidents, human factors, such as expectancy, are important contributors to the behavior of the workers. This is a critical highlight of this study since there are minimal practices that address human factors and ergonomics for such scenarios. OSHA requires workers to use fall protection when working from a height of more than 6 ft. However, most of the fall incidents in this study occurred from heights less than 6 ft. These incidents cause varying types of injuries with varying levels of severity. While OSHA guidance may be effective in preventing more severe incidents at heights greater than 6 ft, the findings of this study suggest a need for addressing the human factors and ergonomics aspect possibly through considerations in the design phase.

Table 2. Overall Top Frequent Incident Reasons

Top frequent incident causes	Frequency
1. Object being lifted or handled	641
2. Falling or flying objects	313
3. Fall from different level	239
4. Hand tool or machine in use	189
5. Pushing or pulling	149
6. Foreign matter in the eye	143
7. Fall on ice or wet floor	141
8. Chemicals, liquids, or vapors	115
9. Vehicle upset	106
10. Animal or insect	99
11. Stationary or sharp objects	62
12. Hot object and temperature extreme	47

Phase III: Identifying Safety Best Practices. As previously mentioned in the literature review, each state develops its own safety programs to address issues in their own unique transportation system. In addition, federal regulations and standards do not offer much to address the needs found in this study. There seems to be a sizable knowledge gap in standardized effective safety practices directly applicable to specific highway maintenance operations, especially for addressing human factors and ergonomics. To compensate for this, the research team reviewed and collected safety practices resources according to the following criteria:

1. Practices addressing specific frequent safety issues (operation specific when possible);
2. Safety resources produced through academic research;
3. Safety practices presented in industry guidance; and
4. Government regulations and standards.

After outlining the above criteria for collecting effective safety practices, several resources were identified including the following:

- 29 CRF 1926 OSHA regulations
- OSHA Ergonomics E-tool
- Safety tool kits from Kansas State University
- Safety tool kits from the University of New Hampshire
- Safety tool kits from the University of Washington
- Roadway Safety training program
- Construction solutions by CPWR
- NIOSH standards
- Other states' employee health manuals
- Others

Once these resources were reviewed and practices were collected, practices were consolidated, and the details were condensed to abbreviate the presentation within the safety tool. In other words, the intent of the safety tool is not to be a training mechanism of the safety practices but to present practices in a quick and fundamentally complete manner such that use of the tool would be effective but not time consuming.

Design of the Safety Tool

The final product and central deliverable of this study is a pre-task safety tool that could be reviewed prior to the workday's planned operations to highlight potential hazards and mitigation measures for maintenance forepersons and supervisors. The forepersons and supervisors could then use the presented information to prepare for a customized pre-task safety talk catered to the workday's operation, specific hazards present, how those hazards have historically injured KYTC maintenance workers, and practices to mitigate those hazards. The purpose of using the tool is to increase workers awareness of the potential hazards related to the work they are preparing to do and suggest safety practices to avoid or minimize the potential risk of these hazards.

The tool was designed with the following considerations in mind:

1. Be simple and intuitive to use and understand;
2. Be quick to complete;
3. Address the top frequent potential hazards;
4. Expandable to all maintenance operations; and
5. Be adjustable so it can be improved based on safety performance.

A Microsoft Excel macro-enabled spreadsheet was used to design the tool since it is easy to use and accessible to KYTC employees. Incompatibility and platform accessibility concerns also steered the research team away from a mobile device platform and web-based applications.

Throughout the design process, different display design and cognitive principles were considered to enhance the usability of the tool. For example, to reduce visual and motor work and to maintain simplicity, the number of buttons was minimized. This in turn is expected to provide an easy-to-use and pleasant interface (*13*). To improve navigation and flow, information and elements on screen were organized rhythmically, guiding the user's eye orientation through the display. Main navigation buttons, such as "Main Menu", "Back", etc. were placed at similar locations throughout the tool to maintain predictability and consistency.

The tool consists of three main sections beside the introductory instructional sheet. The opening sheet includes brief instructions on how to navigate throughout the tool (Figure 1). The first section of the tool is the main selection page that includes the ten maintenance operations along with the overall incidents statistics of KYTC maintenance employees for a ten-year period (Figure 2). The main purpose of this section is to select the desired operations from the list of typical operations. The second section is more specific to the operation and includes a statistical summary of previous injuries of KYTC maintenance employees within each operation and the associated top frequent incident causes (Figure 3). This section helps users to understand the trend and frequent causes of incidents within each operation in order to place more focus on specific safety issues when developing their safety talks. This also helps

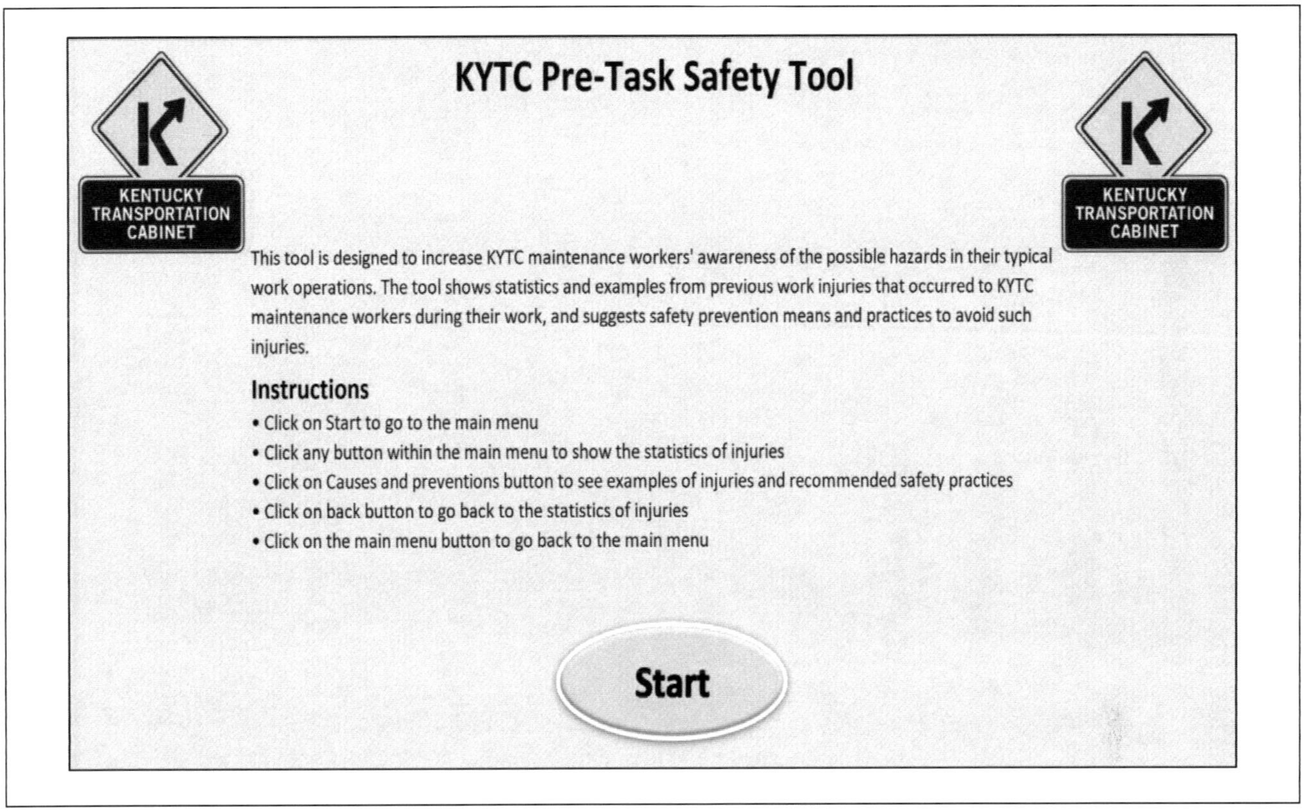

Figure 1. Introductory user form of the tool.

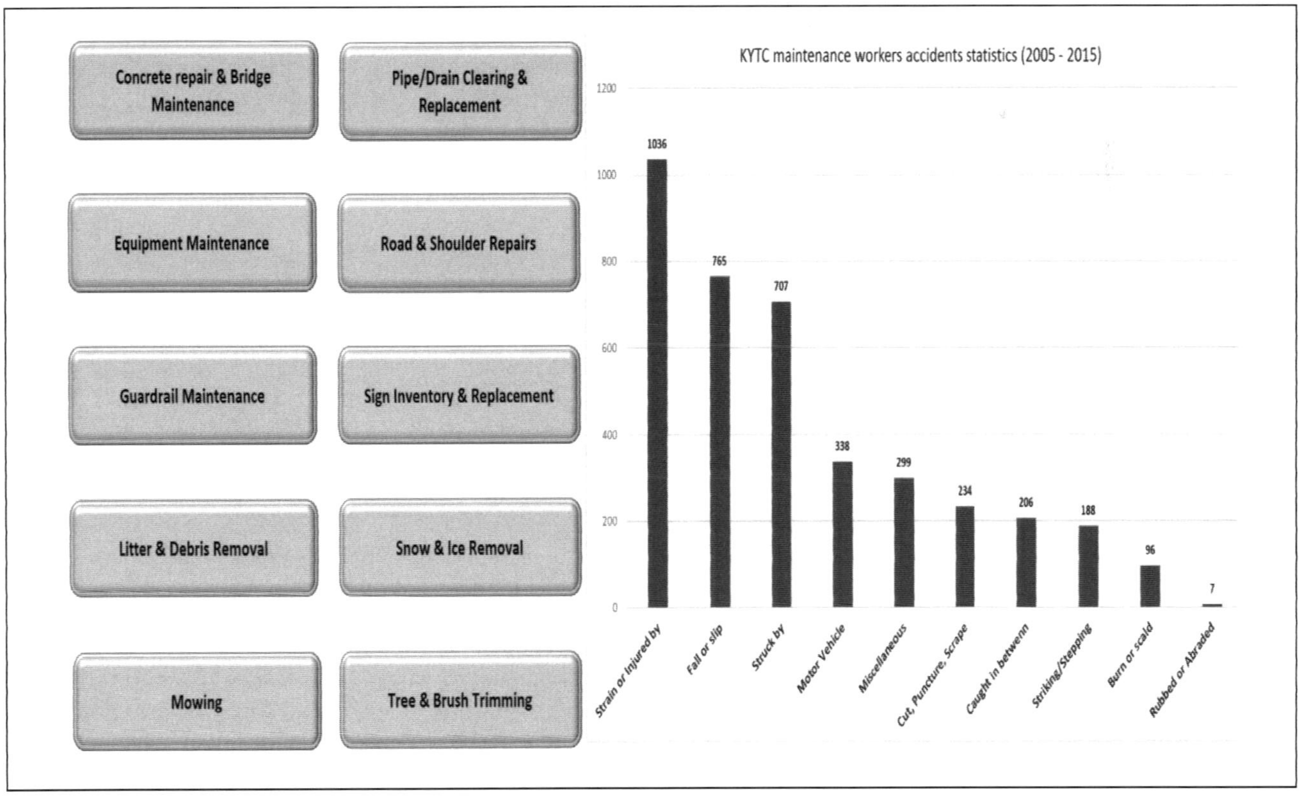

Figure 2. Operation selection and total incident frequency.

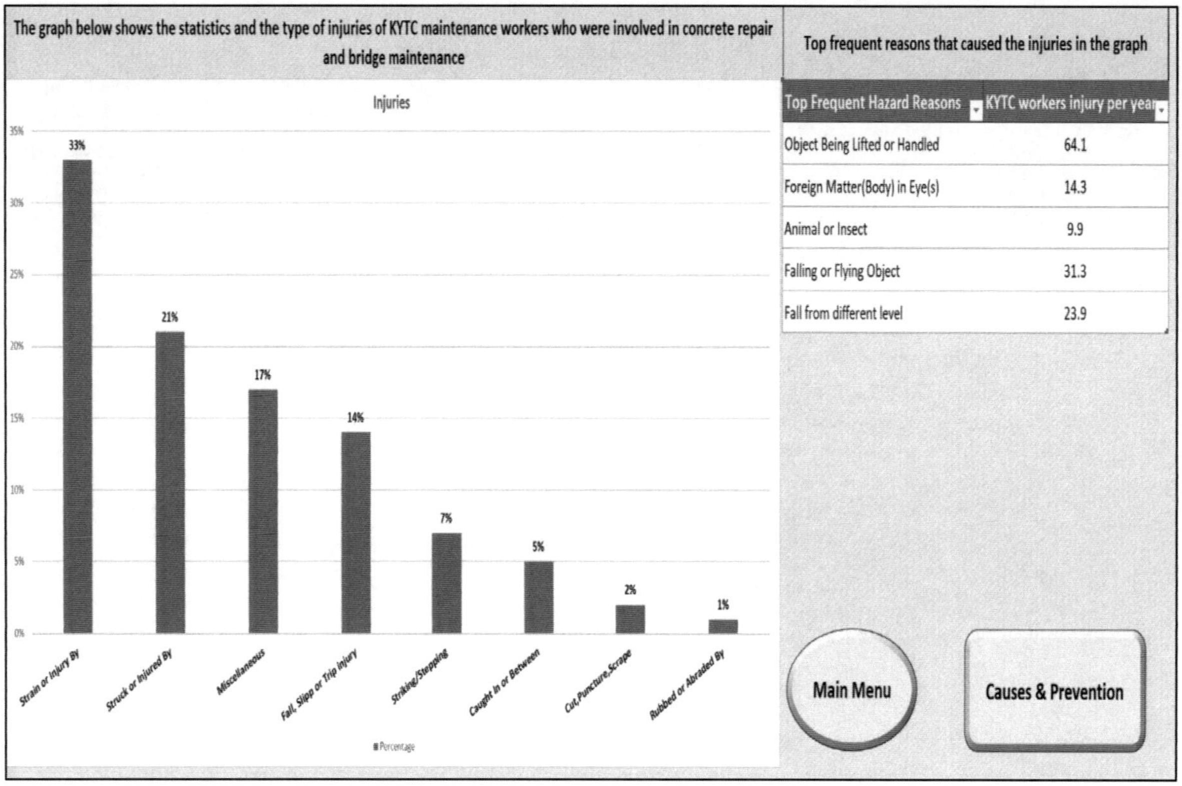

Figure 3. Statistics of injuries and associated reasons for KYTC maintenance employees.

superintendents and forepersons in identifying what to look for at the work site to make it safer. The third section of the tool presents examples of previous work incidents for each of the frequent causes and the suggested safety practices to minimize or eliminate the associated risk (Figure 4). In accordance with OSHA recommendations, the written language of each section is not complex and can be easily understood by users. In addition, the tool includes some pictorial demonstrations for practices potentially requiring additional explanation (Figure 5) with the goal of increasing workers' understanding and reducing the verbiage and time required to use the tool. These sections are formatted to fit an individual's computer screen without the need for scrolling. Coloring and organization of each section are kept similar throughout the tool to maintain consistency, regularity, balance, and unity. Adopting such display design and cognitive principles makes the tool aesthetically pleasing and reduces visual clutters (*13*).

Conclusion

Highway maintenance has unique hazards, but seemingly less directly applicable safety standards, regulations, and programs compared with the whole construction industry. This work is dangerous due to the proximity of work sites to the passing traffic. In addition, this work is normally performed outside and occasionally at nighttime. DOT employees and contractors working in highway maintenance are exposed to a variety of unique hazards specifically associated with their work and not relating to the adjacent traffic. Highway site safety does not receive sufficient attention in terms of safety research and programs. Additionally, the data analysis of the study highlighted that human factors and ergonomics play a role in the injuries related to the subject population, yet there is little guidance or standards for addressing these factors. This study attempted to focus on the work-related hazards present specifically for highway maintenance workers.

By analyzing data associated with incidents involving KYTC maintenance employees over a period of ten years (2005–2015), this study was able to develop a pre-task safety tool applicable to ten different operations these employees perform. Using this tool will assist supervisors in communicating safety concerns and increasing KYTC maintenance workers' awareness of the potential hazards in their work environment. Additionally, the tool suggests safety practices specific to their work task to prevent or minimize the risk of such hazards. The design of

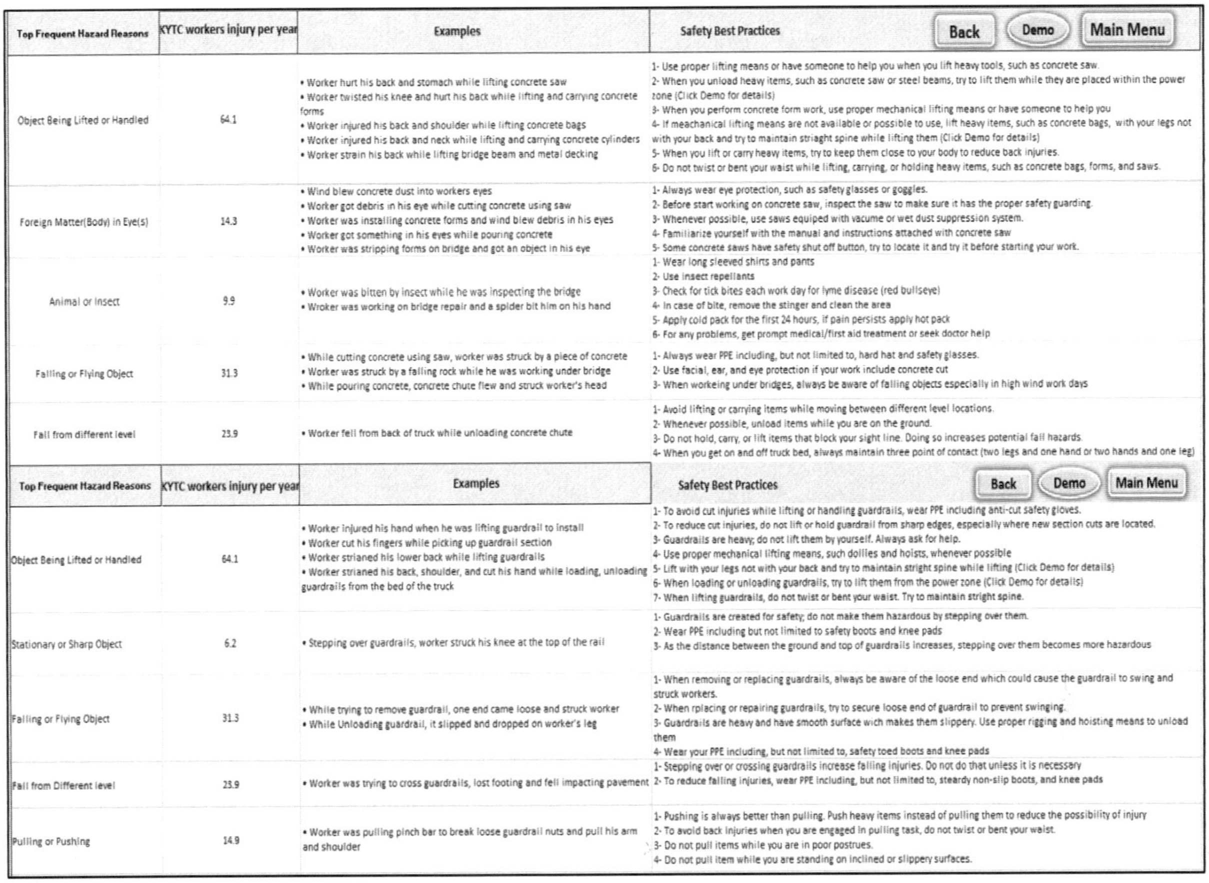

Top Frequent Hazard Reasons	KYTC workers injury per year	Examples	Safety Best Practices Back Demo Main Menu
Object Being Lifted or Handled	64.1	• Worker hurt his back and stomach while lifting concrete saw • Worker twisted his knee and hurt his back while lifting and carrying concrete forms • Worker injured his back and shoulder while lifting concrete bags • Worker injured his back and neck while lifting and carrying concrete cylinders • Worker strain his back while lifting bridge beam and metal decking	1- Use proper lifting means or have someone to help you when you lift heavy tools, such as concrete saw. 2- When you unload heavy items, such as concrete saw or steel beams, try to lift them while they are placed within the power zone (Click Demo for details) 3- When you perform concrete form work, use proper mechanical lifting means or have someone to help you 4- If meachanical lifting means are not available or possible to use, lift heavy items, such as concrete bags, with your legs not with your back and try to maintain striaght spine while lifting them (Click Demo for details) 5- When you lift or carry heavy items, try to keep them close to your body to reduce back injuries. 6- Do not twist or bent your waist while lifting, carrying, or holding heavy items, such as concrete bags, forms, and saws.
Foreign Matter(Body) in Eye(s)	14.3	• Wind blew concrete dust into workers eyes • Worker got debris in his eye while cutting concrete using saw • Worker was installing concrete forms and wind blew debris in his eyes • Worker got something in his eyes while pouring concrete • Worker was stripping forms on bridge and got an object in his eye	1- Always wear eye protection, such as safety glasses or goggles. 2- Before start working on concrete saw, inspect the saw to make sure it has the proper safety guarding. 3- Whenever possible, use saws equiped with vacume or wet dust suppression system. 4- Familiarize yourself with the manual and instructions attached with concrete saw 5- Some concrete saws have safety shut off button, try to locate it and try it before starting your work.
Animal or Insect	9.9	• Worker was bitten by insect while he was inspecting the bridge • Wroker was working on bridge repair and a spider bit him on his hand	1- Wear long sleeved shirts and pants 2- Use insect repellants 3- Check for tick bites each work day for lyme disease (red bullseye) 4- In case of bite, remove the stinger and clean the area 5- Apply cold pack for the first 24 hours, if pain persists apply hot pack 6- For any problems, get prompt medical/first aid treatment or seek doctor help
Falling or Flying Object	31.3	• While cutting concrete using saw, worker was struck by a piece of concrete • Worker was struck by a falling rock while he was working under bridge • While pouring concrete, concrete chute flew and struck worker's head	1- Always wear PPE including, but not limited to, hard hat and safety glasses. 2- Use facial, ear, and eye protection if your work include concrete cut 3- When workeing under bridges, always be aware of falling objects especially in high wind work days
Fall from different level	23.9	• Worker fell from back of truck while unloading concrete chute	1- Avoid lifting or carrying items while moving between different level locations. 2- Whenever possible, unload items while you are on the ground. 3- Do not hold, carry, or lift items that block your sight line. Doing so increases potential fall hazards. 4- When you get on and off truck bed, always maintain three point of contact (two legs and one hand or two hands and one leg)

Top Frequent Hazard Reasons	KYTC workers injury per year	Examples	Safety Best Practices Back Demo Main Menu
Object Being Lifted or Handled	64.1	• Worker injured his hand when he was lifting guardrail to install • Worker cut his fingers while picking up guardrail section • Worker strianed his lower back while lifting guardrails • Worker strianed his back, shoulder, and cut his hand while loading, unloading guardrails from the bed of the truck	1- To avoid cut injuries while lifting or handling guardrails, wear PPE including anti-cut safety gloves. 2- To reduce cut injuries, do not lift or hold guardrail from sharp edges, especially where new section cuts are located. 3- Guardrails are heavy; do not lift them by yourself. Always ask for help. 4- Use proper mechanical lifting means, such dollies and hoists, whenever possible 5- Lift with your legs not with your back and try to maintain stright spine while lifting (Click Demo for details) 6- When loading or unloading guardrails, try to lift them from the power zone (Click Demo for details) 7- When lifting guardrails, do not twist or bent your waist. Try to maintain stright spine.
Stationary or Sharp Object	6.2	• Stepping over guardrails, worker struck his knee at the top of the rail	1- Guardrails are created for safety; do not make them hazardous by stepping over them. 2- Wear PPE including but not limited to safety boots and knee pads 3- As the distance between the ground and top of guardrails increases, stepping over them becomes more hazardous
Falling or Flying Object	31.3	• While trying to remove guardrail, one end came loose and struck worker • While Unloading guardrail, it slipped and dropped on worker's leg	1- When removing or replacing guardrails, always be aware of the loose end which could cause the guardrail to swing and struck workers. 2- When rplacing or repairing guardrails, try to secure loose end of guardrail to prevent swinging. 3- Guardrails are heavy and have smooth surface wich makes them slippery. Use proper rigging and hoisting means to unload them 4- Wear your PPE including, but not limited to, safety toed boots and knee pads
Fall from Different level	23.9	• Worker was trying to cross guardrails, lost footing and fell impacting pavement	1- Stepping over or crossing guardrails increase falling injuries. Do not do that unless it is necessary 2- To reduce falling injuries, wear PPE including, but not limited to, steardy non-slip boots, and knee pads
Pulling or Pushing	14.9	• Worker was pulling pinch bar to break loose guardrail nuts and pull his arm and shoulder	1- Pushing is always better than pulling. Push heavy items instead of pulling them to reduce the possibility of injury 2- To avoid back injuries when you are engaged in pulling task, do not twist or bent your waist. 3- Do not pull items while you are in poor postrues. 4- Do not pull item while you are standing on inclined or slippery surfaces.

Figure 4. Examples of previous incidents and recommended best practices.

the tool was based on careful examination and analysis of incident data. After determining the leading causes of incidents within each maintenance operation and identifying the best practices to address these hazards, the final product of this study is an electronic tool that can be used by KYTC maintenance crews prior to any workday. The tool is simple to use and would ideally help prepare forepersons or supervisors for a pre-task safety talk specific to the workday's activities, their associated hazards, and specific measures for mitigating these hazards. This tool can be expanded to include all KYTC maintenance operations once data is available and can be improved based on safety performance.

Recommendations and Future Work

Due to incomplete or unavailable data, the tool does not cover all 17 typical maintenance operations as mentioned in the KYTC Maintenance Field Operations Guide. Therefore, one recommendation is to develop data collection standards that may report work operations as standardized within KYTC process manuals. Also, additional description of incidents should be considered such that a sufficient description of claim is available to adequately understand the happenings of the incident and the hazards that may need future mitigation. One possible area for investigation is to examine what other state DOTs use for data collection and archiving. This will help the KYTC in expanding and improving the tool and in evaluating safety performance as part of using the tool. Having accurate data will help in establishing a robust benchmarking and evaluation system for future safety programs. It also helps decision makers in KYTC to make more informed data-driven decisions.

Another area of recommended future effort is an assessment of the impact of the safety tool through a pre-test and post-test of maintenance crews using the tool. This study would highlight the hazard recognition abilities of the maintenance staff and if those abilities were improved through using the safety tool. The assessment

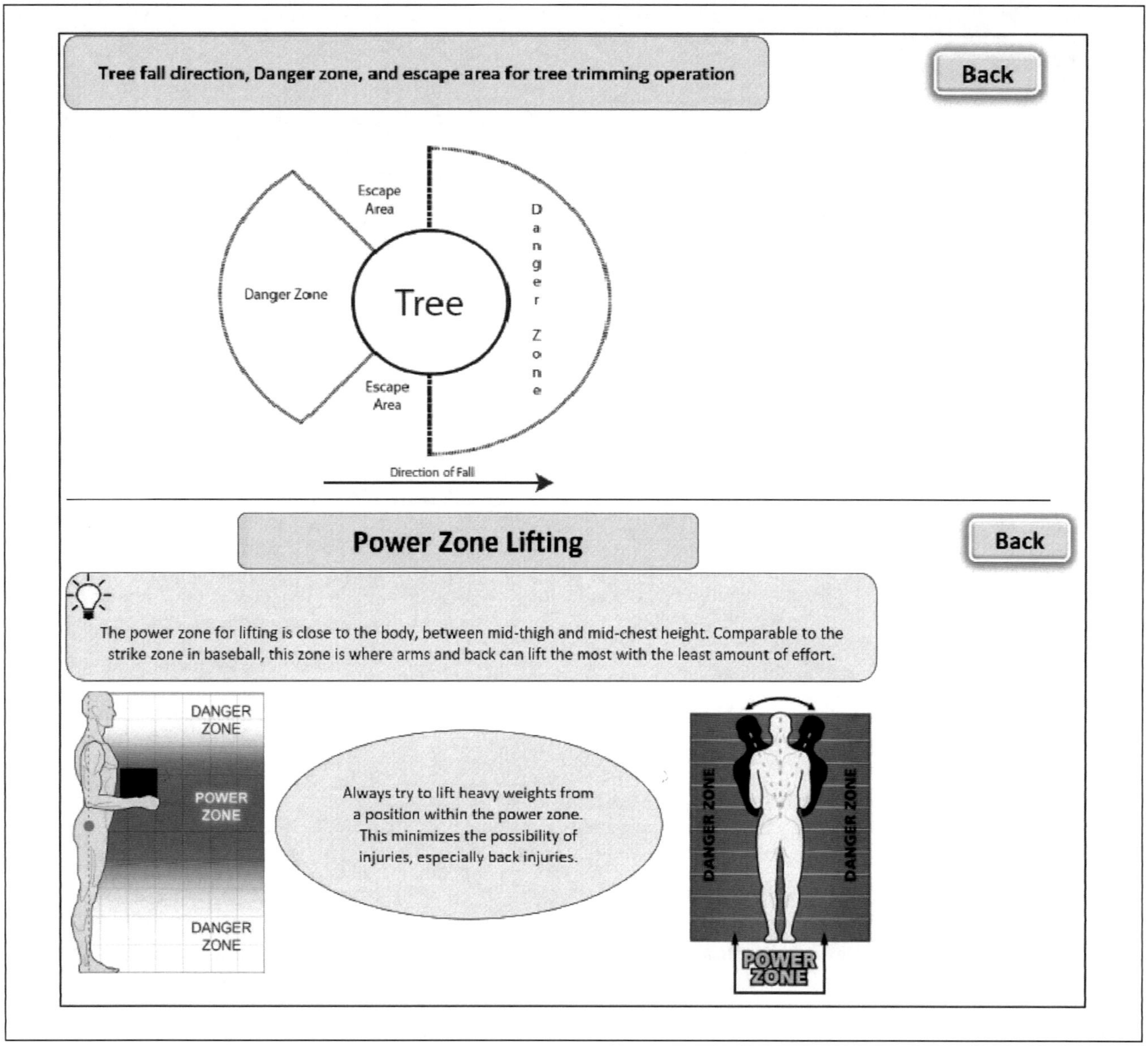

Figure 5. Demonstration of recommended lifting and tree-trimming practices.

would also help in improving and evaluating the effectiveness of the tool. As part of this study, the research could examine the relationship between the hazard recognition ability of maintenance crews and associated implementation of safe practices. This aspect of the study would help in determining the impact of safety programs on maintenance crews and the areas which need to be addressed. It also indicates safety value and strength with the specific audience of workers and reveals their priorities and behavior drivers. The culmination of this study would result in an evaluation of KYTC's maintenance safety culture and safety citizenship allowing for performance measurement and improvement.

Acknowledgments

The authors wish to thank the Kentucky Transportation Cabinet for their participation and support in this study. In addition, study advisory committee members for KYSPR 15-508 were critical to the project effort.

Author Contributions

The authors confirm contribution to the paper as follows: study conception and design: Sturgill, Dadi; data collection: Sturgill; analysis and interpretation of results: Al-Shabbani, Sturgill, Dadi; draft manuscript preparation: Al-Shabbani. All authors reviewed the results and approved the final version of the manuscript.

References

1. Pegula, S. Fatal Occupational Injuries at Road Construction Sites. Monthly Labor Review. *U.S. Bureau of Labor Statistics*, Vol. 127, No. 43, 2004.
2. Pegula, S. An Analysis of Fatal Occupational Injuries at Road Construction Sites, 2003–2010. Monthly Labor Review. *U.S. Bureau of Labor Statistics*, Vol. 136, No. 11, 2013. https://doi.org/10.21916/mlr.2013.36.
3. Bureau of Labor Statistics, U.S. Department of Labor. Fatal Injuries at Road Work Zones. *The Economics Daily*, 2017. http://www.bls.gov/opub/ted/2017/fatal-injuries-at-road-work-zones.htm. Accessed July 10, 2017.
4. Federal Highway Administration. *Work Zone Management Program*. Federal Highway Administration, Washington, D.C., 2016. https://ops.fhwa.dot.gov/wz/resources/facts_stats/safety.htm. Accessed July 5, 2017.
5. Hecker, R. *KYTC Annual Employee Safety Report and Recommendations*. Employee Safety and Health, Kentucky Transportation Cabinet, 2016.
6. Hopwood, T., and S. Palle. *Safety and Health Concerns for KYTC and Contractor Personnel*. Kentucky Transportation Center, University of Kentucky, 2004. http://uknowledge.uky.edu/ktc_researchreports/199. Accessed June 25, 2017.
7. Hancher, D. E., K. Bussey, R. Meagher, J. Ross, and K. Smith. *Improve Safety of Workers During Highway Construction and Maintenance*. Kentucky Transportation Center, University of Kentucky, 2007. http://uknowledge.uky.edu/ktc_researchreports/149. Accessed May 20, 2017.
8. Kentucky Transportation Cabinet. *Maintenance Field Operations Guide*. Kentucky Transportation Cabinet Policy Manuals, 2016.
9. Abdulhamid, T. S., and J. G. Everette. Identifying Root Causes of Construction Accidents. *Journal of Construction Engineering and Management*, Vol. 126, No. 1, 2000, pp. 52–60.
10. Federal Highway Administration. *Manual on Uniform Traffic Control Devices (MUTCD)*. Federal Highway Administration, Washington, D.C., 2009.
11. National Academies of Sciences, Engineering, and Medicine. *Highway Worker Safety*. The National Academies Press, Washington, D.C., 2017. https://doi.org/10.17226/24776.
12. Caltrans. *Roadside Management Toolbox*. California Department of Transportation (Caltrans), Sacramento, 2014. http://www.dot.ca.gov/hq/LandArch/16_la_design/guidance/roadside_safety_tb/index.htm. Accessed May 17, 2017.
13. Galitz, W. O. *The Essential Guide to User Interface Design: An Introduction on GUI Design Principles and Techniques*. John Wiley & Sons, Indianapolis, Indiana, 2007.

The Standing Committee on Maintenance and Operations Personnel (AHD15) peer-reviewed this paper (18-04190).

Article

TRR
JOURNAL OF THE TRANSPORTATION RESEARCH BOARD

Transportation Research Record
2018, Vol. 2672(12) 198–208
© National Academy of Sciences:
Transportation Research Board 2018
Article reuse guidelines:
sagepub.com/journals-permissions
DOI: 10.1177/0361198118797458
journals.sagepub.com/home/trr
SAGE

Accommodation of Saltwater Temporary Pipelines on the Roadside

William Holik[1], Cesar Quiroga[1], Ioannis Tsapakis[1], and Jing Li[1]

Abstract

Drilling and completing oil and gas wells, particularly when using horizontal and hydraulic fracturing techniques, requires enormous amounts of water. Generally, it is cheaper for the industry to move fluids by pipeline than by truck, hence the interest in using permanent and/or temporary pipelines to transport water in areas where oil and gas developments take place. This paper describes temporary pipeline installation and operation practices and how they impact roadside maintenance activities as well as offering guidelines on how to install and operate temporary pipelines. A GIS database of temporary pipeline locations was developed from permits issued by the Texas Department of Transportation between July 2011 and August 2016. General trends indicate that temporary pipelines are typically 3, 4, 8, or 10 inches in diameter. Operators tend to favor certain highway segments to install temporary pipelines within the right of way. When multiple temporary pipelines are installed on segments repeatedly, this can affect maintenance operations. Several trends were observed that necessitate the development of guidelines for temporary pipelines. Many temporary pipelines were placed away from the right of way line, which creates conflicts with maintenance operations and results in some temporary pipelines being in the clear zone. Many temporary pipelines are not anchored in place and roll into the bottom of ditches or do not maintain a uniform alignment, which affects roadside maintenance. The percentage reduction in overtopping flow rate due to installing temporary pipelines through culverts is also analyzed.

One of the effects of using horizontal drilling and hydraulic fracturing techniques to extract oil and gas resources has been a significant increase in traffic volumes. These traffic increases have been particularly noticeable in the form of large numbers of heavy loaded trucks providing transportation services to develop and operate energy sites. Most well development-related traffic occurs on rural roads that were never designed to carry the huge amount of truck traffic associated with energy developments. Most of those roads were built decades ago to serve local low-volume traffic needs, and not repetitive heavy truckloads. The result has been accelerated degradation of pavements and roadside infrastructure, as well as increases in congestion and crash and fatality rates (*1–4*).

The number of trucks needed to develop and operate oil and gas wells varies depending on the region where the energy development takes place. Based on several estimates gathered over the last few years, a high-level depiction of the number of heavy trucks needed to develop a typical horizontal well in the Barnett Shale, Eagle Ford Shale, Permian Basin, Marcellus, and Niobrara regions is provided (Table 1) (*1, 3, 5*).

Water is used for many oil and gas activities, including enhanced recovery applications, drilling, and completing oil and gas wells. Water is also one of the byproducts of the operation of a well. Drilling a well can require anywhere from 65,000 to 600,000 gallons of water. Hydraulic fracturing involves pumping into the formation large volumes of water which includes components such as a friction reducer, surfactant and clay stabilizer, and sand. Hydraulic fracturing of a horizontal well can require 2 million to 6 million gallons of water (*3*). Moving these enormous amounts of water and other fluids requires considerable resources. As Table 1 shows, carrying water for hydraulic fracturing accounts for a significant percentage of all the heavy truckloads typically needed to develop a well.

Generally, it is cheaper for the industry to move fluids by pipeline than by truck, hence the interest by the industry in using pipelines to transport water in areas where oil and gas developments take place. Some of those pipelines are located within the right of way of public roads. One interest in temporary pipelines is the reduction in

[1]Texas A&M Transportation Institute (TTI), Texas A&M University System, San Antonio, TX

Corresponding Author:
Address correspondence to William Holik: w-holik@tti.tamu.edu

Table 1. Number of Trucks Needed to Develop a Well

Well development	Number of trucks				
	Barnett Shale	Eagle Ford Shale	Permian Basin	Marcellus formation	Niobrara formation
Drilling pad and construction equipment	70	70	70	45	23
Drilling rig, equipment, materials, and fluid	117	117	117	87	203
Fracking equipment: pump trucks, tanks	74	74	74	206	320
Fracking water	533	1,021	527	1,667	1,050
Fracking sand	57	147	66	77	48
Other additives and fluids	4	24	11	53	43
Flowback water removal	133	255	132	383	1,400
Total	988	1,708	997	2,518	3,087

pavement degradation by removing hundreds or thousands of truck journeys from the roads per well. The Texas Department of Transportation (TxDOT) recently completed a study to determine the impacts of temporary pipelines on roadside maintenance activities, determine how temporary pipelines could be safely accommodated, and develop guidelines for temporary pipelines on the roadside (*6*). This paper details the results of this study.

While crude oil and natural gas prices have dropped from their peaks in recent years, the U.S. Energy Information Administration forecast increasing crude oil and natural gas prices for 2018 (*7*). Oil and gas operators will respond to increased prices by drilling and completing more wells, meaning the use of temporary pipelines is likely to increase. Advances in technology relating to horizontal drilling and hydraulic fracturing have made it economically feasible to produce crude oil and natural gas in regions previously thought unprofitable. As a result, states with little history of energy developments are seeing increased activity.

Water is a critical component of drilling and completing wells, and due to lower costs and reduced impacts, water will continue to be transported in pipelines. Since only a short amount of time (weeks to months) is needed to drill and complete a well, operators prefer to use aboveground temporary pipelines to transport water for energy developments. As the following section shows, several issues can arise from the use of temporary pipelines, creating the need for guidelines to accommodate temporary pipelines safely and effectively in the right of way.

Impacts on Roadside Maintenance

Although the use of temporary pipelines has brought considerable benefits because fewer trucks are needed for energy developments, many issues related to the installation and operation of temporary pipelines exist. The conditions warrant implementation of the developed guidelines that clarify roles and responsibilities, improve installation and operation practices, and minimize the level of risk and exposure to transportation agencies and the public. Examples of issues observed in Texas include the following:

- Temporary pipelines that occupy the right of way without proper authorization.
- Temporary pipelines that use highway culverts incorrectly or driveway culverts illegally.
- Water with a much higher salt content than what TxDOT envisioned when temporary pipelines were first permitted on the state right of way.
- Temporary pipelines that are placed all over the roadside, including at the bottom of ditches and near the edge of pavement, therefore affecting the clear zone.
- Perception by temporary pipeline operators that they follow effective practices, not realizing that their pipelines do interfere with TxDOT roadside maintenance activities.

On occasion, temporary pipelines are placed near the edge of a roadway with tall fence posts used to anchor the temporary pipeline (Figure 1). This could present safety issues depending on clear zone requirements and the height of the stake. A temporary pipeline leaking from a joint between two pipeline segments into a river was discovered (Figure 2). Depending on the water being transported this could have environmental impacts.

Data and Descriptive Statistics

Copies of applications for temporary pipeline permits were obtained from TxDOT to develop a geographic information system (GIS)-based database of the locations of temporary pipelines. The database includes data attributes such as district, county, and road location; size and length of temporary pipeline; location of temporary

Figure 1. Metal fence posts used to anchor temporary pipelines in place.

Figure 2. Water leaking from a temporary pipeline into a river.

pipeline with respect to the road; and number of pipelines included in the permit. In total, 1,091 permits for 1,426 temporary pipelines were included in the database. The permits covered dates ranging from July 2011 to August 2016. In most cases, permit applicants provided a summary of where the pipeline was to be located, including distance from nearby intersections and which culverts would be used if crossing under a road. The application also included coordinates where the temporary pipelines would enter and exit the state right of way, as well as a map depicting the proposed location of the temporary pipeline. This information allowed the permits to be transcribed geographically in the database.

The GIS database of historical temporary pipelines facilitated the completion of a variety of analyses. For example, a map was created showing temporary pipeline locations in the Eagle Ford region as well as the locations of wells that were completed in the region from 2011 to 2016 (Figure 3). Notice that temporary pipelines were used throughout the region, but the geographic distribution was not uniform.

In Figure 4 a map of the Eagle Ford region shows where temporary pipelines were installed. The roadway segments are color-coded depending on the amount of time that temporary pipelines occupied each roadway segment during the entire study period (i.e., July 2011 through August 2016) (Figure 4*a*). The map also shows a zoomed-in view in Karnes County and part of DeWitt County which experienced significant temporary pipeline installation activity during the study period (Figure 4*b*). Notice that certain corridors were used extensively, for example, FM 1144 and FM 882, which experienced temporary pipelines for more than 36 months (i.e., three years). Other corridors experienced temporary pipelines for more than a year, e.g., State Highway (SH) 80, SH 119, FM 99, FM 238, FM 626, FM 1353, and FM 2102.

The researchers conducted a more disaggregated analysis that involved dividing each roadway segment into sub-segments at break points that corresponded to the start and end points of each temporary pipeline. The end result was a list of smaller segments, all of which were occupied by temporary pipelines during the period of analysis. A summary of trends derived from this analysis was created (Figure 5). A description for each of the columns in this table is as follows:

- Roadway Segments by Number of Temporary Pipeline Permits: This column shows a grouping of different segment types according to the total number of temporary pipeline permits that were issued for each segment during the study period. The range in the number of temporary pipeline permits per segment was 1 through 24.
- Total Length (miles): This column shows the total length of the segments per number of permits. Overall, 415 roadway miles were occupied by one or more pipelines during the study period.
- Length (%): This column shows the total length of the segments as a percentage of the total length of the examined network. For example, segments that involved only one permit covered 51.5% of the 415 miles of roadways where there were permits.
- Cumulative Length (%): This column shows the cumulative length percentage for each segment group. For example, segments that involved up to five permits covered close to 90% of the 415 miles of roadways where there were permits. Only 10% of the 415 miles of roadways where there were permits involved more than five permits.
- Average Duration (months): This column shows the average time period (months) during which segments where occupied by pipelines. For

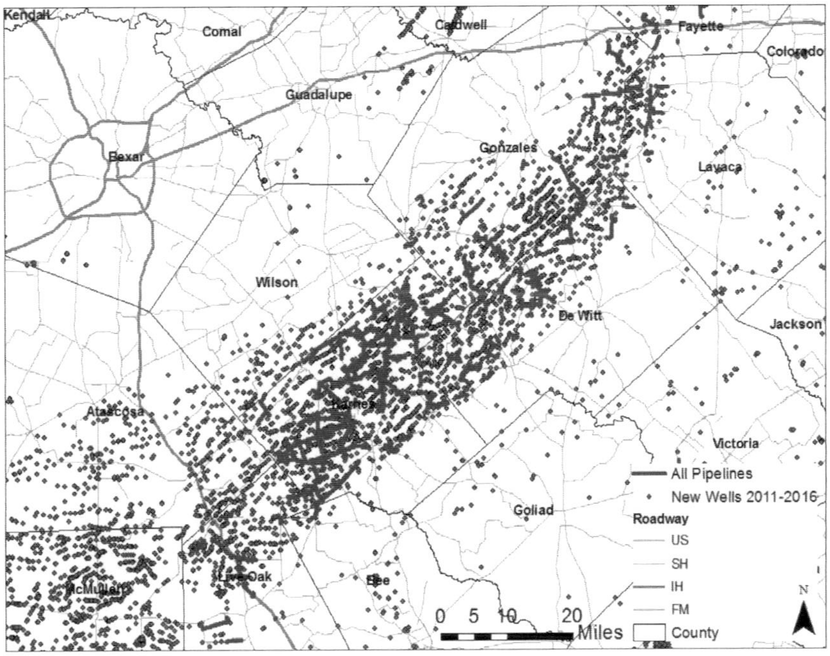

Figure 3. Locations of temporary pipelines in the Eagle Ford Shale region.

example, segments that had one permit were occupied for three months on average. Segments that had two permits were occupied by pipelines for 5.6 months on average. With a few exceptions, the average duration of pipeline occupancy increased with the number of permits.

- Duration as Percent of Study Period: This column shows the average occupancy duration as a percentage of the total study period (5.1 years).
- Average Number of Operators: This column shows the average number of operators per segment group. Not surprisingly, the average number of operators increased with the number of permits per segment group.

Accommodation Guidelines for Temporary Pipelines

The following section describes recommended practices for the accommodation, installation, operation, and maintenance of temporary pipelines within the right of way. Several topics are covered, including temporary pipeline specifications, hydraulic analysis of temporary pipelines occupying culverts, anchoring devices to keep pipelines in place, and inspection protocols based on the type of water being transported. The topics covered herein are a subset of the guidelines developed for TxDOT. However, the entire guidebook of recommendations developed for TxDOT can be applied to other states as well (*8*).

Temporary Pipeline Specifications

Water Characteristics. It was assumed that the water transported for energy developments was predominately fresh water. However, after meetings with pipeline operators and TxDOT officials it became clear that the water being transported in temporary pipelines has a wide range of total dissolved solids (TDS). TDS is a measure of both dissolved inorganic and organic matters, and is widely used to express salinity.

Water that can be transported in aboveground temporary pipelines could fall under one of the following categories:

- Fresh water (i.e., TDS up to 1,000 mg/L): This is the preferred type of water to be transported by aboveground temporary pipelines.
- Slightly saline water (i.e., TDS greater than 1,000 and up to 3,000 mg/L): This type of water is acceptable to be transported by aboveground temporary pipelines.
- Moderately saline water (i.e., TDS greater than 3,000 and up to 10,000 mg/L): This type of water is not recommended to be transported by temporary pipelines, but exceptions could be made under certain circumstances at the discretion of the DOT.

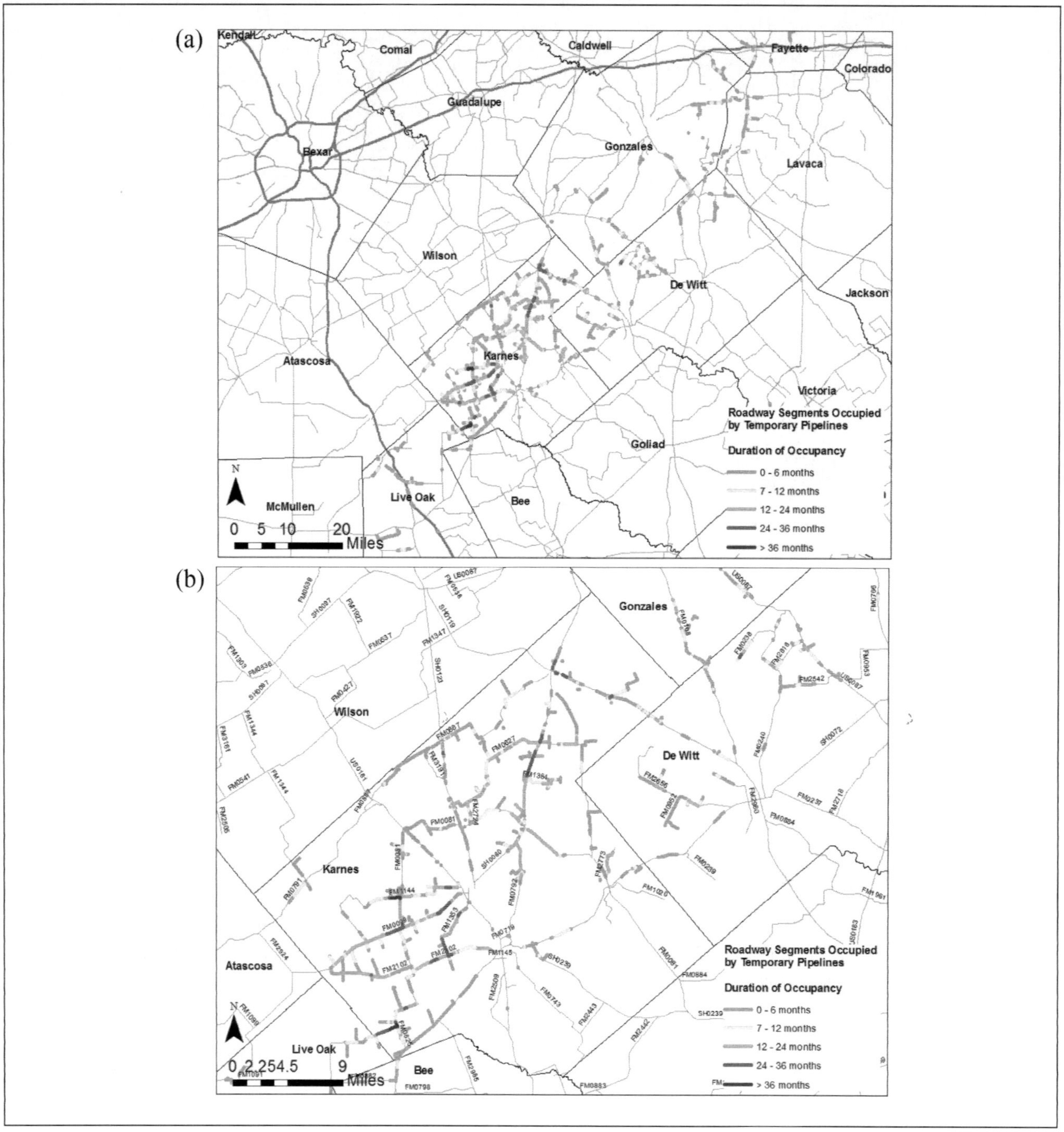

Figure 4. (*a*) Roadway segments with temporary pipelines in the Eagle Ford Shale region; (*b*) zoom-in view focusing on Karnes and De Witt Counties.

Very saline water (i.e., TDS greater than 10,000 mg/L) should not be allowed to be transported in aboveground temporary pipelines.

Temporary Pipeline Size and Material. From the database of temporary pipeline permits, temporary pipelines were found to range in size from 2 in. to 12 in. However, the most common sizes are 3 and 4 in. (which are used during the drilling phase of well development) and 8 and 10 in. (which are used during the hydraulic fracturing or completion stage of well development).

Common materials include lay-flat and high-density polyethylene (HDPE) pipelines. When operators first began transferring water in temporary pipelines for

Roadway Segments by Number of Temporary Pipeline Permits	Roadway Length			Duration of Occupancy		Average Number of Operators
	Total Length (miles)	Length (%)	Cumulative Length (%)	Average Duration (months)	Duration as Percent of Study Period*	
Segments with 1 permit	213.9	51.5%	51.5%	3.0	5%	1
Segments with 2 permits	82.5	19.9%	71.4%	5.6	9%	1-2
Segments with 3 permits	37.2	9.0%	80.3%	8.1	13%	1-2
Segments with 4 permits	16.5	4.0%	84.3%	9.9	16%	2-3
Segments with 5 permits	19.4	4.7%	89.0%	13	21%	2-3
Segments with 6 permits	9.2	2.2%	91.2%	15	24%	2-3
Segments with 7 permits	6.9	1.7%	92.9%	17	28%	2-3
Segments with 8 permits	8.3	2.0%	94.9%	18	30%	3-4
Segments with 9 permits	4.8	1.2%	96.0%	21	34%	3-4
Segments with 10 permits	2.4	0.6%	96.6%	21	35%	3-4
Segments with 11 permits	2.3	0.6%	97.2%	22	36%	3-4
Segments with 12 permits	2.6	0.6%	97.8%	25	41%	4-5
Segments with 13 permits	1.9	0.5%	98.3%	28	45%	4-5
Segments with 14 permits	1.0	0.2%	98.5%	26	43%	4-5
Segments with 15 permits	2.4	0.6%	99.1%	28	45%	4-5
Segments with 16 permits	0.7	0.2%	99.3%	33	54%	4-5
Segments with 17 permits	0.7	0.2%	99.4%	30	49%	4-5
Segments with 18 permits	0.5	0.1%	99.5%	31	51%	4-5
Segments with 19 permits	0.6	0.1%	99.7%	34	56%	5-6
Segments with 20 permits	0.3	0.1%	99.8%	38	62%	5-6
Segments with 21 permits	0.1	0.0%	99.8%	36	58%	5-6
Segments with 22 permits	0.5	0.1%	99.9%	41	67%	5-6
Segments with 23 permits	0.3	0.1%	100.0%	36	58%	6-7
Segments with 24 permits	0.2	0.0%	100.0%	42	68%	6-7
Segments with 1-24 permits (TOTAL)	**415.1**	**100%**				

* Note: The total duration of the study period was from 7/22/2011 to 8/11/2016.

Figure 5. Descriptive statistics of roadway segments occupied by temporary pipelines.

energy developments they used aluminum pipelines. These pipelines tend to leak more frequently and are harder to install and remove than lay-flat and HDPE temporary pipelines. Due to safety concerns if a vehicle were to strike a temporary pipeline and issues with leaks and installation/removal, it is not recommended to use metal temporary pipelines.

Hydraulic Analysis

A common practice for temporary pipeline operators is to install temporary pipelines through culverts to cross roads. Occupying culverts is cheaper than boring, does not damage the pavement structure like open cutting the road, and reduces impacts to vehicular traffic as if a road crossing were used. The primary purpose of culverts is to convey water; however, the capacity of culverts is reduced by installing temporary pipelines. Hydraulic analysis was conducted to determine the impacts of installing temporary pipelines through culverts to provide guidelines on the size of temporary pipelines that should be allowed in certain sizes of culverts.

To conduct the hydraulic analysis, several geometric data elements are necessary including culvert size, culvert inlet and outlet elevations, roadway elevation, and drainage channel invert elevation. These data elements were captured for several culverts as part of a field data collection campaign. For those culverts, the researchers found that the average difference in elevation between the top of the culvert and the edge of the pavement was about 10 in. When accounting for a typical cross slope of 2% to the centerline of the road and an average combined shoulder and lane width of 13.9 ft, the elevation at the road crest would be 1 3 in. higher than the top of the culvert. One important criterion for hydraulic analysis is the overtopping flow rate, which is based partly on the difference in elevation between the top of the culvert and the crest of the road.

The researchers analyzed pipe culverts and box culverts, while assuming all culverts to be straight single-barrel concrete culverts. The pipe culverts analyzed ranged in diameter from 24 in. to 60 in. and the box culverts ranged in width and height from 2 ft by 2 ft to 6 ft by 6 ft. The temporary pipeline sizes considered had

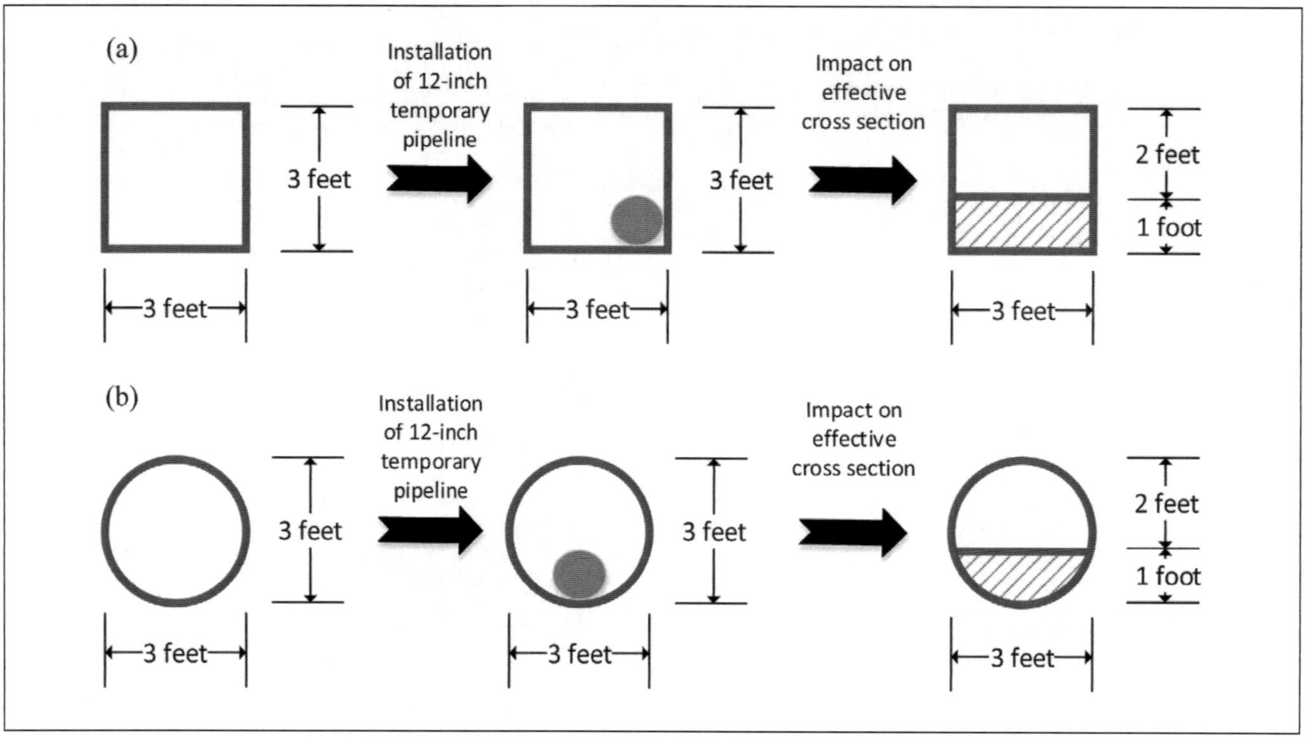

Figure 6. Reduction of effective cross section after inserting pipeline in culvert: (*a*) box culvert; (*b*) pipe culvert.

diameters of 3, 4, 6, 8, 10, and 12 in. One and two temporary pipelines of the same size were modeled in the analysis.

Typical design parameters for culverts include minimum, maximum, and design flow rates. These values can be measured from discharge data collected in the field or calculated using conceptual methods such as the rational method or the unit hydrograph method. For this analysis, the researchers were not designing culverts but quantifying the effects of installing temporary pipelines in existing culverts. To this end, the researchers compared the overtopping flow rate of culverts without temporary pipelines with the overtopping flow rate of culverts of the same size with temporary pipelines. The overtopping flow rate is calculated as part of the hydraulic analysis, therefore it was not necessary to determine exact minimum, maximum, and design flow rates. The researchers used reasonable estimates for flow rate inputs and compared the resulting overtopping flow rates of culverts with and without temporary pipelines.

The researchers used FHWA's hydraulic analysis software HY-8, version 7.50, for the hydraulic analysis (*9, 10*). This software is a culvert analysis tool, not a water surface computation program, which made it suitable for the needs of the analysis. The HY-8 software does not account directly for flow obstructions such as temporary pipelines inside of culverts. To model culverts with temporary pipelines, the researchers reduced the effective cross section of the culvert. In the field, the researchers noticed several cases where a single temporary pipeline caused additional sedimentation which had the effect of raising the bottom of the cross section inside the culvert to roughly match the top of the pipeline. For example, a 3 ft × 3 ft culvert with a 12-in. temporary pipeline would be reduced to a 2 ft × 3 ft cross section with its bottom raised by 12 in. (Figure 6*a*). Similarly, inserting a 12-in. pipeline in a 36-in. pipe culvert would raise the effective invert by 12-in. and reduce the effective cross section accordingly (Figure 6*b*).

The researchers calculated the relative reduction in overtopping flow rate for the various combinations of temporary pipeline sizes and culvert types and sizes described previously (Table 2). For example, the impact of inserting a 6-in. pipeline in a 3 ft × 3 ft box culvert is a 20% reduction in the overtopping flow rate of the box culvert. As the size of the temporary pipeline increases, the reduction in overtopping flow rate increases. For example, the impact of inserting a 10-in. pipeline would be a 39% reduction in the overtopping flow rate of the box culvert. Notice that the impact of inserting two temporary pipelines of equal diameter is the same as that associated with a single pipeline.

It is assumed that there is a level of risk associated with obstructing a culvert, due to reduction of the effective cross-sectional area and hydraulic capacity of the culvert. Depending on the functional class and traffic volume of the roadway, DOTs will design culverts for a specific return period. For example, TxDOT designs

Table 2. Reduction in Overtopping Flow Rate Caused by Temporary Pipelines in Box and Pipe Culverts

	Temporary pipeline diameter (inches) – single and double pipeline occupancy											
	3"	4"	6"	8"	10"	12"	3" + 3"	4" + 4"	6" + 6"	8" + 8"	10" + 10"	12" + 12"
Box culvert span × rise (feet)	Reduction in overtopping flow rate for box culverts (%)											
2×2	14	19	28	37	46	55	14	19	28	37	46	55
2×3	10	13	20	26	33	39	10	13	20	26	33	39
3×2	14	19	28	38	47	56	14	19	28	38	47	56
3×3	10	13	20	26	33	39	10	13	20	26	33	39
3×4	7	10	14	20	25	29	7	10	14	20	25	29
4×2	4	18	32	36	45	58	4	18	32	36	45	58
4×3	12	13	19	25	31	38	12	13	19	25	31	38
4×4	6	8	15	18	25	28	6	8	15	18	25	28
4×5	9	9	12	15	23	24	9	9	12	15	23	24
5×2	7	12	22	38	41	50	7	12	22	38	41	50
5×3	16	18	20	24	30	41	16	18	20	24	30	41
5×4	9	11	14	24	25	27	9	11	14	24	25	27
5×5	2	4	12	13	15	19	2	4	12	13	15	19
5×6	4	5	10	11	14	20	4	5	10	11	14	20
6×3	9	12	23	24	30	42	9	12	23	24	30	42
6×4	12	13	14	18	19	27	12	13	14	18	19	27
6×5	5	7	11	18	18	21	5	7	11	18	18	21
6×6	1	3	6	9	12	15	1	3	6	9	12	15
Pipe culvert diameter (inches)	Reduction in overtopping flow rate for pipe culverts (%)											
24	8	12	21	31	42	53	10	15	27	40	53	65
30	6	9	16	23	32	40	7	11	20	30	40	51
36	1	3	9	15	22	29	2	4	12	19	28	37
42	2	3	7	15	18	24	3	4	9	17	22	27
48	1	2	3	6	9	16	2	2	4	8	12	18
60	1	1	2	3	6	11	1	1	2	4	10	11

culverts for a 25-year return period for principal arterials and for a 10-year return period on minor arterials and collectors. Using design rainfall intensities (*11*) for each county in oil and gas producing regions in Texas, the percentage reductions in flow rate for corresponding reductions in storm return periods are calculated. Aggregated results are calculated for the Barnett Shale, Eagle Ford Shale, and Permian Basin regions (Table 3). In general:

- For a design return period of 25 years, a 15% reduction in capacity would reduce the return period to 10 years, a 28% reduction in capacity would reduce the return period to 5 years, and a 35% reduction in capacity would reduce the return period to 2 years.
- For a design return period of 10 years, a 15% reduction in capacity would reduce the return period to 5 years, and a 35% reduction in capacity would reduce the return period to 2 years.
- When analyzing overtopping flow for a return period of 100 years, a 12% reduction in capacity would reduce the return period to 50 years, a 23% reduction in capacity would reduce the return period to 25 years, and a 35% reduction in capacity would reduce the return period to 10 years.

Based on the hydraulic analysis and feedback from various stakeholders, a 25% maximum reduction in capacity is used as the criterion for whether or not a certain sized temporary pipeline should be allowed in a culvert. The maximum allowable reduction in capacity is a policy decision, there is nothing intrinsically wrong with selecting a different threshold, such as a more stringent 10% reduction. The recommended number of temporary pipelines that should be allowed in box and pipe culverts is developed from a maximum reduction in capacity of 25% and the limit of only allowing two temporary pipelines in a culvert at a time (Table 4).

Anchoring Devices

Unless temporary pipelines are anchored, they tend to move, either because of changes in the operating pressure inside the pipeline or because of where the pipes are placed on the ground. Pipeline operators should ensure their pipelines do not shift laterally so as to not interfere with maintenance operations. This means that, under

Table 3. Reduction in Flow Rate Caused by Temporary Pipelines in Culverts by Oil and Gas Producing Regions of Texas

Reduction in return period (Years)	Region: Barnett Shale	Region: Eagle Ford Shale	Region: Permian Basin
	Reduction in flow rate for corresponding reduction in return period (%)		
100 to 50	12	12	12
100 to 25	23	23	23
100 to 10	35	35	36
100 to 5	44	44	45
100 to 2	57	57	59
50 to 25	13	13	13
50 to 10	26	26	27
50 to 5	37	37	38
50 to 2	52	52	54
25 to 10	15	15	16
25 to 5	28	28	29
25 to 2	45	45	47
10 to 5	15	15	15
10 to 2	35	35	37

Table 4. Recommended Maximum Number of Temporary Pipelines That Can Be Installed in Box and Pipe Culverts

	Temporary pipeline diameter (inches)					
	3	4	6	8	10	12
Box culvert span×rise (feet)	Maximum number of temporary pipelines in box culvert					
2×2	2	2	2	1	0	0
2×3	2	2	2	1	1	0
3×2	2	2	2	2	1	0
3×3	2	2	2	2	1	0
3×4	2	2	2	2	2	1
4×2	2	2	2	2	2	1
4×3	2	2	2	2	2	1
≥4×4	2	2	2	2	2	2
Pipe culvert diameter (inches)	Maximum number of temporary pipelines in pipe culvert					
24	2	2	1	0	0	0
30	2	2	2	1	0	0
36	2	2	2	2	1	0
42	2	2	2	2	2	1
≥48	2	2	2	2	2	2

normal circumstances, temporary pipelines must be anchored. When anchoring temporary pipelines, pipeline operators must be aware of the following:

- The method used to secure temporary pipelines should not interfere with any underground utilities and must comply with damage prevention laws and regulations (*12*).
- The method used to secure temporary pipelines must be crashworthy in accordance with the requirements included in tests 70, 71, and 72 in the AASHTO Manual for Assessing Safety Hardware (MASH) (*13*).
- The method used to secure temporary pipelines should not damage the right of way or right of way fence.
- The temporary pipeline should not deviate from its centerline by greater than 6 in.

The pipeline operator must use devices to keep temporary pipelines from deviating from their centerline or moving to the bottom of the slope. The operator may use stakes, pipes, or other acceptable devices, provided the devices meet the following requirements:

- The device must not exceed 4 ft in length and 4.5 in. in diameter or width.

- The height of the device above ground must not exceed 3 ft.
- The depth of the device below ground must not exceed 16 in., or the maximum allowed depth per One Call requirements (*12*).
- The transportation agency may request proof of compliance with AASHTO MASH if the device is a stake or pipe. For any other device, the operator must submit proof of compliance with AASHTO MASH as part of the installation approval process.
- When using a stake, the operator must use a protective sleeve or similar appurtenance to protect the temporary pipeline from rubbing against the stake. The operator must secure the protective sleeve to the stake to prevent vertical movement of the sleeve, which might cause the temporary pipeline to rub against the stake.

Inspection Protocols

While the temporary pipeline is in operation, inspections should be conducted at least twice daily using a method that ensures early, effective detection of anomalies such as leaks and spills, damage to the structural integrity of the pipeline, and excessive shifts in pipeline alignment. The use of aboveground temporary pipelines is limited to water with TDS up to 10,000 mg/L. Produced water (including flowback water) or any water with TDS greater than 10,000 mg/L should use a different mode of transportation, for example, truck or underground pipelines. The protocols apply in response to leaks or spills affecting temporary pipelines (Table 5).

Conclusions

The water transfer industry has evolved over time as the oil and gas industry has increased the use of temporary pipelines to transport water needed for drilling and completing oil and gas wells. From initially using rigid aluminum pipelines, temporary pipeline operators have largely migrated to polyethylene and lay-flat pipelines. These newer temporary pipelines are lighter and more durable than aluminum temporary pipelines making them more economical to install, operate, and maintain.

A GIS database of temporary pipeline locations was developed from permit applications submitted to TxDOT between July 2011 and August 2016. General trends indicate temporary pipelines are typically 3, 4, 8, or 10 in. in diameter. Operators tend to favor certain highway segments to install temporary pipelines within the right of way. When multiple temporary pipelines are installed on segments repeatedly, this can impact maintenance operations.

Several trends were observed that necessitate the development of guidelines for temporary pipelines. Many temporary pipelines were placed away from the right of way line, which can interfere with maintenance operations and also means some temporary pipelines are installed in the clear zone. Many temporary pipelines are not anchored and roll into the bottom of ditches or do not maintain a uniform alignment and impact vegetation management.

To conduct the hydraulic analysis, the researchers combined field geometric data about culverts and roadways with estimated flow data in a simulation software environment to arrive at typical discharge flow rates through various types and sizes of culverts. The

Table 5. Leak and Spill Response Protocol for Aboveground Temporary Pipelines

Type of water	What to do after leaks/spills
Fresh water (TDS $\leq$ 1,000 mg/L)	• The pipeline operator must repair the pipeline within 24 hours of detection.
Slightly saline water (1,000 $<$ TDS $\leq$ 3,000 mg/L)	• The pipeline operator must repair the pipeline within 24 hours of detection.
Moderately saline water (3,000 $<$ TDS $\leq$ 10,000 mg/L)	• The pipeline operator must repair the pipeline within 24 hours of detection. • The pipeline operator must estimate the amount of water leaked or spilled and the potential environmental impact resulting from the leak or spill. • The pipeline operator must report the incident to the transportation agency within 24 hours of detection of the incident so that the transportation agency can document and begin to monitor the spill location. Long-term monitoring is likely to be expected. • If the impact of the leak or spill on vegetation in the affected area becomes evident, the pipeline operator must conduct a soil assessment and implement appropriate remediation solutions to restore the soil productivity and structure to the conditions prior to the incident. This includes restoring the vegetation in the affected area to its original condition. If the transportation agency makes such restorations because of inability of the pipeline operator to complete this task, the pipeline operator should reimburse the transportation agency.

researchers made comparisons between the output flow rate with and without temporary pipelines occupying culverts to estimate potential reductions in capacity. Based on the hydraulic analysis and district stakeholder feedback, the researchers used a 25% maximum reduction in capacity as the basic criterion for whether or not a certain sized temporary pipeline should be allowed in a culvert. The researchers then combined this requirement with a limit of no more than two temporary pipelines in any culvert to arrive at a tabulation of the maximum number of temporary pipelines (and their size) to allow inside culverts of any size. A reduction in capacity would have the effect of reducing the effective return period. For a design return period of 25 years, a 15% reduction in capacity would reduce the return period to 10 years, a 25% reduction in capacity would reduce the return period to 5 years, and a 40% reduction in capacity would reduce the return period to 2 years.

Acknowledgments

This work was completed as part of the research project "Engineering Guidelines for Installing Temporary Pipelines within the Right of Way."

Authors' Note

Jing Li is now with TJKM Transportation Consultants in Pleasanton, CA.

Author Contributions

The authors confirm contribution to the paper as follows: study conception and design: WH and CQ; data collection: WH, CQ, JL; analysis and interpretation of results: WH, CQ, IT, JL; draft manuscript presentation: WH, CQ, IT. All authors reviewed the results and approved the final version of the manuscript.

References

1. Quiroga, C. A., I. Tsapakis, J. Li, W. Holik, and E. Kraus. *Truck Traffic and Truck Loads Associated with Unconventional Oil and Gas Developments in Texas.* 2016 Update. Report RR-16-01. Texas A&M Transportation Institute, Texas Department of Transportation, College Station, TX, August 2016.
2. Quiroga, C. A., I. Tsapakis, E. Kraus, J. Le, and A. Wimsatt *Oil and Gas Energy Developments and Changes in Pavement Conditions in Texas.* Final Report. Report PRC 14-35F. Transportation Policy Center, Texas A&M Transportation Institute, College Station, TX, March 2015. http://tti.tamu.edu/documents/PRC-14-35F.pdf. Accessed July 11, 2017.
3. Quiroga, C. A., E. G. Fernando, and J. H. Oh. *Energy Developments and the Transportation Infrastructure in Texas: Impacts and Strategies.* Report FHWA/TX-12/0-6498-1. Texas Transportation Institute, Texas Department of Transportation, College Station, TX, March 2012. http://tti.tamu.edu/documents/0-6498-1.pdf. Accessed July 11, 2017.
4. Quiroga, C. A., and I. Tsapakis. *Oil and Gas Energy Developments and Changes in Crash Trends in Texas.* Final Report. Report PRC 15-35 F. Transportation Policy Center, Texas A&M Transportation Institute, College Station, TX, October 2015. http://tti.tamu.edu/documents/PRC-15-35-F.pdf. Accessed July 11, 2017.
5. Quiroga, C. A., E. Kraus, I. Tsapakis, J. Li, and W. Holik. *Truck Traffic and Truck Loads Associated with Unconventional Oil and Gas Developments in Texas.* Report RR-15-01. Texas A&M Transportation Institute, Texas Department of Transportation, College Station, TX, August 2015.
6. Quiroga, C. A., W. Holik, J. Li, I. Tsapakis, and D. Bierling. *Temporary Saltwater Pipelines: Current Practice and Recommendations.* Report FHWA/TX-17/0-6886-1. Texas Transportation Institute, Texas Department of Transportation, College Station, TX, August 2017.
7. U.S. Energy Information Administration. *Short-Term Energy Outlook –July 2017.* U.S. Energy Information Administration, Washington, D.C., 2017. https://www.eia.gov/outlooks/steo/tables/pdf/2tab.pdf. Accessed July 22, 2017.
8. Quiroga, C. A., W. Holik, and J. Li. *Guidelines for Installing, Operating, and Maintaining Temporary Saltwater Pipelines.* Texas A&M Transportation Institute, Texas Department of Transportation, Austin, TX, August 2017.
9. Federal Highway Administration. *HY-8 Culvert Hydraulic Analysis Program.* Federal Highway Administration, Washington, D.C., 2016. https://www.fhwa.dot.gov/engineering/hydraulics/software/hy8/. Accessed July 11, 2017.
10. Federal Highway Administration. *Hydraulic Design of Highway Culverts.* Hydraulic Design Series Number 5. Federal Highway Administration, Washington, D.C., 2012. https://www.fhwa.dot.gov/engineering/hydraulics/pubs/12026/hif12026.pdf. Accessed July 11, 2017.
11. Cleveland, T. G., G. R. Herrmann, C. C. Tay, C. M. Neale, M. R. Schwarz, and W. H. Asquith. *New Rainfall Coefficients – Including Tools for Estimation of Intensity and Hyetographs in Texas.* Report FHWA/TX-15/0-6824-1. Texas Tech University, Lubbock, TX, August 2015. http://onlinemanuals.txdot.gov/txdotmanuals/hyd/0-6824-fpr-main.pdf. Accessed July 11, 2017.
12. State of Texas, *Texas Utilities Code*, Title 5, Chapter 251. http://www.statutes.legis.state.tx.us/Docs/UT/htm/UT.251.htm. Accessed July 11, 2017.
13. AASHTO. *Manual for Assessing Safety Hardware.* American Association of State Highway and Transportation Officials, Washington, D.C., 2016.

The Standing Committee on Roadside Maintenance Operations (AHD50) peer-reviewed this paper (18-05033).

The contents of and opinions expressed in this paper are those of the authors and are not the official views or policies of TxDOT or the Federal Highway Administration.

Article

TRR
JOURNAL OF THE TRANSPORTATION RESEARCH BOARD

Transportation Research Record
2018, Vol. 2672(12) 209–219

DOI: 10.1177/0361198118787081
journals.sagepub.com/home/trr
SAGE

Development and Implementation of a Pavement Marking Management System in Florida

Bouzid Choubane[1], Joshua Sevearance[2], Charles Holzschuher[1], James Fletcher[2], and Chieh (Ross) Wang[3]

Abstract

The visibility of pavement markings is an important aspect of a safe transportation system as the markings convey vital roadway warnings and guidance information to the traveling public. Therefore, it is beneficial to maintain acceptable visibility levels of markings on pavements under all weather and lighting conditions. To ensure the intended in-service visibility level is adequately maintained, the reflectivity must be monitored and quantified accordingly. Historically, visibility or retroreflectivity of in-service pavement markings has been measured with handheld devices and visual inspections. However, visual surveys are considered subjective and the handheld measurements are tedious and potentially hazardous. Consequently, the Florida Department of Transportation (FDOT) has focused on the use of a non-contact technology capable of assessing pavement markings continuously at highway speeds with improved safety and efficiency. The use of mobile technology for measuring reflectivity has allowed FDOT to develop and, subsequently, implement a Pavement Marking Management System (PMMS) to improve the safety and nighttime visibility of its roadways. Implementation of such a system provides an efficient and less subjective methodology to identify conditions that are detrimental to roadway safety, and strategize mitigating solutions including the selection of appropriate materials and application techniques. The system will ultimately result in an effective use of state funds while ensuring the safety of the traveling public. This paper presents a description of the Florida Pavement Markings Management System and its subsequent implementation including FDOT's effort to ensure the quality, consistency, repeatability, and accessibility of statewide pavement marking retroreflectivity data.

The importance of pavement markings to the driving public is well recognized. The level of retroreflectivity provided by pavement markings significantly impacts the nighttime visibility and comfort to those traveling. According to the Federal Highway Administration (FHWA) Roadway Departure Safety "Nighttime Visibility" web page, over half of all traffic fatalities occur at night even though three quarters of travel occurs in daylight (*1*). Thus, when factoring in miles driven, the fatality rate at night is triple that of the daytime rate. Although other factors such as alcohol and fatigue cannot be minimized, pavement markings have been shown to reduce crashes when providing appropriate visibility levels to drivers at night and to facilitate safe nighttime driving (*1*).

Section 406 of the 1993 United States Department of Transportation Appropriation Act requires that FHWA specify a minimum level of retroreflectance that must be maintained for pavement markings and traffic signs (*2*). As a response to this act, FHWA released a notice of proposed amendments (NPA) in April 2010, which proposed that the Manual on Uniform Traffic Control Devices (MUTCD) include standards, guidance, options, and supporting information needed for maintaining the minimum levels of pavement marking retroreflectivity (*3*). Correspondingly, the proposed revisions to the MUTCD, including the minimum levels of pavement marking retroreflectivity, have been designated as Revision 1 to the 2009 Edition of the MUTCD (*4*). FHWA has recently released a supplemental notice of proposed amendments (SNPA) in January 2017, to amend the 2009 MUTCD and the 2010 NPA (*5*).

Retroreflectivity is created by reflecting light from the vehicle's headlamps directly back to the driver's eyes as illustrated in Figure 1. It is quantified as the ratio of the luminance (or brightness detected by a human eye) of an object to the illuminance of the object by a light source (amount of light illuminating the object) and is expressed in units of millicandelas per meter squared per lux ($mcd/m^2/lux$) (*6*).

[1]Florida Department of Transportation, Materials Research Park, Gainesville, FL
[2]University of North Florida (UNF), Mechanical Engineering, Jacksonville, FL
[3]Applied Research Associates, Inc., Gainesville, FL

Corresponding Author:
Address correspondence to Bouzid Choubane:
bouzid.choubane@dot.state.fl.us

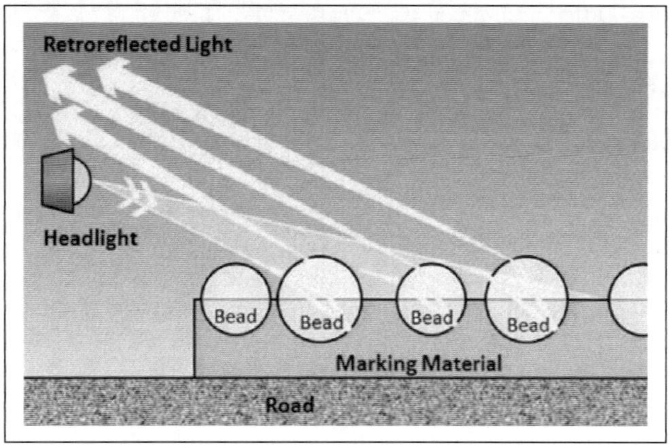

Figure 1. Illustration of the retroreflective effect using glass beads (7).

Pavement markings typically provide retroreflectivity through the application of small glass spheres that are partially embedded into the pavement marking material.

There are many pavement marking materials available, including paints, thermoplastics, and tapes, with many utilizing glass beads to enhance retroreflectivity as shown in Figure 1 (*7*). However, the intended retroreflectivity level is not always achieved given the inherent non-uniformity of the material such as that created by the random dispersion of glass beads. In addition, the environmental conditions such as rain, fog, snow, ultraviolet light, and heat as well as the constant action of tires, debris, and dirt can all affect retroreflective properties. Thus, to ensure the intended in-service visibility level is adequately maintained, the retroreflectivity must be measured and quantified accordingly.

Historically, the visibility or retroreflectivity of in-service pavement markings has been measured with handheld devices (see Figure 2) and visual inspections. However, visual surveys are considered subjective and handheld measurements are tedious and potentially hazardous. In addition, the Florida Department of Transportation (FDOT) has always searched for more innovative ways to monitor the performance, condition, and safety of its roadway infrastructure. It has carried out several initiatives focusing on the appropriate implementation of faster and safer testing methods and technologies to minimize maintenance of traffic, expedite projects acceptance, and minimize impact to the traveling public. FDOT has particularly focused on the use of non-contact sensor-based technology capable of assessing pavement markings continuously at highway speeds as it provides for versatility, ease, and speed of use. Inventorying stripe retroreflectivity using this mobile technology (mobile retroreflectivity unit or MRU) eliminates the need for maintenance of traffic and provides improved safety and efficiency over the traditional site-specific handheld measurements. FDOT initiated a program to evaluate, optimize, and subsequently develop a comprehensive implementation plan for MRU as an alternative for measuring pavement marking retroreflectivity (*8*, *9*). The plan included the use of MRU to identify pavement markings that are reaching the minimum allowable retroreflectivity, to monitor the retroreflective characteristics of various materials/applications of pavement markings, and to provide inventory for maintenance to assess restriping strategies. This has also provided FDOT with an opportunity to develop and, subsequently, implement a Pavement Marking Management System (PMMS).

Florida Mobile Retroreflectivity Unit (MRU)

FDOT's mobile retroreflectometers measure the retroreflectivity by applying the "30-meter geometry" described in ASTM E 1710 (*10*). The 30-meter geometry consists of the following assumptions: a typical passenger vehicle headlamp

Figure 2. Use of handheld retroreflectometers.

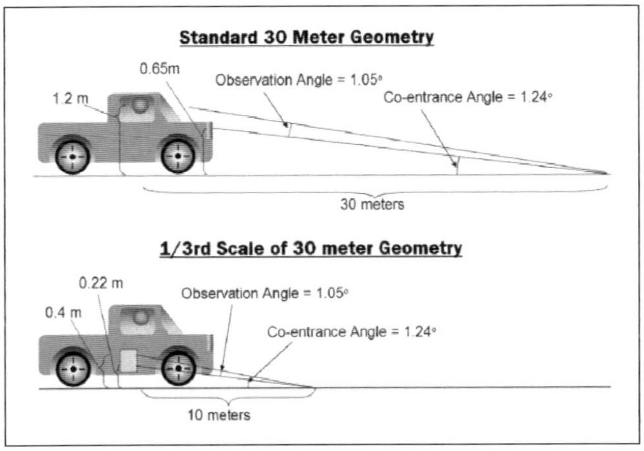

Figure 3. Standard 30-meter geometry and the 1/3 scale used in MRU.

height of 0.65 m (2.1 Ft), a driver eye height of 1.2 m (3.9 Ft), and a distance of 30 m (98 Ft) between the headlamps and the ground-based retroreflectance target. To reduce the size of the measuring device, the MRU uses a proportional scale (e.g., one-third or one-fifth) of the 30-meter geometry shown in Figure 3.

FDOT's MRU is mounted on full-size passenger vehicle equipped with all the mechanical and electrical power supplies for evaluating pavement markings while conforming to ASTM E 1710. In addition, the vehicles include a data acquisition system for collecting and storing information. A distance-measuring instrument (DMI) is provided to determine the position along the roadway. The longitudinal distance measurement is critical in associating the precise location for each 0.16 km (0.1 mile), the interval at which the MRU is reporting the data. FDOT's MRU is shown in Figure 4.

Equipment Optimization and Calibration

The MRU is a sensitive instrument. Thus, FDOT initiated a program to evaluate, develop mitigation strategies, and optimize the technology accordingly (*8*, *9*). For instance, initial testing showed that measurements taken within a short period of time were repeatable, but the results varied considerably throughout the day. Significant temperature sensitivity in the interference filters used to eliminate background lighting was identified. A thermoelectric cooler was therefore installed to maintain the operating temperature, and temperature compensation was added to the software to mitigate this issue (*8*).

Moreover, the precision of the MRU measurements is highly dependent on the calibration of the equipment. It is important that accurate geometry and an appropriate calibration panel be used during MRU calibration to reduce uncertainty in pavement marking measurements. Any errors introduced during the calibration process will be transferred to errors in pavement marking evaluation. The goal of the calibration process is to establish a baseline or reference by scanning a calibration panel of known retroreflectivity value. The consistency of the calibration panel is crucial to the success of the MRU calibration and should result in reproducible calibration factors. Much effort has been spent to improve and optimize the calibration procedure and alternative calibration tools (*11*, *12*). Since there is no standard procedure broadly accepted for the calibration and operation of the MRU, FDOT has researched and implemented a new calibration panel consisting of vinyl sign sheeting that is a more uniform material than the typical beaded panel and provides a reliable measurement when assigning a retroreflectivity value to the material using a photometric range testing equipment. The new vinyl calibration panel has provided more reproducible calibrations for the MRU equipment (*11*, *12*).

Pavement Marking Management System (PMMS)

In 2012, FDOT implemented the PMMS including documentation, quality assurance procedures, and a database. Its primary objective was to evaluate and manage statewide pavement marking retroreflectivity in a timely and cost-effective manner.

Figure 4. FDOT's mobile retroreflectivity units.

Table 1. Pavement Stripe Nomenclature

Purpose	Code	Description
Direction	L	Descending mileposts
	R	Ascending mileposts
Stripe type	EL	White edge line
	CL	Yellow center line. All yellow stripes including yellow skip lines are considered as center lines. When collecting yellow center line pavement markings, white pavement marking such as turn lanes must not influence the yellow center line's reflectivity.
	SL	White skip line. A lane designation number is placed between the Direction and the Stripe Type to indicate the order of the skip lines starting with the one closest to the center line.

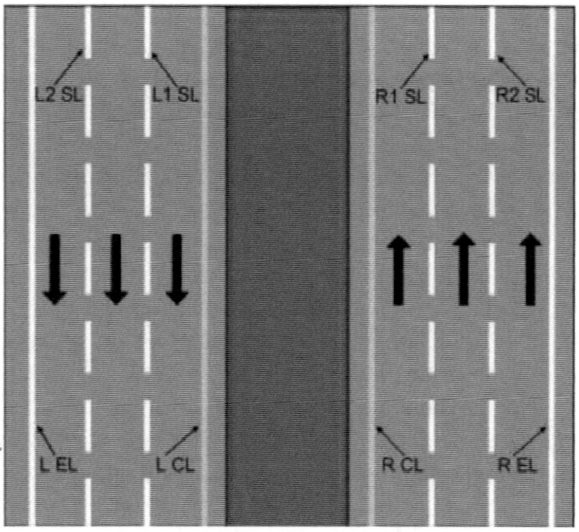

Figure 5. Pavement stripe nomenclature for longitudinal markings.

Annually, approximately 25,000 lane-miles of pavement markings are surveyed. The 25,000 lane-miles consist of 12,000 lane-miles of center lines (yellow solid line), 5,000 lane-miles of white skip lines, 4,000 lane-miles of white edge lines, and 4,000 lane-miles of other special requests from Districts. To ensure robust data collection efforts, a comprehensive process for standardized MRU testing was developed to cover the operation, data worksheets, and quality assurance processes. A database for PMMS was also developed to store and display the MRU data.

Florida Test Method for Retroreflectivity Measurement

To standardize the MRU testing equipment including the data acquisition system, the FDOT developed the "*Florida Test Method for Measuring Retroreflectivity of Pavement Marking Materials Using a Mobile Retroreflectivity Unit*" (*13*). This test method covers measurement of the retroreflective properties of longitudinal pavement markings, using an MRU operated at highway speed at a prescribed geometry that corresponds to the European Committee for Standardization (CEN) geometry and is the standard geometry adopted by ASTM E 1710 (*10*). This test method ensures that the nighttime retroreflectivity provided by pavement markings is adequate.

A standard naming convention was developed to identify each pavement stripe regardless of the number of lanes. The pavement stripe nomenclature used in Florida, which is shown in Table 1, includes the direction of travel followed by the stripe type. For skip lines, a lane designation number is inserted between the direction and the stripe type.

For example, as depicted in Figure 5, the northbound skip line closest to the center line would be identified as R1SL; the edge line in the same direction would be identified as REL. It must be noted, these directional codes only indicate the travel direction of the MRU, not the individual stripe (left or right stripe) nor the direction of stripe application.

The test method also normalizes the data acquisition system, calibration, and standardization (including quality assurance and frequency of check and calibration procedures), procedure (warm up, calibration, and minimum vehicle dynamics), interference and environmental influence (temperature, speed, RPMs, Event Codes), report (MRU Worksheet), and equipment precision. References of this normalization process are described in the following sections.

MRU Operations

In the process of researching the MRU equipment, the FDOT has developed the "*Operations Manual for Mobile Retroreflectivity Testing*" which includes calibration of the equipment, collection and post processing of the data, equipment maintenance, and troubleshooting procedures in great detail (*14*).

Data Worksheets

All pavement marking data is required in a standardized format, referred to as a worksheet, allowing the data to be

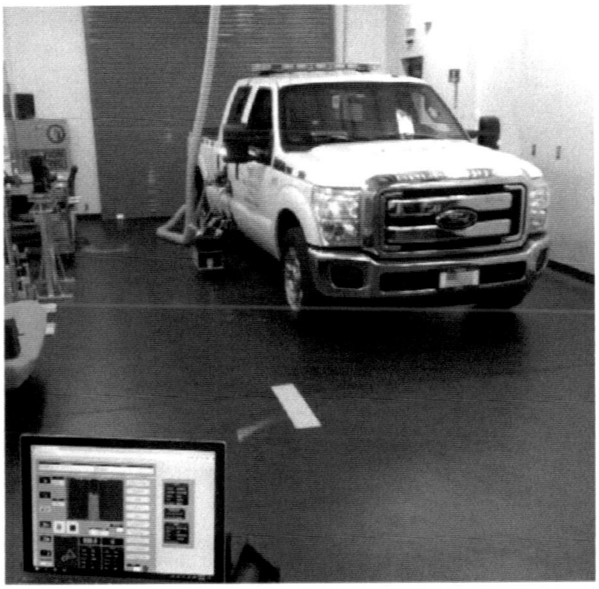

Figure 6. FDOT's calibration bay for mobile retroreflectivity units.

uploaded into the database. Information in the worksheets includes details of the county section roadway, pavement marking information, and retroreflectivity measurements.

Retroreflectivity data collected with an MRU is typically reported at 0.1-mile intervals. The FDOT *MRU Worksheet* (*15*) standardizes the format of the pavement marking data with information including details of the county section roadway, calibration information, environmental conditions, retroreflectivity measurements on both the left and right stripe with its location on the roadway, and event codes that identify pavement marking characteristics, such as pavement changes, no line, turn lane, etc.

Retroreflectivity data of new construction or restriping are typically collected with handheld retroreflectometers and reported in the FDOT *Traffic Marking Certification Worksheet*, which requires the following material details, depending on the type of pavement marking contract: if it is a temporary or permanent pavement marking, Approved Product List (APL) material number, APL bead number(s), pay item, description, and batch number. The *Traffic Marking Certification Worksheet* also provides the contract number, Financial Project Number (FIN), Federal Aid Number, contractor, subcontractor, name of individual collecting the data, date of testing, and period represented by certification.

Quality Assurance

As part of the PMMS, the FDOT has developed a "*Quality Assurance for Mobile Retroreflectivity Units*" document for MRU equipment and operators (*16*). This document establishes a series of laboratory and field tests at the State Materials Office (SMO) located in Gainesville, Florida. The laboratory test consists of a series of experiments that are conducted in a controlled environment (i.e., FDOT's MRU calibration bay as shown in Figure 6); field evaluations, including DMI/GPS accuracy and MRU precisions and stabilities, are conducted on a series of roadways in Alachua County, Florida, with various pavement markings and pavement types.

Before providing MRU services to FDOT, the contractor must demonstrate that their MRU(s) and operator(s) satisfy all of the quality assurance criteria described above. To continue providing MRU services, the contractor must undergo the quality assurance tests at the frequency designated for each respective test. Any time a quality assurance criterion is not met, the contractor must inform FDOT and take remedial action. The contractor must also inform FDOT after remedial action has been implemented and again demonstrate that all quality assurance criteria are satisfied. The quality assurance tests that need to be repeated shall be determined by FDOT after reviewing the problem and the remedial action taken by the contractor. The requirements to be followed by MRU equipment and operators for collecting data for the PMMS are found in Table 2.

In addition to all laboratory and field tests required for contractors, FDOT further conducts a separate, independent quality assurance sampling of the stripes evaluated by the contractor to ensure that the retroreflectivity data provided by the contractor is reliable. Quality assurance testing is performed in close proximity to the contractor on at least 20% of the stripes evaluated by the contractor. The contractor's data are then compared with FDOT's data in a 1-mile running average. If the two datasets differ by more than FDOT's dictated retest criteria, the mile sections that failed must be retested by the contractor.

PMMS Database

A comprehensive PMMS database, accessible to FDOT personnel via intranet and a web-based interface, was developed to store and display pavement marking data. Information on the Traffic Marking Certification and MRU worksheets is uploaded to the PMMS database to link available pavement marking data with characteristics of the roadway, such as average daily traffic, speed limit, and pavement type, from the Roadway Characteristics Inventory (RCI). The PMMS database contains tables with raw and summarized pavement marking retroreflectivity data, roadway sections information, equipment and operator information, as well as spatial and temporal information (e.g., date and time of the test and corresponding GPS coordinates). Some of the additional features include the ability to query data using a user-friendly, web-based interface that filters pavement marking retroreflectivity data based on the criteria found in Figure 7.

As depicted in Figure 8, the results screen shows a summary table of the average retroreflectivity of all roadway sections that met the query criteria. This table contains

Table 2. Frequency of Quality Assurance Testing

	Testing frequency		
Type of test	35 days	65 days	365 days
Laboratory testing			
Set-up and visual inspection		X	
Verification testing		X	
Background noise testing			X
Equipment dynamics		X	
Field testing			
DMI and GPS accuracy	X		
Precision test		X	
Reflectivity stability with speed test			X
Reflectivity stability with temperature test			X
Operational competency			X

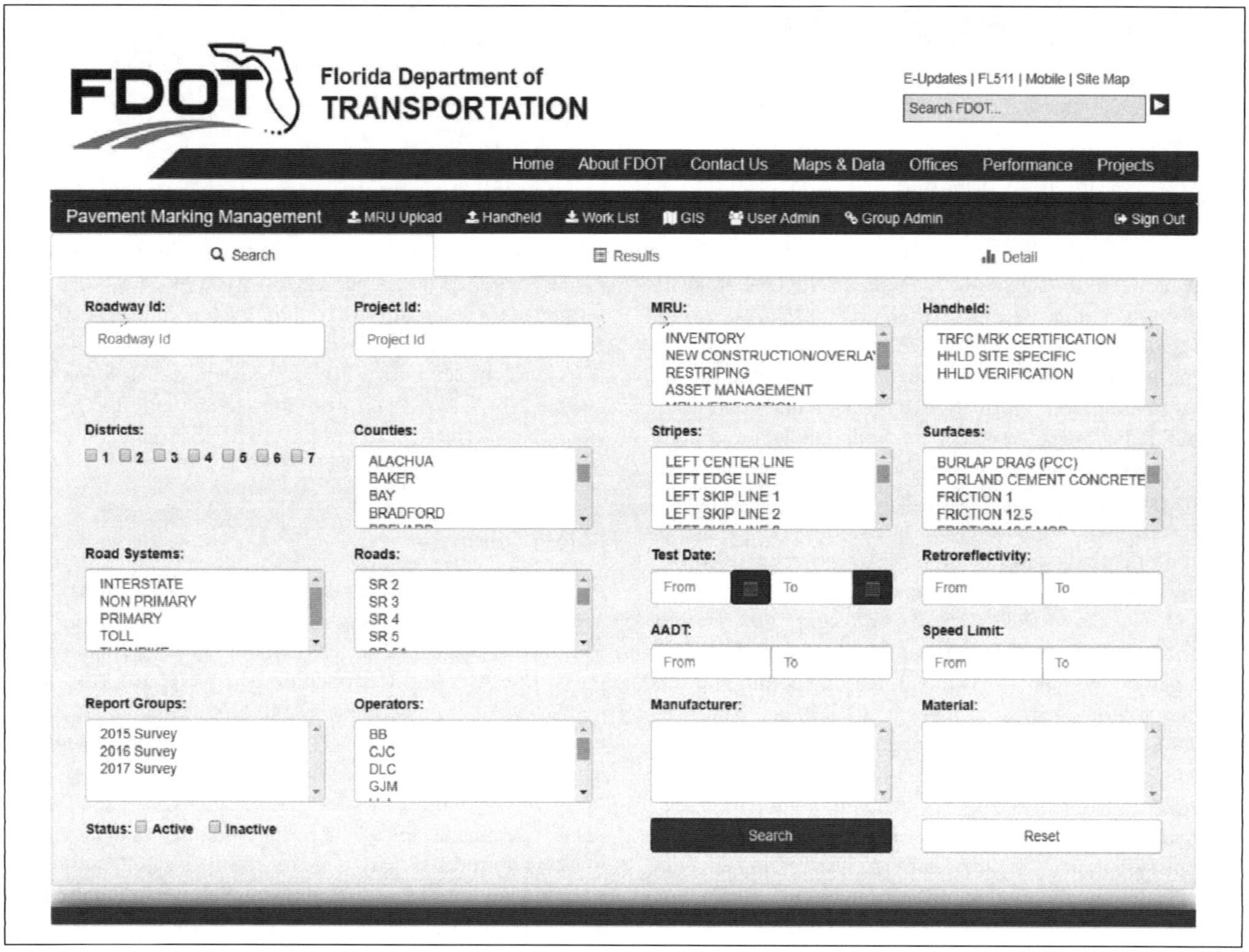

Figure 7. FDOT's Pavement Marking Management System database search screen.

information such as the District, the roadway ID, County, state road number, report group (i.e., survey year), beginning and ending mileposts, date of data collection, test type, different stripes with corresponding retroreflectivity average, and weather and status of the roadway (active or inactive). Depending on the average retroreflectivity of the

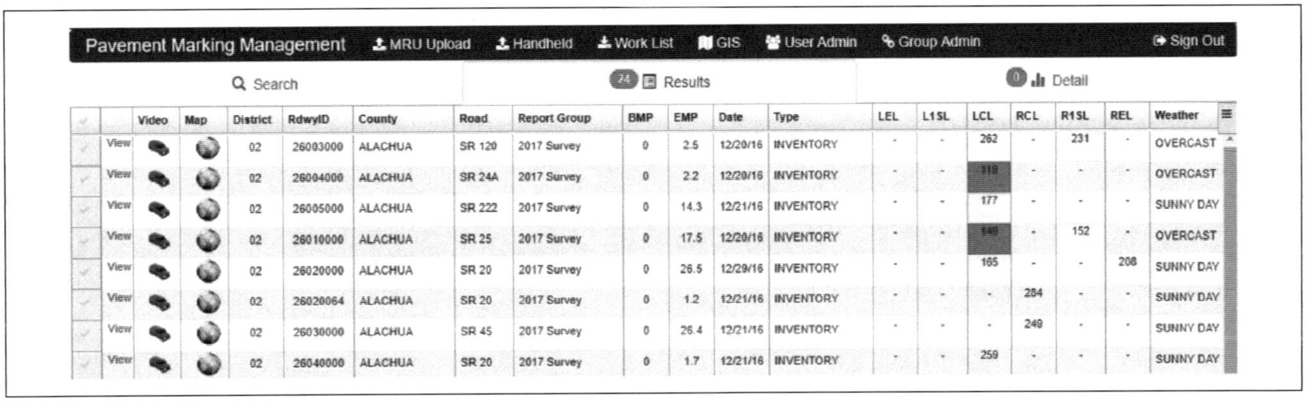

		Video	Map	District	RdwyID	County	Road	Report Group	BMP	EMP	Date	Type	LEL	L1SL	LCL	RCL	R1SL	REL	Weather
	View			02	26003000	ALACHUA	SR 120	2017 Survey	0	2.5	12/20/16	INVENTORY	-	-	262	-	231	-	OVERCAST
	View			02	26004000	ALACHUA	SR 24A	2017 Survey	0	2.2	12/20/16	INVENTORY	-	-	[illegible]	-	-	-	OVERCAST
	View			02	26005000	ALACHUA	SR 222	2017 Survey	0	14.3	12/21/16	INVENTORY	-	-	177	-	-	-	SUNNY DAY
	View			02	26010000	ALACHUA	SR 25	2017 Survey	0	17.5	12/20/16	INVENTORY	-	-	[illegible]	-	152	-	OVERCAST
	View			02	26020000	ALACHUA	SR 20	2017 Survey	0	26.5	12/29/16	INVENTORY	-	-	165	-	-	208	SUNNY DAY
	View			02	26020064	ALACHUA	SR 20	2017 Survey	0	1.2	12/21/16	INVENTORY	-	-	-	284	-	-	SUNNY DAY
	View			02	26030000	ALACHUA	SR 45	2017 Survey	0	26.4	12/21/16	INVENTORY	-	-	-	249	-	-	SUNNY DAY
	View			02	26040000	ALACHUA	SR 20	2017 Survey	0	1.7	12/21/16	INVENTORY	-	-	259	-	-	-	SUNNY DAY

Figure 8. Search results by road sections.

stripe, highlights may be employed to allow the end user to visually sort through the data. Red indicates that the stripe has an average retroreflectivity value of 150 mcd/m^2/lux or below, and yellow represents an average retroreflectivity between 150 and 250 mcd/m^2/lux. Stripes with an average retroreflectivity above 250 mcd/m^2/lux are not highlighted.

In addition, the results tab includes the ability to: 1) view any roadway section in 0.1-mile detail; 2) show video log images of the selected roadway section via FDOT's Video Log Viewer; and 3) visualize the data on FDOT's enterprise GIS application.

Figure 9 shows the detailed roadway section data and the graph from the "View" option. This option provides similar information as the results tab but at 0.1-mile intervals. Some additional information and features, such as speed limit, Average Annual Daily Traffic (AADT), roadway surface type, material type, and an interactive graph of the roadway section, are included.

The video link to the Video Log Viewer Application provides an image to visually understand characteristics of the roadway (e.g., intersections, bridges, dirt roads, etc.). An example image of this Video Log Viewer Application is shown in Figure 10.

The GIS Application link provides a map on which the average retroreflectivity is displayed at the District, County, Roadway section, or 0.1-mile segment levels depending on the zooming scales. The GIS Application also provides the ability to select a feature on the map to display additional details. An example map of this GIS Application is shown in Figure 11.

The PMMS database also includes the ability to export results tables as Excel files. These features will allow districts to better understand the performance of in-service pavement markings and make informed and timely maintenance decisions.

Completion Notice and Annual Reporting

In addition to providing data on the PMMS database to allow immediate access, a county completion notice is sent on completing the MRU survey in a county. This notifies the County and the District Offices that the data collection as part of the PMMS has been completed. They are asked to verify the data and notify any discrepancies within 45 days from the receipt of the notice.

Moreover, at the end of survey year, FDOT provides an annual report with essential information regarding conditions of pavement markings on the Florida State Highway System as collected during that year. The comprehensive database, county completion notification, and annual facts and figures reports for statewide pavement marking performance are examples of the rich information the PMMS provides for Districts to make data-driven decisions and manage statewide pavement markings in a timely and cost-effective manner.

Conclusion

This paper describes the development and implementation of FDOT Pavement Marking Management System using Mobile Retroreflectivity Units to monitor the retroreflective characteristics of in-service pavement markings in Florida. The required foundational efforts, including documentation, aggressive quality assurance metrics, and a comprehensive database were successfully defined and developed, and a robust PMMS for FDOT was implemented. The PMMS is in its fifth year and it has provided valuable lessons and quality data to the State of Florida for pavement marking performance evaluation and management. The collected data are used to analyze long-term performance of pavement marking materials and inform best practices for pavement marking management and contracting. Some key conclusions of the PMMS implementation are summarized in the following:

- MRU technology collects pavement marking retroreflectivity at highway speeds in a more efficient and objective way than manual and visual inspections.
- A comprehensive procedure, including the operation, precision, standardized test method of the testing

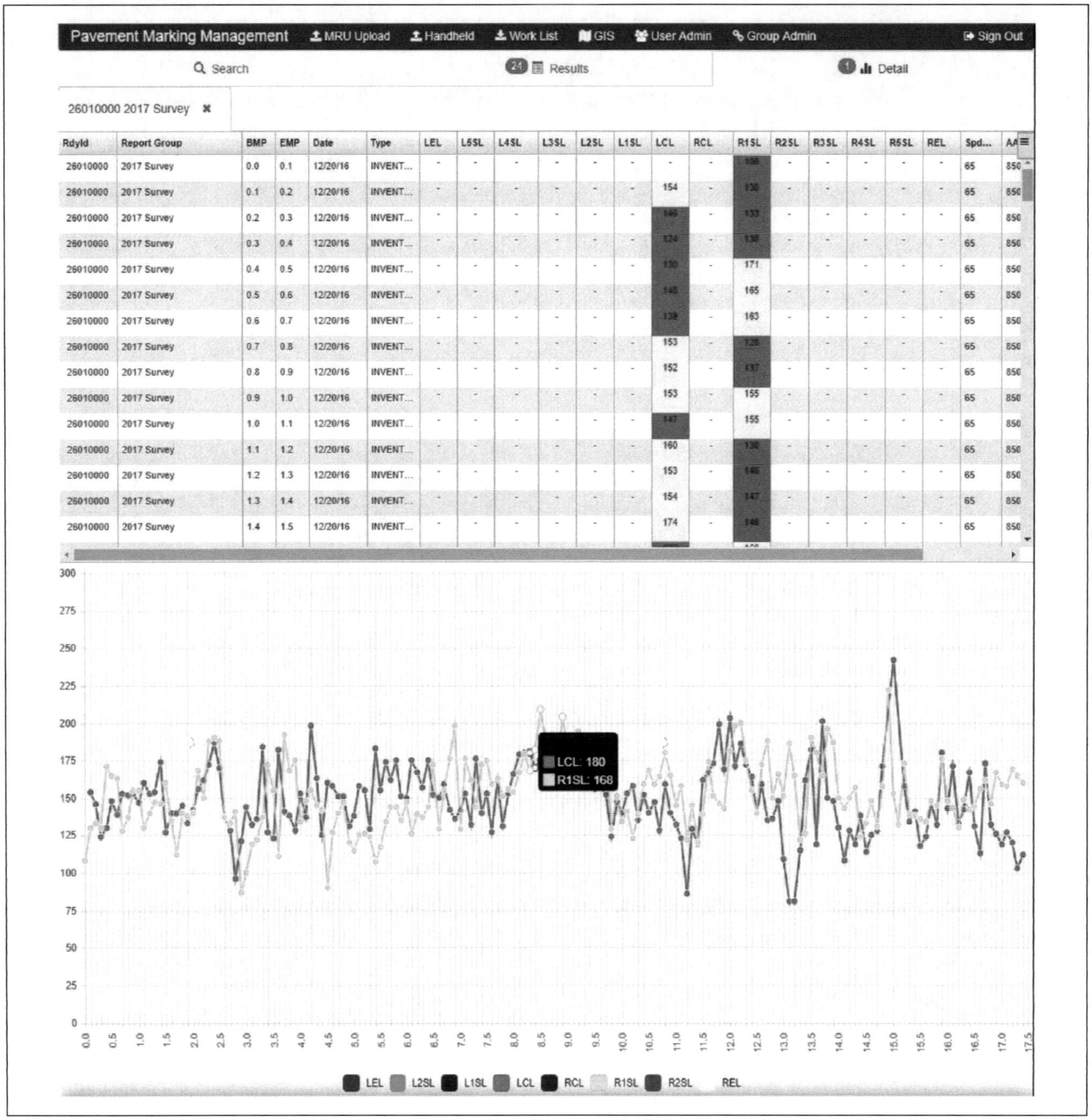

Figure 9. Detailed roadway section result table and chart.

equipment, and data worksheets of the retroreflectivity measurements, was developed to ensure robust and consistent data collection.

- A unique, comprehensive quality assurance procedure was initiated to conduct a series of laboratory and field tests and verify 20% of the stripes surveyed by the contractor to ensure the reliability and quality of retroreflectivity data.
- A state-of-the-art database was developed to manage pavement marking data and made accessible to FDOT personnel via intranet and a web-based interface with the ability to query and visualize data in a GIS application.

The PMMS provides an efficient and less subjective methodology than manual and visual inspections to monitor and assess the safety and nighttime visibility of our roadway system. It will result in a more effective/strategic use of state funds through informed decision-making processes while ensuring the safety of the traveling public. The

Figure 10. Video Log Viewer Application of the roadway section.

Florida TaxWatch organization has recently honored this development with a Productivity Award and highlighted that this work "significantly and measurably increases productivity and promotes innovation to improve the delivery of state services and save money for Florida taxpayers and businesses."

Acknowledgments

The work represented here was the result of a team effort. The authors would like to acknowledge the staff of FDOT's State Materials Office and Office of Information Technology for their assistance.

Author Contributions

The authors confirm contribution to the paper as follows: study conception and design: Bouzid Choubane; data collection: Joshua Sevearance and Charles Holzschuher; analysis and interpretation of results: Bouzid Choubane, Joshua Sevearance, Charles Holzschuher, James Fletcher, and Chieh (Ross) Wang; draft manuscript preparation: Bouzid Choubane, Joshua Sevearance, Charles Holzschuher, James Fletcher, and Chieh (Ross) Wang. All authors reviewed the results and approved the final version of the manuscript.

References

1. Federal Highway Administration. *Nighttime Visibility – Safety*. Federal Highway Administration. https://safety.fhwa.dot.gov/roadway_dept/night_visib/. Accessed July 27, 2017.
2. Public Law. *Department of Transportation and Related Agencies Appropriations Act*, Congress Public Law, 1993.
3. Federal Highway Administration. *Notice of Proposed Amendments*. FHWA Docket No. FHWA-2009-0139. Federal Highway Administration, Washington, D.C., 2010.
4. Federal Highway Administration. Manual on Uniform Traffic Control Devices (MUTCD). Federal Highway Administration. https://mutcd.fhwa.dot.gov/. Accessed July 27, 2017.
5. Federal Highway Administration. *Supplemental Notice of Proposed Amendments*. FHWA Docket No. FHWA-2009-0139. Federal Highway Administration, Washington, D.C., 2017.
6. ASTM International. *ASTM Standard E808, Standard Practice for Describing Retroreflection*. ASTM International, West Conshohocken, Pa., 2001.
7. Choubane, B., J. Sevearance, H. Lee, P. Upshaw, and J. Fletcher. Repeatability and Reproducibility of Mobile Retroreflectivity Units for Measurement of Pavement Markings. *Transportation Research Record: Journal of the Transportation Research Board*, 2013. 2337: 74–82.
8. Fletcher, J., A. Philpott, B. Choubane, and C. Holzschuher. Characterization and Mitigation of Temperature Sensitivity within Mobile Retroreflectometer Unit. *Transportation Research Record: Journal of the Transportation Research Board*, 2007. 2015: 91–102.
9. Holzschuher, C., B. Choubane, J. Fletcher, J. Sevearance, and H. Lee. Repeatability of Mobile Retroreflectometer Unit for Measurement of Pavement Markings. *Transportation Research Record: Journal of the Transportation Research Board*, 2010. 2169: 95–106.
10. ASTM International. *ASTM Standard E1710, Standard Test Method for Measurement of Retroreflective Pavement Marking Materials with CEN-Prescribed Geometry Using a Portable Retroreflectometer*. ASTM International, West Conshohocken, Pa., 2011.

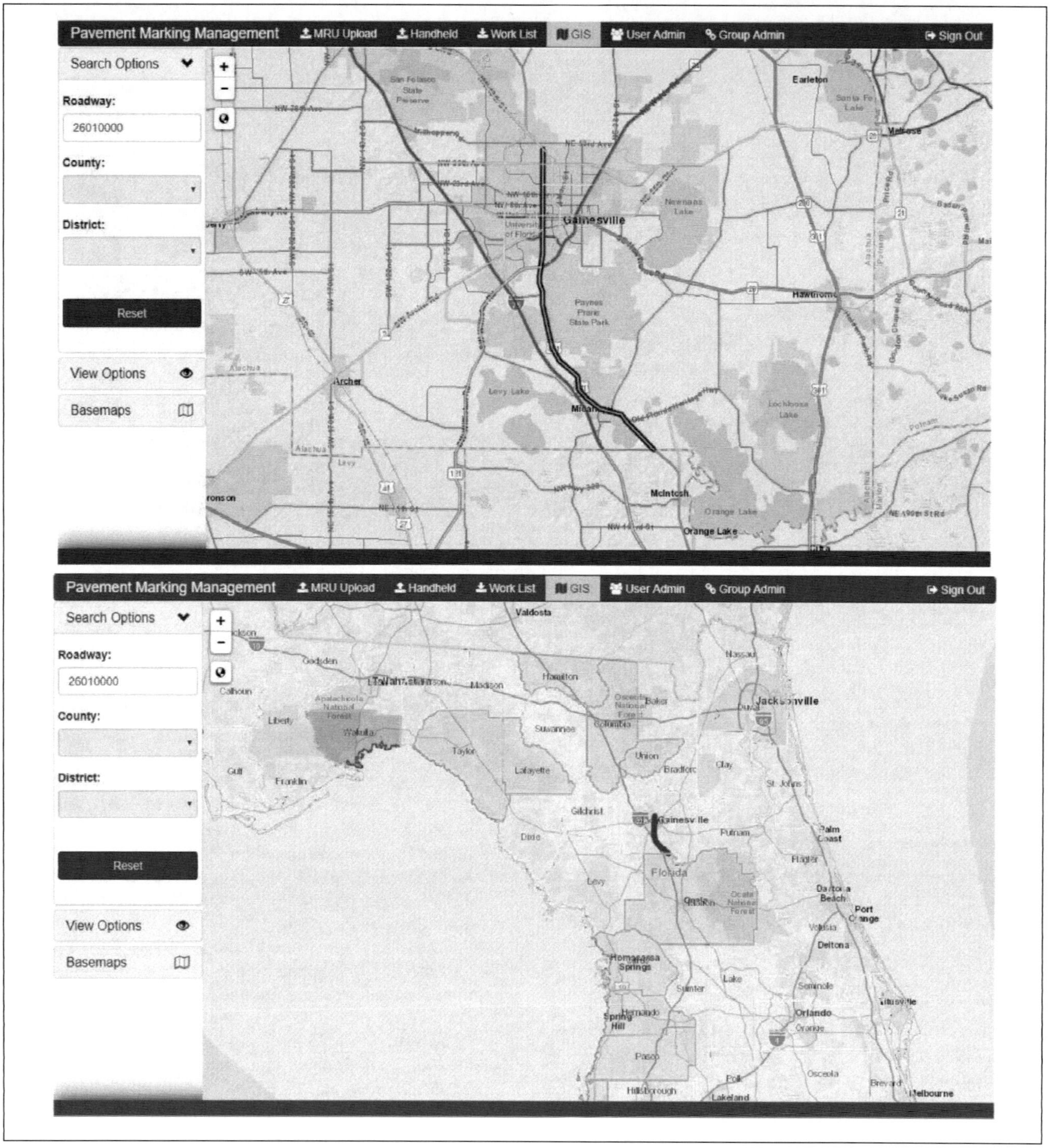

Figure 11. A GIS application for PMMS.

11. Fletcher, J., J. Sevearance, B. Choubane, and H. Charles. Characteristics of Calibration Standard for Mobile Retroreflectometer Unit. Presented at 88th Annual Meeting of the Transportation Research Board, Washington, D.C., 2009.
12. Gonos, J. S., J. Fletcher, H. Holzschuher, and B. Choubane. Calibration Standards for Mobile Retroreflectometer Units. Presented at 87th Annual Meeting of the Transportation Research Board, Washington, D.C., 2008.
13. Florida Department of Transportation. *Florida Test Method for Measuring Retroreflectivity of Pavement Marking Materials Using a Mobile Retroreflectivity Unit (FM5-600)*. Florida Department of Transportation. www.fdot.gov/materials/

administration/resources/library/publications/fstm/methods/fm5-600.pdf. Accessed January 12, 2016.
14. Upshaw, P., J. Sevearance, and H. Lee. *Operations Manual for Mobile Retroreflectivity Units*. Florida Department of Transportation. www.fdot.gov/materials/pavement/performance/ndt/documents/operationsmanual.pdf. Accessed February, 2013.
15. Florida Department of Transportation. *Mobile Retroreflectivity Unit (MRU) Worksheet (Form 675-060-15)*. Florida Department of Transportation. https://fms.fdot.gov/Anonymous/SendDocumentToClient?documentId=1208. Accessed January, 2016.
16. Sevearance, J., and C. Holzschuher. *Quality Assurance for Mobile Retroreflectivity Units*. Florida Department of Transportation. www.fdot.gov/materials/pavement/performance/ndt/documents/qualityassurance.pdf. Accessed June, 2012.

The Standing Committee on Signing and Marking Materials (AHD55) peer-reviewed this paper (18-01510).

The content of this paper reflects the views of the authors who are solely responsible for the facts and accuracy of the data as well as for the opinions, findings, and conclusions presented here. The contents do not necessarily reflect the official views or policies of the Florida Department of Transportation. This paper does not constitute a standard, specification, or regulation. In addition, the above listed agency assumes no liability for its contents or use thereof.

TRR
JOURNAL OF THE TRANSPORTATION RESEARCH BOARD

Article

Transportation Research Record
2018, Vol. 2672(12) 220–231

DOI: 10.1177/0361198118772959
journals.sagepub.com/home/trr
SAGE

Modeling Surface Temperatures for Snow-Covered Roads: A Case Study for a Heated Pavement in Bærum, Norway

Anne D. W. Nuijten[1], Inge Hoff[1], and Knut V. Høyland[2]

Abstract

Heated pavements are used as an alternative to removing snow and ice mechanically and chemically. Usually a heated pavement system is automatically switched on when snowfall starts or when there is a risk of ice formation. Ideally, these systems run based on accurate predictions of surface conditions a couple of hours ahead of time, for which both weather forecasts and reliable surface temperature predictions are needed. The effective thermal conductivity of the snow layer is often described as a function of its density. However the thermal conductivity of a snow layer can vary considerably, not only for snow samples with a different density, but also for snow samples with the same density, but with a variation in the liquid water content. In this paper a physical temperature and surface condition model is described for snow-covered roads. The model is validated for an entire winter season on a heated pavement in Norway. Two different models to describe the thermal conductivity through the snow layer were compared. Results show that the thermal conductivity of the snow layer can be best described as a function of the density for snow with a low liquid water content. For snow with a high water content, the thermal conductivity can be best described as a function of the volume fractions and thermal conductivity of ice, water, and air, in which air and ice are modeled as a series system and water and air/ice in parallel.

During winter conditions, snow and frost formation on roads can cause slippery conditions decreasing mobility and having a detrimental effect on safety. To prevent slippery conditions, winter maintenance measures are taken including snow plowing, sanding, or applying de-icing agents. An alternative to removing snow and ice mechanically and chemically is the use of heating. In Nordic countries, heated pavements are frequently used to keep the pavement free of snow and ice in pedestrian areas, on bicycle paths, and in other places where snow removal is difficult.

Heating of pavements can be an effective way of removing snow and ice but often has high operational costs. Currently it is difficult to predict accurately the energy demands of a heated pavement system at a specific location. It is also challenging to determine how the surface condition (layer of snow, slush, ice on top of the pavement) changes, since not only is the heating changing the surface condition, but the surface condition is also impacted by the weather and traffic. In addition, it is not clear what is the most efficient way to control such a system. This results in higher costs than necessary. Numerous efforts have been undertaken to optimize such systems by: increasing the thermal conductivity of the asphalt layer, optimizing the location and type of heating used, or optimizing the control system (*1, 2*).

Usually a heated pavement system is automatically switched on when snowfall starts or when there is a risk of ice forming. Many systems run on measured surface temperatures and react when the pavement is covered with snow or ice. Ideally, these systems run based on accurate predictions of the surface condition a couple of hours ahead of time, for which both weather forecasts and reliable surface temperature predictions are needed. A prediction of meteorological data such as the snowfall is usually available. It is therefore of interest to predict the surface temperature also, from two to three hours up to a day in advance of snowfall.

[1]Department of Civil and Environmental Engineering, Norwegian University of Science and Technology, Trondheim, Norway
[2]Norwegian University of Science and Technology, Sustainable Arctic Marine and Coastal Technology (SAMCoT), Centre for Research-based Innovation (CRI), Trondheim, Norway

Corresponding Author:
Address correspondence to Anne D. W. Nuijten: annenuijten@gmail.com

Pavement surface temperature and condition prediction models have been developed and described in a number of studies (*3–9*). Most of these models do not include the effect of the surface layer (the layer of snow, slush, ice on top of the pavement) on the thermal balance of the system. When the pavement is covered with dry uncompressed snow, the snow layer behaves as an ideal insulator and has a notable effect on the thermal balance.

Some of the prediction models which include winter conditions have been developed for the modeling of snow melting systems (*10–13*). Most of the previously developed models of snow melting systems are steady-state systems. One of the earliest models of this kind was developed by Chapman (*10*). Later models were developed based on transient conditions, such as those developed by Liu et al. (*11*), Rees et al. (*12*) and Xu and Tan (*13*). In the model developed by Liu et al. (*11*), which was a modification of the model of Rees et al. (*12*), a snow layer was included which was either a dry snow layer, a slush layer, or a dry snow layer on top of a slush layer. The dry snow layer was assumed to be homogeneous and porous, the slush layer was assumed to be isothermal. The thermal conductivity of the snow layer (k_s) was described as $k_s = 2.22362 \cdot \rho_s^{1.885}$ as described by Yen (*14*), in which ρ_s is the density of the snow layer.

The effective thermal conductivity of a snow layer is often described as a function of the density (*14–17*). Most relationships between the thermal conductivity and density are based on data retrieved from thermal conductivity measurements of various types of seasonal snow with densities up to 600 kg m^{-3} and for snow with a low liquid water content. However the thermal conductivity of a snow layer can vary considerably, not only for snow samples with different densities but also for snow samples with the same density but with variation in the liquid water content.

In this paper a physical temperature and surface condition model is described for snow-covered roads and heated pavements, which includes the thermal conductivity through the snow layer. The model can be used to identify risk of slippery conditions and to determine how much heat is needed to maintain usable and safe roads in cold climates. The model is validated for an entire winter season on a heated pavement in Norway, but can also be applied to unheated roads. Two different models to describe the thermal conductivity through the snow layer were compared. Results show that the thermal conductivity of the snow layer can be best described as a function of the density, as found by Sturm et al. (*17*) for snow with a low liquid water content. For snow with a high water content, the thermal conductivity can be best described as a function of the volume fractions and thermal conductivity of ice, water, and air as proposed by Nuijten and Høyland (*18*).

Surface Temperature Prediction Model

Outline of the Model

Figure 1 gives a schematic overview of the surface temperature and condition prediction model for heated pavements. The pavement surface temperatures are predicted based on the heat fluxes at the pavement surface. Some of these fluxes are affected by the surface condition.

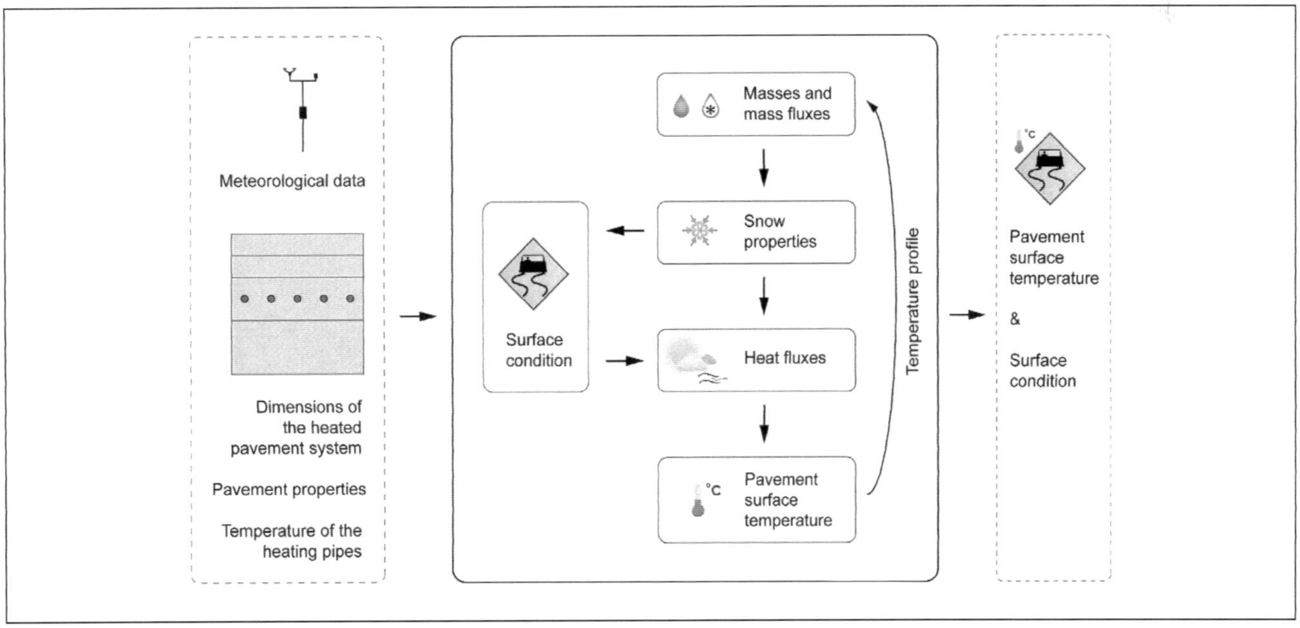

Figure 1. Overview of the surface temperature and condition prediction model for heated pavements.

The surface condition is calculated based on the mass fluxes of ice and water. Input parameters are:

- meteorological data: air temperature, relative humidity, wind speed, the amount of precipitation, and the incoming and outgoing longwave and shortwave radiation,
- dimensions of the pavement and heated pavement system,
- thermophysical pavement properties: density, specific heat capacity and thermal conductivity, and
- temperature on the outside of the heating pipes.

Mass Fluxes

The mass fluxes included in the model are the mass flux of water and ice.

Mass Balance of Water. The mass flux of water $\dot{m}_w$ (kg $m^{-2}s^{-1}$) is determined by adding up the mass flux of rain $\dot{m}_{wp}$ (kg m^{-2} s^{-1}), the mass flux of melting and freezing $\dot{m}_m$ (kg $m^{-2}s^{-1}$) and the mass flux due to runoff $\dot{m}_{wr}$ (kg $m^{-2}s^{-1}$):

$$\dot{m}_w = \dot{m}_{wp} + \dot{m}_m - \dot{m}_{wr} \tag{1}$$

$\dot{m}_m$ is given by

$$\dot{m}_m = \frac{q_{tot}}{L_f} \tag{2}$$

where q_{tot} is the total heat flux at the pavement surface (W m^{-2}) and L_f is the latent heat of fusion ($334 \cdot 10^3$ J kg^{-1}). If the total heat flux toward the pavement surface is positive, ice or snow is present at the pavement surface and if the pavement surface temperature is at the melting temperature then melting occurs. $\dot{m}_{wr}$ is given by

$$\dot{m}_{wr} = (h_w - h_{w,min}) \cdot \rho_w / \Delta t \tag{3}$$

where h_w is the height of the water level (m) calculated based on the mass of water on the surface and the density, $h_{w,min}$ is the moisture level below which drainage does not occur (taken as 0.5 mm), ρ_w is the density of the water (kg m^{-3}) and Δt is the time step (s). For water levels below $h_{w,min}$ $\dot{m}_{wr}$ is 0.

Mass Balance of Ice. The mass flux of ice $\dot{m}_i$ (kg $m^{-2}s^{-1}$) is determined by adding up the mass flux of snowfall $\dot{m}_{ip}$ (kg $m^{-2}s^{-1}$) and the mass flux of melting and freezing $\dot{m}_m$ (kg $m^{-2}s^{-1}$):

$$\dot{m}_i = \dot{m}_{ip} - \dot{m}_m \tag{4}$$

The mass fluxes of sublimation, deposition, evaporation, and condensation are not included in the model.

Snow Properties

The properties of the snow layer are assumed to be uniform throughout the layer. Two different models to describe the thermal conductivity through the snow layer were compared. The first is the formula by Sturm et al. (*17*) in which the thermal conductivity is described as a function of the density:

$$\lambda_s = 10^{(2.650\rho_s - 1.652)} \tag{5}$$

This formula is based on data retrieved from thermal conductivity measurements of various types of seasonal snow with densities up to 600 kg m^{-3} and for snow with a low liquid water content.

The second is the formula by Nuijten and Høyland (*18*) which is developed specifically to determine the thermal conductivity of a melting snow layer, and validated with measurements of a melting uncompressed and compressed snow layer. The thermal conductivity of the snow layer λ_s is calculated based on the thermal conductivity and volume fractions of ice, water, and air as follows (*18*):

$$\lambda_s = \frac{\theta_i + \theta_a}{\left(\frac{\theta_i}{\lambda_i} + \frac{\theta_a}{\lambda_a}\right)} + \lambda_w \cdot \theta_w \tag{6}$$

where λ_i, λ_w, and λ_a are the thermal conductivities of ice, water, and air and θ_i, θ_w, and θ_a are the volume fractions of ice, water, and air. θ_i and θ_w are given as

$$\theta_i = \frac{M_i \cdot \rho_s}{M_s \cdot \rho_i} \tag{7}$$

$$\theta_w = \frac{M_w \cdot \rho_s}{M_s \cdot \rho_w} \tag{8}$$

where M_i, M_w, and M_s, are the masses of ice, water, and the snow layer (kg m^{-2}), ρ_s is the density of the snow layer (kg m^3), and ρ_i and ρ_w are the densities of ice and water. M_s is calculated as the sum of the mass of ice M_i and water M_w. The density of the snow layer ρ_s is calculated as follows:

$$\rho_s = \frac{M_s}{h_s} \tag{9}$$

where h_s is the height of the snow layer which is calculated as follows:

$$h_s^{t+1} = h_s^t + \dot{h}_{ip} - \dot{h}_m \tag{10}$$

where h_s^t and h_s^{t+1} are the heights of the snow layer during the current and next time step, $\dot{h}_{ip}$ is the height of the freshly fallen snow during the current time step and $\dot{h}_m$ is the height of the layer of snow which is melted. The heights $\dot{h}_{ip}$ and $\dot{h}_m$ are calculated based on the mass flux

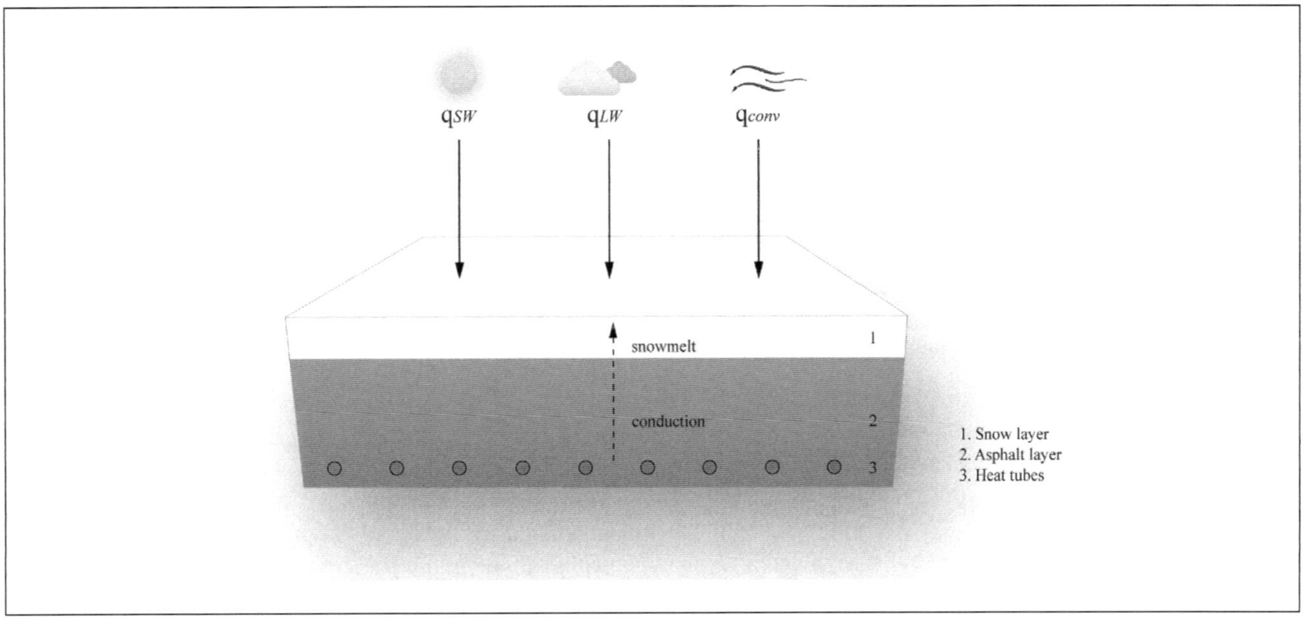

Figure 2. Overview of the heat fluxes included in the model.

of snowfall and melting respectively and the density of the snow. The density of freshly fallen snow is assumed to be 100 kg m^{-3}. Due to traffic the snow will be compressed. Since no traffic data is available it is assumed that the density of the freshly fallen snow is increased to 500 kg m^{-3} after it reaches the ground.

Surface Condition

The surface condition prediction is based on the following surface condition categories: dry, wet, dry snow, wet snow, slush, hoarfrost, and ice. For the surface conditions "dry snow", "wet snow," and "slush" the ICSI classification system is used to determine the condition based on the water content (*19*). Snow with a liquid water content of 0% is identified as dry snow, snow with a liquid water content up to 15% is identified as wet snow, and snow with a liquid water content above 15% is identified as slush.

Heat Fluxes

The following heat fluxes are included in the model:

- q_{conv}: convective heat flux
- q_{LW}: longwave radiation
- q_{SW}: shortwave radiation
- q_{pav}: conductive heat flux through the pavement
- q_{snow}: conductive heat flux through the snow layer
- q_m: melting flux

As described by Nuijten (*20*), the heat fluxes of evaporation, condensation, sublimation, and deposition are relatively small compared with other heat fluxes such as those for shortwave radiation, longwave radiation, convection, and conduction. These fluxes are therefore not taken into account in the model.

The model includes the conduction through the snow when a snow layer is present. When there is a snow layer on the pavement, the shortwave radiative heat flux, longwave radiative heat flux, and convective heat flux are expected to affect the temperature at the top of the snow layer. An overview of the heat balance system is shown in Figure 2.

Convection. The convective heat flux q_{conv} (W m^{-2}) is given by (*21*)

$$q_{conv} = -h_c \cdot \left(T_{p0} - T_a\right) \tag{11}$$

where h_c is the convective heat transfer coefficient (in W m^{-2} K^{-1}) and T_a is the air temperature (°C). Formulas for the convective heat transfer coefficient have been described by a number of authors (*5, 7, 22–24*). Preliminary testing of the formulas showed that the formula for the sensible heat flux due to natural wind as described by Fujimoto (*23*) gave the best results. The formula is

$$h_c = 10.4 \cdot v_w^{0.7} + 2. \tag{12}$$

where v_w is the wind speed.

Longwave and Shortwave Radiation. The incoming and outgoing longwave radiative heat flux and the incoming shortwave radiative heat flux are measured. The absorbed shortwave radiation depends on the surface condition. The shortwave or solar radiation absorbed by the pavement q_{SW} (W m^{-2}) is given by

$$q_{SW} = q_{SW,in} \cdot (1 - \propto) \tag{13}$$

where $q_{SW,in}$ is the incoming shortwave radiation (W m^{-2}) and α is the albedo. The albedo values used in the model are 0.15 for dry asphalt, 0.20 for a wet surface, and 0.60 for a road covered with a dry snow layer. These values are based on (*22, 25–27*), and take into account that the pavement is only covered with a thin layer of snow.

Conduction through the Pavement. The conductive heat flux at the pavement surface q_{cond_p} (W m^{-2}) is given by (*21*)

$$q_{cond\_p} = \lambda_p \frac{\Delta T_p}{\Delta z} \tag{14}$$

where λ_p is the thermal conductivity of the pavement (W m^{-1}K^{-1}) and ΔT_p is the temperature difference between the pavement surface and the pavement temperature at a distance Δz from the surface (°C).

Conduction through the Snow. The conduction through the snow layer is calculated as

$$q_{cond\_s} = \lambda_s \frac{\Delta T_s}{h_s} \tag{15}$$

Where λ_s is the thermal conductivity through the snow layer, ΔT_s is the temperature difference between the top and bottom of the snow layer, and h_s is the height of the snow layer.

Melting Flux. When the pavement surface is covered with a snow or ice layer and the pavement surface temperature is above freezing point there is a melting flux which is calculated as follows:

$$q_m = q_{cond\_p} + q_{SW} - q_{cond\_s} \tag{16}$$

In all other cases the melting flux is 0.

Calculating the Pavement and Snow Temperatures

The pavement and snow temperatures are predicted based on the temperature diffusion equation:

$$\lambda \frac{\partial^2 T}{\partial z^2} = \rho c_p \frac{\partial T}{\partial t} \tag{17}$$

where λ is the thermal conductivity (W m^{-1}K^{-1}), ∂T is the change in pavement temperature (°C), ∂z is the node distance (m), ∂t is the time step (s), ρ is the density of the pavement or snow (kg m^{-3}), and c_p is the specific heat capacity (J kg^{-1}K^{-1}). An explicit finite difference method is used to determine the pavement and snow temperature at various depths. The pavement and snow layers are divided into layers with a vertical distance of 5 mm, for which at each time step the temperature is determined based on the temperature during the previous time step, the surface heat flux, and the thermal properties of the pavement and snow layer. The height of the snow layer is rounded up to a multiple of 5 mm, whereby snow layers smaller than 1 mm are taken as zero. A time step of one minute was used when the model was run for the entire winter season. When the model was run for a shorter period in January, a time step of 0.1 second was used to meet the stability conditions. The heating pipes are at a depth of 120 mm. The temperature at that depth is measured and used as the input value for the model. During snow melting the temperature on top of the pavement layer is set at 0°C.

The temperature on top of the snow layer is calculated as follows:

$$\rho_s c_s \frac{\Delta T_s}{\Delta t} = \frac{q_{LW} + q_{conv} + q_{SW} + q_{cond\_s}}{0.5 \cdot \Delta z} \tag{18}$$

where ρ_s is the density of the snow layer, c_s is the specific heat capacity of the snow, ΔT is the change in temperature of the snow layer, Δt is the time step, and Δz is the node distance. If there is no snow layer the same formula is used to calculate the pavement surface temperature whereby ρ_s, c_s, and q_{cond_s} are replaced by ρ_p, c_p, and q_{cond_p}.

Model Input Data

Overview Heated Pavement System

Validation of the model is based on measured meteorological data from the winter of 2015–2016 and estimated pavement properties of the newly installed heated pavement system at the Willy Greiners Vei (road), next to the Sandvika train station. Sandvika is a town in the municipality of Bærum which is located close to Oslo, Norway, at 59°53′34″N, 10°31′32″E at 12 m above sea level (Figure 3). The heated pavement system was built during the winter of 2014–2015. The heated pavement area consists of a pedestrian area and a road with a roundabout. The road consists of the following layers: 40 mm surface course, 40 mm binder course, 180 mm base course, and 300 mm subbase. At a depth of 120 mm heating pipes are placed with a horizontal distance between the pipes of 50 mm. Warm water is distributed over the area through loops of pipe. At two locations within one of these loops, one at the beginning of the loop and one at the end, the temperature is measured at various depths. The

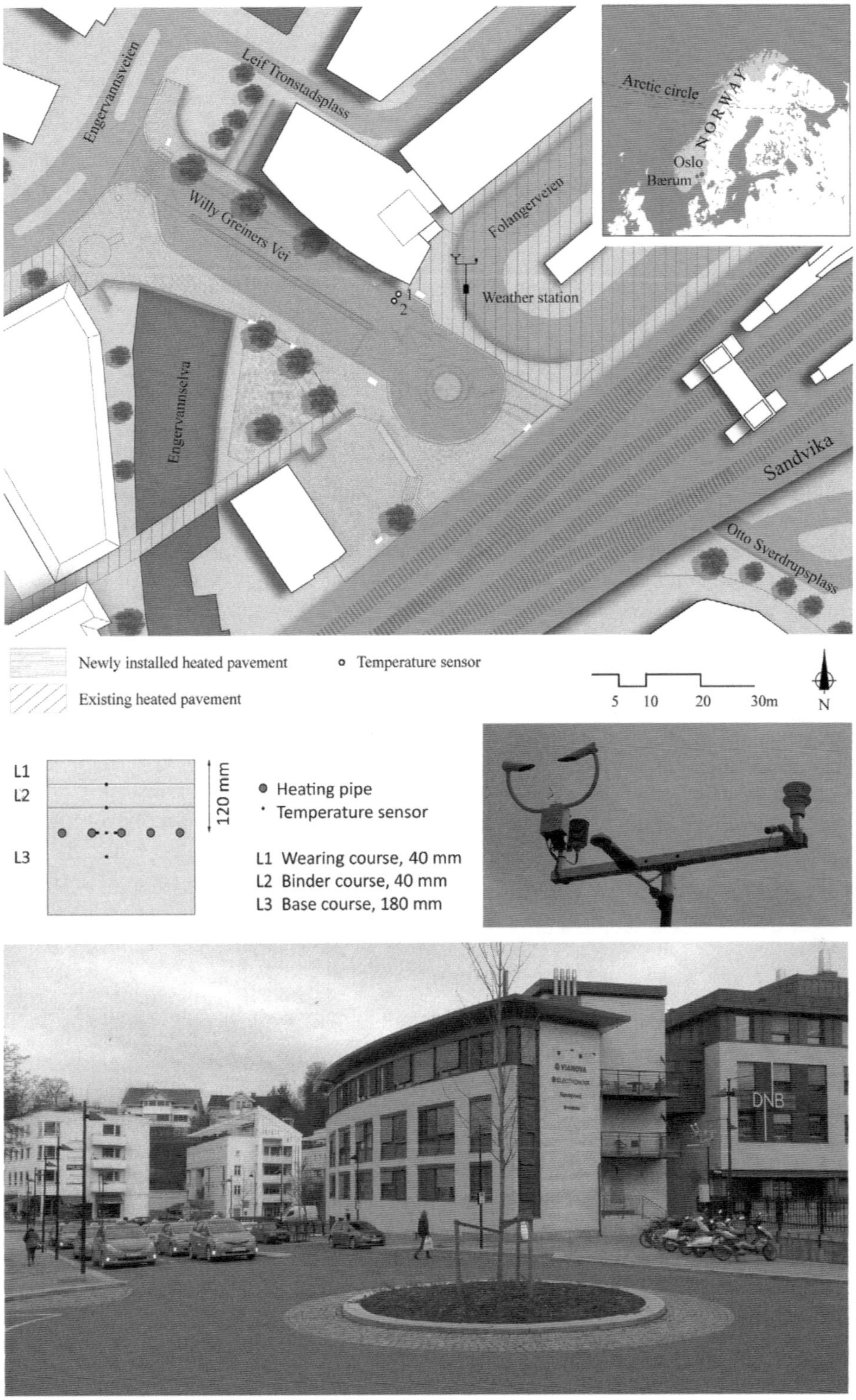

Figure 3. Overview and cross section of the heated pavement system and location of the weather station and temperature sensors. Middle right: Weather station located next to the heated pavement system.

Table 1. Overview of the Measured Data, Sensors, and Measuring Frequency

No.	Measured data	Type of sensor/logger	Frequency
1	Air temperature (°C)	CS215	1 min
2	Precipitation (mm)	CS125	1 min
3	Surface temperatures (°C)	Apogee SI-411	1 min
4	Pavement temperatures (°C)	Pt100 1/10 DIN	1 min
5	Incoming and outgoing shortwave radiation ($W\ m^{-2}$)	CNR4 NET radiometer	1 min
6	Incoming and outgoing longwave radiation ($W\ m^{-2}$)	CNR4 NET radiometer	1 min
8	Wind speed ($m\ s^{-1}$) and direction	V200A Ultrasonic wind sensor	10 min
9	Relative humidity	CS215	60 min

temperatures measured on top of the heating pipes are higher at the beginning of the loop than at the end. The temperature is measured with Pt-100 sensors at a depth of 40, 80, 120, and 160 mm and on top of and in between two of the heating pipes. The properties of the asphalt layers are based on values found in the literature: a dry density (ρ) of 2100 $kg\ m^{-3}$, a specific heat capacity (c_p) of 0.92 $kJ\ kg^{-1}K^{-1}$, and a thermal conductivity (λ) of 0.74–2.89 $W\ m^{-1}K^{-1}$ (*26*).

Meteorological Data

A weather station is located on the sidewalk on the north side of the heated pavement system, about 10 m from where the pavement temperatures are measured. The surface temperature is measured with an Apogee SI-411 sensor close to the weather station. The weather station also stores the data of the pavement temperatures in the locations indicated in Figure 3. An overview of the measured data, the sensors, and their measuring frequency is given in Table 1.

Results

Temperature Prediction

The pavement surface temperature was predicted for two locations of the heated pavement system in Bærum (see Figure 3), during the period between November 1, 2015 and April 1, 2016. The model was run in long-term prediction mode using one-minute data. Measured meteorological data is used instead of forecasts. In this way, the performance only reflects the deviations of the model and not the errors in the forecasted input data. In this mode the thermal conductivity through the snow layer was not measured.

Figure 4*a* shows the predicted surface temperatures for the two locations and the measured surface temperatures close to the weather station. The figure shows that the model is stable and errors only occur during a short time period and do not have a long-term effect on the results. The average error for location 1 ($T_{predicted}-T_{measured}$) is 0.85°C and the root mean square error (RMSE) is 1.73°C. The average error and RMSE for location 2 are 0.35°C and 1.59°C respectively. The predicted surface temperatures are higher at location 1 than at location 2. This is due to the higher temperature of the heating pipes used as input for the model. The model tends to overestimate the surface temperature for surface temperature above −5°C and underestimate the surface temperature for temperatures below −5°C.

The effect of the accuracy of the following input variables on the prediction error and RMSE was investigated: air temperature, temperature on the outside of the heating pipes, longwave radiation, shortwave radiation, wind speed, and thermal conductivity of the asphalt. The variables having most effect on the RMSE are longwave and shortwave radiation. By increasing the longwave radiation by 10% or decreasing the shortwave radiation by 5%, the RMSE is reduced by 4–5%. Changing the value for the thermal conductivity of the asphalt layer from 1.0 to 1.5 $W\ m^{-1}K^{-1}$ increases the RMSE by 5%. A change of 0.5°C in air or heating pipe temperature affects the RMSE by 2.5%.

Figure 4*b* shows the temperatures measured on the outside of the heating pipes at the start and end of one of the loops. The peak temperatures show when the system is switched on. Figure 4*c* shows the measured surface temperature and dew point temperature and Figure 4*d* shows the amount of snowfall and rainfall throughout the winter of 2015–2016 near the heated pavement system. Figure 4*b* shows that the heating system is switched on during snowfall, but also during rainfall with pavement surface temperatures below 0°C.

Effect of Including the Thermal Conductivity through the Snow Layer

The highest snow precipitation of the winter season 2015–2016 was measured in January. For the period of January 7 to 14, 2016, a multiple day temperature and condition prediction was completed. The thermal conductivity through the snow layer was included in the

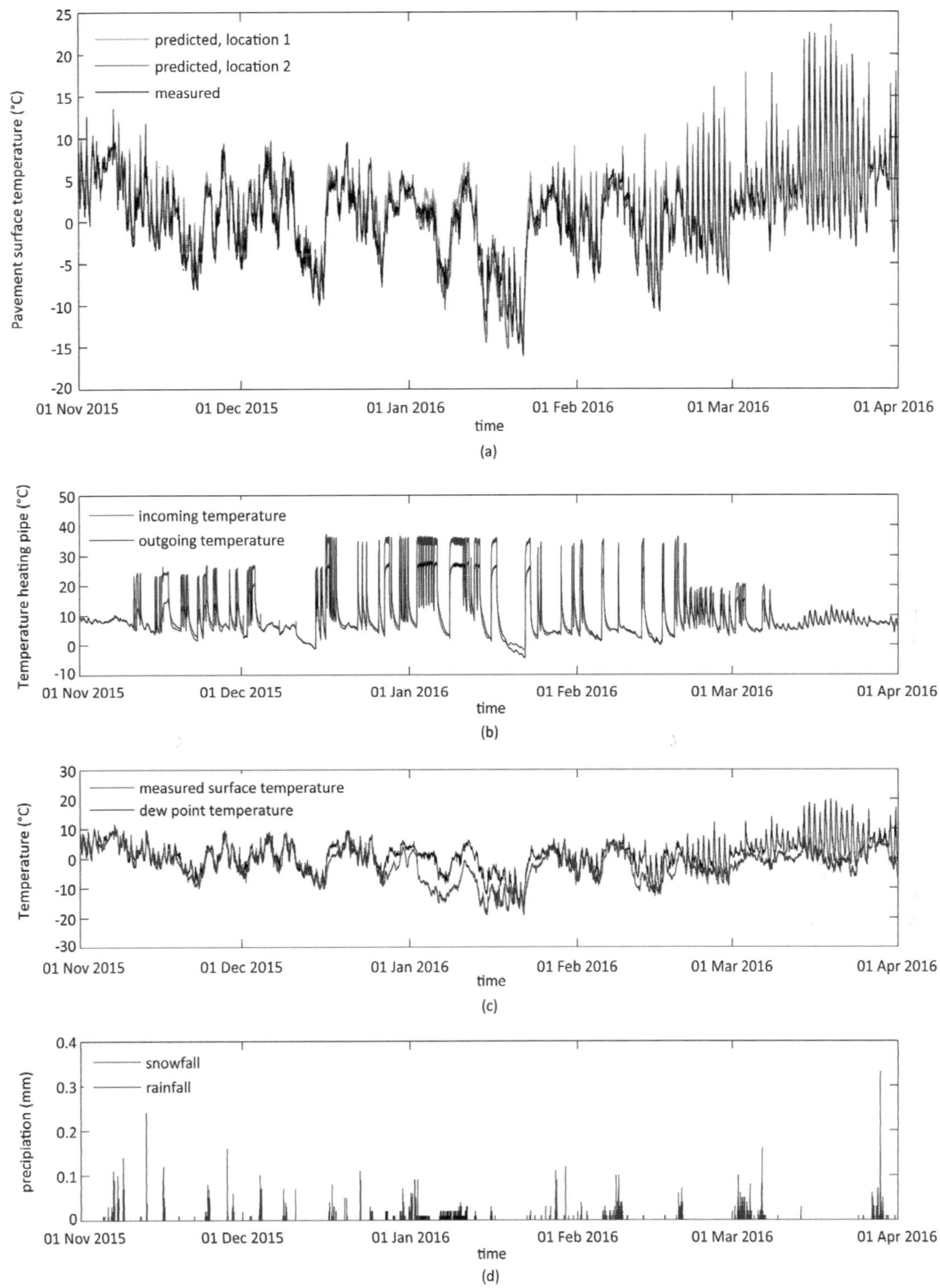

Figure 4. (*a*) Predicted and measured pavement surface temperatures of the heated pavement system in Sandvika during the winter season 2015–2016, (*b*) Incoming and outgoing temperature measured on top of the heating pipes, (*c*) Measured surface temperature and dew point temperature, and (*d*) Amount of precipitation during the winter season of 2015–2016.

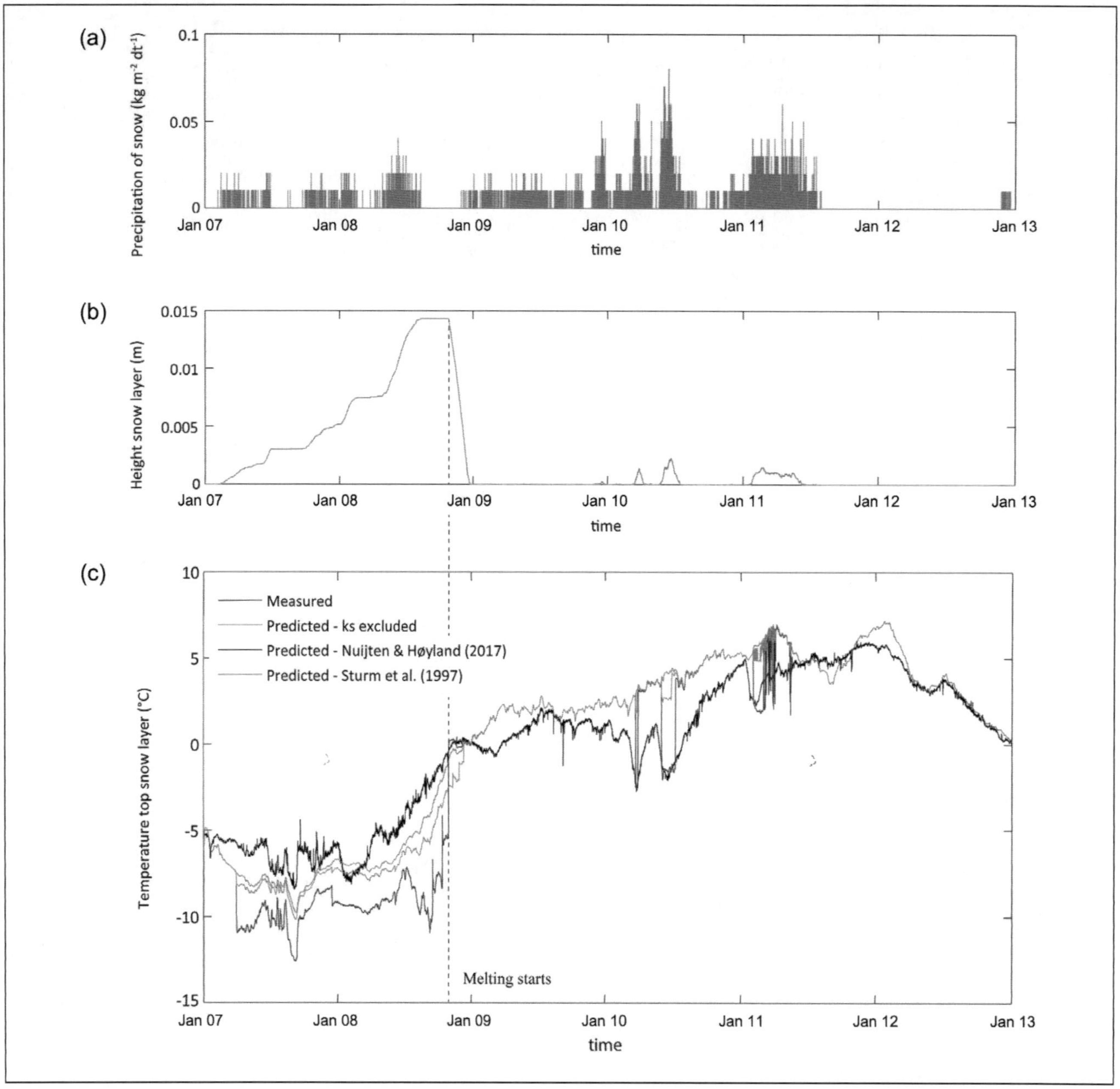

Figure 5. Results for January 2016: (*a*) Precipitation of snow, (*b*) Predicted height of the snow layer, and (*c*) Measured and predicted temperature on top of the snow layer. The dark blue line shows the predicted temperatures for the model including the thermal conductivity through the snow layer. The light blue line shows the prediction excluding the thermal conductivity through the snow layer.

model. Figure 5 shows the predicted height of the snow layer and the measured and predicted temperature on top of the snow layer. The results are shown for one scenario where the thermal conductivity through the snow layer was not included in the model and two scenarios where it was. The thermal conductivity through the snow layer was calculated based on the formula by Sturm et al. (*17*), which relates the thermal conductivity to the density of the snow, and based on the formulas developed by Nuijten and Høyland (*18*), which is used for a melting snow layer. Excluding the thermal conductivity results in a RMSE of 1.74°C, the RMSE based on the formula by Sturm et al. (*17*) and Nuijten and Høyland (*18*) are 1.81°C and 2.43°C respectively.

On January 8 the model in which the thermal conductivity through the snow layer is excluded gives a better prediction than the models in which the thermal conductivity through the snow is included. On January 10 and 11, however, the predictive performance of the model is greatly improved by including the thermal conductivity through the snow layer. As can be seen in Figure 5, the predictive performance of the formula by Sturm et al. (*17*) gave the best results during the first period, but still slightly worse results than the model in which the thermal conductivity through the snow layer was excluded. The formula by Nuijten and Høyland (*18*) gave the best results during the period starting on January 9, with a RMSE of 1.71°C compared with a RMSE of 2.02°C using the formula of Sturm et al. (*17*) or a RMSE of 2.16°C by excluding the thermal conductivity through the snow layer.

Discussion

Predicted Surface Temperatures

The model presented in this paper is stable and gives reasonably accurate results using historical metrological data. The quality of the weather forecast will obviously have an influence on the performance of the system. Differences in the measured and predicted temperatures can be explained by the difference in location of the measured surface temperature and pavement temperatures. The surface temperature is measured near the weather station, which is not exactly the same location as the location for which the predictions are made (Figure 3). Also it is possible that the surface temperature is measured partly on the pedestrian area instead of the road. The pavement materials used in this area and possibly also the depth of the heating pipes varies from what is used for a road, and surface temperatures can therefore also vary.

There are differences in the predicted temperatures for locations 1 and 2. A possible explanation for this difference is the location of the sensors measuring the temperature on top of the heating pipes. This data is used as input data for the model. The heated pavement consists of several loops of pipes. At the beginning of such a loop, the temperature is about 10°C higher than at the end. One of the sensors is located close to the beginning of such a loop and the other close to the end.

Thermal Conductivity through the Snow Layer

From January 7 to 14, 2016, the road was covered with snow. Two different models for thermal conductivity through the snow layer were compared. The formula by Sturm et al. (*17*) gave the best results during the first part and the formula by Nuijten and Høyland (*18*) gave the best results during the second period.

During the first period the liquid water content was zero. Starting at 7.37 p.m. on January 8 the asphalt surface temperature reached 0°C and the snow started to melt. The formula by Sturm et al. (*17*) and other formulas in which the thermal conductivity is described as a function of the density (*14–16*) are based on data retrieved from thermal conductivity measurements of various types of seasonal snow with densities up to 600 kg m^{-3} and for snow with a low liquid water content. The thermal conductivity can be increased by either an increase in liquid water content, which finally results in the thermal conductivity of water (0.58 W $m^{-1}K^{-1}$), or by a decrease of the volume fraction of air, which results in the thermal conductivity of ice (2.1 W $m^{-1}K^{-1}$). Since the formulas by Sturm et al. (*17*), Abels (*15*), Aggarwal et al. (*16*), and Yen (*14*) are based on snow with a low liquid water content, they will give too large values for snow with increased densities due to an increase in liquid water content.

The formula by Nuijten and Høyland (*18*) is developed specifically to determine the thermal conductivity of a melting snow layer, and validated with measurements of a melting uncompressed and compressed snow layer. In this formula, air and ice are modeled as a series system and water and the combination of air/ice in parallel. As can be seen in Figure 5, the model improves the results for a melting snow layer, but gives worse results for dry snow conditions. When the liquid water content of the snow is very low, the model only consists of ice and water parts which are modeled in series. As shown by Nuijten and Høyland (*18*), a series system will give comparable thermal conductivity values for low densities, but for increasing densities due to compression, such as is caused by traffic, the series system will give values which are too low.

Use of the Model for Optimal Regulation of the System

When snowfall is expected or when slippery conditions are predicted to occur, the temperature of the input water for the heating pipes could be adjusted to increase the pavement surface temperature. Since the system has a delayed response it is necessary to adjust the water in the heating pipes about two hours before a predicted weather change. In order to keep the pavement and sidewalks free of snow and ice at all times and during all weather conditions, the surface temperature must be kept quite high, which means much energy is needed to heat up the pavement surface temperature. If the system is allowed to have short periods during which the road is covered with snow, such as during hours when the traffic intensity is very low, less energy is needed. Also during extreme weather events longer periods with slippery conditions could possibly be allowed. The public will not expect perfect conditions during a heavy snowstorm and will take other measures to be safe.

Conclusion

A physical model for temperature and surface condition prediction is built which can, in combination with weather forecasts, be used to predict slippery conditions for snow-covered roads and heated pavements. The model presented is stable and can accurately predict the surface temperature during most of the winter season. When run with measured weather data the average error of the model is 0.85°C and 0.35°C for the two measured locations and the RMSE is 1.73°C and 1.59°C respectively.

When the surface of the road is covered with snow the thermal conductivity through the snow layer is included in the model. Two different models to describe the thermal conductivity through the snow layer were compared. Results show that when the liquid water content of the snow is very low the thermal conductivity through the snow layer can best be determined based on the formula by Sturm et al. (*17*), in which the thermal conductivity is related to the density. For a liquid water content above zero (when the snow starts to melt) the thermal conductivity can best be determined based on the formula by Nuijten and Høyland (*18*), in which the thermal conductivity is calculated as a function of the mass fractions and the thermal conductivities of ice, air, and water. In the model, air and ice are modeled as a series system and water and the combination of air/ice in parallel.

When the liquid water content of the snow is very low, the model by Nuijten and Høyland (*18*) only consists of an ice and water part, modeled in series. For increasing densities due to compression the model gives values which are too low. The formula by Sturm et al. (*17*) is based on snow with densities up to 600 kg m^{-3} and a low liquid water content. For increasing densities the thermal conductivity approaches the value of ice instead of water. Better results are expected when these thermal conductivity formulas are applied on a uncompressed melting snow layer with no runoff. Further research needs to be completed to examine the effect of runoff and compression due to traffic on the thermal conductivity of a melting snow layer. Even though this model is developed for melting snow layers on heated pavement, it might also be used for snow melting due to other sources, for example, solar radiation or traffic heat.

Acknowledgments

This work has been made possible by financial support from the Norwegian Public Roads Administration. The authors are grateful for their support. The authors also thank the Bærum Kommune for providing the data for this analysis by equipping a heated pavement system with temperature sensors and installing a weather station. The authors thank Ed McCormack for his comments on the paper.

Author Contributions

The authors confirm contribution to the paper as follows: study conception and design: AN, IH, KH; data collection: AN; analysis and interpretation of results: AN, IH, KH; draft manuscript preparation: AN. All authors reviewed the results and approved the final version of the manuscript.

References

1. van Bijsterveld, W. T., and A. H. de Bondt. *Structural Aspects of Asphalt Pavement - Heating and Cooling Systems.* Third International Symposium on 3D Finite Element for Pavement Analysis, Design and Research, Amsterdam, 2002.
2. Xu, H., and Y. Tan. Modeling and Operation Strategy of Pavement Snow Melting Systems Utilizing Low-Temperature Heating Fluids. *Energy*, Vol. 80, 2015, pp. 666–676. https://doi.org/10.1016/j.energy.2014.12.022
3. Chapman, L., J. E. Thornes, and A. V. Bradley. Modelling of Road Surface Temperature from a Geographical Parameter Database. Part 2: Numerical. *Meteorological Applications*, Vol. 8, No. 4, 2001, pp. 421–436.
4. Crevier, L.-P, and Y. Delage. METRo: A New Model for Road-Condition Forecasting in Canada. *Journal of Applied Meteorology*, Vol. 40, No. 11, 2001, pp. 2026–2037.
5. Denby, B. R., I. Sundvor, C. Johansson, L. Pirjola, M. Ketzel, M. Norman, K. Kupiainen, M. Gustafsson, G. Blomqvist, M. Kauhaniemi, and G. Omstedt. A Coupled Road Dust and Surface Moisture Model to Predict Non-Exhaust Road Traffic Induced Particle Emissions (NORTRIP). Part 2: Surface Moisture and Salt Impact Modelling. *Atmospheric Environment*, Vol. 81, 2013 pp. 485–503. https://doi.org/10.1016/j.atmosenv.2013.09.003
6. Greenfield, T. M., and E. S. Takle. Bridge Frost Prediction by Heat and Mass Transfer Methods. *Journal of Applied Meteorology and Climatology*, Vol. 45, No. 3, 2006, pp. 517–525.
7. Hermansson, Å. Mathematical Model for Paved Surface Summer and Winter Temperature: Comparison of Calculated and Measured Temperatures. *Cold Regions Science and Technology*, Vol. 40, No. 1–2, 2004, pp. 1–17. https://doi.org/10.1016/j.coldregions.2004.03.001
8. Kangas, M., M. Heikinheimo, and M. Hippi. RoadSurf: A Modelling System for Predicting Road Weather and Road Surface Conditions. *Meteorological Applications*, Vol. 22, No. 3, 2015, pp. 544–553.
9. Shao, J., and P. J. Lister. An Automated Nowcasting Model of Road Surface Temperature and State for Winter Road Maintenance. *Journal of Applied Meteorology*, Vol. 35, No. 8, 1996, pp. 1352–1361. https://doi.org/10.1175/1520-0450(1996)035<1352:AANMOR>2.0.CO;2
10. Chapman, W. P. Design of Snow Melting Systems. *Heating and Ventilating*, Vol. 49, 1952, pp. 88–95.
11. Liu, X., S. J. Rees, and J. D. Spitler. Modeling Snow Melting on Heated Pavement Surfaces. Part I: Model Development. *Applied Thermal Engineering*, Vol. 27, No. 5–6, 2007, pp. 1115–1124. https://doi.org/10.1016/j.applthermaleng.2006.06.017

12. Rees, S. J., J. D. Spitler, and X. Xiao. Transient Analysis of Snow-Melting System Performance. *ASHRAE Transactions*, Vol. 108, No. 2, 2002, pp. 406–423.
13. Xu, H., and Y. Tan. Development and Testing of Heat- and Mass-Coupled Model of Snow Melting for Hydronically Heated Pavement. *Transportation Research Record: Journal of the Transportation Research Board*, 2012. 2282: 14–21.
14. Yen, Y. C. *Review of the Thermal Properties of Snow, Ice and Sea Ice*. CRREL Report 81-10. U.S. Army Cold Regions Research and Engineering Laboratory, Hanover, N.H., 1981.
15. Abels, H. Beobachtungen der Täglichen Periode der Temperatur im Schnee und Bestimmung des Warmeleitungsvermogens des Schnees als Function Seiner Dichtigkeit (Observations of Daily Temperature Variations in a Snow Cover) (in German). *Repert. Meteorol. Herausgegeben Kaiserlichen Akademie der Wissenschaften*, Vol. 16, 1892, pp. 1–53.
16. Aggarwal, R. K., P. S. Negi, and P. K. Satyawali. New Density-based Thermal Conductivity Equation for Snow. *Defence Science Journal*, Vol. 59, No. 2, 2009, pp. 126–130.
17. Sturm, M., J. Holmgren, M. König, and K. Morris. The Thermal Conductivity of Seasonal Snow. *Journal of Glaciology*, Vol. 43, No. 143, 1997, pp. 26–41.
18. Nuijten, A. D. W, and K. V. Høyland. Modelling the Thermal Conductivity of a Melting Snow Layer on a Heated Pavement. *Cold Regions Science and Technology*, Vol. 140, 2017, pp. 20–29. https://doi.org/10.1016/j.coldregions.2017.04.008
19. Colbeck, S., E. Akitaya, R. Armstrong, H. Gubler, J. Lafeuille, K. Lied, D. McClung, and E. Morris. *The International Classification for Seasonal Snow on the Ground*. The International Commission on Snow and Ice of the International Association of Scientific Hydrology, Boulder, Colo., 1990.
20. Nuijten, A. D. W. Runway Temperature Prediction, A Case Study for Oslo Airport, Norway. *Cold Regions Science and Technology*, Vol. 125, 2016, pp. 72–84. https://doi.org/10.1016/j.coldregions.2016.02.004
21. F. P. Incropera, T. L. Bergman, D. P. DeWitt, and A. S. Lavine. *Principles of Heat and Mass Transfer*. John Wiley & Sons, New York, 2013.
22. Solaimanian, M. K., and T. W. Kennedy. Predicting Maximum Pavement Surface Temperature Using Maximum Air Temperature and Hourly Solar Radiation. *Transportation Research Record: Journal of the Transportation Research Board*, 1993. 1417: 1–11.
23. Fujimoto, A., R. A. Tokunaga, M. Kiriishi, Y. Kawabata, N. Takahashi, T. Ishida, and T. Fukuhara. A Road Surface Freezing Model Using Heat, Water and Salt Balance and Its Validation by Field Experiments. *Cold Regions Science and Technology*, Vol. 106–107, 2014, pp. 1–10. https://doi.org/10.1016/j.coldregions.2014.06.001
24. Chiasson, A., J. D. Spitler, S. J. Rees, and M. D. Smith. A Model for Simulating the Performance of a Pavement Heating System as a Supplemental Heat Rejecter with Closed-Loop Ground-Source Heat Pump Systems. *Journal of Solar Energy Engineering*, Vol. 122, No. 4, 2000, pp. 183–191.
25. Hermansson, Å. Simulation Model for Calculating Pavement Temperatures Including Maximum Temperature. *Transportation Research Record: Journal of the Transportation Research Board*, 2000. 1699: 134–141.
26. Andersland, O. B, and B. Ladanyi. *Frozen Ground Engineering*. John Wiley & Sons, New York, 2004.
27. Oke, T. R. *Boundary Layer Climates*, 2nd ed. Methuen Publishers, Lagos, 1987.

The Standing Committee on Winter Maintenance (AHD65) peer-reviewed this paper (18-03966).

Article

TRR
JOURNAL OF THE TRANSPORTATION RESEARCH BOARD

Transportation Research Record
2018, Vol. 2672(12) 232–242

DOI: 10.1177/0361198118796354
journals.sagepub.com/home/trr
SAGE

Role of Route Optimization in Benefiting Winter Maintenance Operations

Tyler Miller[1], Ben Gleichert[1], Hiram Crabtree[2], Jamie Hendershot[2], Rod Nuveman[2], and William Schneider[1]

Abstract

For winter maintenance purposes, the Ohio Department of Transportation (ODOT) deploys a fleet of approximately 1,600 snowplow trucks that maintain 43,000 lane miles of roadway. These trucks are based out of 200 garages, yards, and outposts that also house a combined total of 650,000 tons of salt. The deployment of such a large number of trucks over a vast maintenance area creates an operational problem in determining the optimal maintenance routes and fleet size. Traditionally, ODOT has used county borders as maintenance boundaries for ODOT garages. However, by removing these borders and optimizing the snowplow routes, ODOT may benefit from significant time and cost savings. The results of this project provide ODOT a tool to determine the minimum number of trucks needed to maintain the necessary roadways within Districts 1, 2, and 10. In addition, the project recommends to ODOT a tool to assign assets to specific facilities and the most optimal routes for each truck in a district. This research indicates the possibility of reducing the fleet size within Districts 1, 2, and 10 by a total of 29 trucks while maintaining the same level of service. Apart from fleet size reduction, the route optimization can decrease the amount of time in which to treat each road in the three districts by 17 hours for one iteration.

For winter maintenance purposes, the Ohio Department of Transportation (ODOT) deploys a fleet of approximately 1,600 snowplow trucks that maintain 43,000 lane miles of roadway. These trucks are based out of 200 garages, yards, and outposts that also house a combined total of 650,000 tons of salt (*1*). The deployment of such a large number of trucks over a vast maintenance area creates an operational problem in determining the optimal maintenance routes and fleet size. ODOT recently through the use of ArcGIS is trying to advance their route optimization. This optimization will help future planning in determining the necessary number of trucks that may maintain state highways. Traditionally, ODOT has used county borders as maintenance boundaries for ODOT garages. However, by removing these borders and optimizing the snowplow routes, ODOT may realize significant time and cost savings.

Objectives

The major objective of this research project is to optimize snow and ice routes for ODOT's snowplow trucks in Districts 1, 2, and 10 while eliminating county border restrictions. There are numerous benefits expected from the outcome of this project. One important benefit will be the economization of the fleet size in three of ODOT's twelve districts which ensures ODOT's efficiency in maintaining all the required roadways in these districts. In addition, each facility within Districts 1, 2, and 10 will know the specific roadways that it must maintain, regardless of the amount of resources available during winter maintenance operations. Another benefit of this research is the route optimization model (ROM) which serves as a tool to analyze new equipment technology and new operational considerations.

Literature Review

In determining the potential cost savings from the optimization of winter maintenance operations, a literature review was conducted on how optimization models were created and implemented in other agencies and organizations. The literature review consists of articles published academically and a look at the optimization of fleets and

[1]Department of Civil Engineering, The University of Akron, Akron, OH
[2]The Ohio Department of Transportation, Columbus, OH

Corresponding Author:
Address correspondence to William Schneider: whs4@uakron.edu

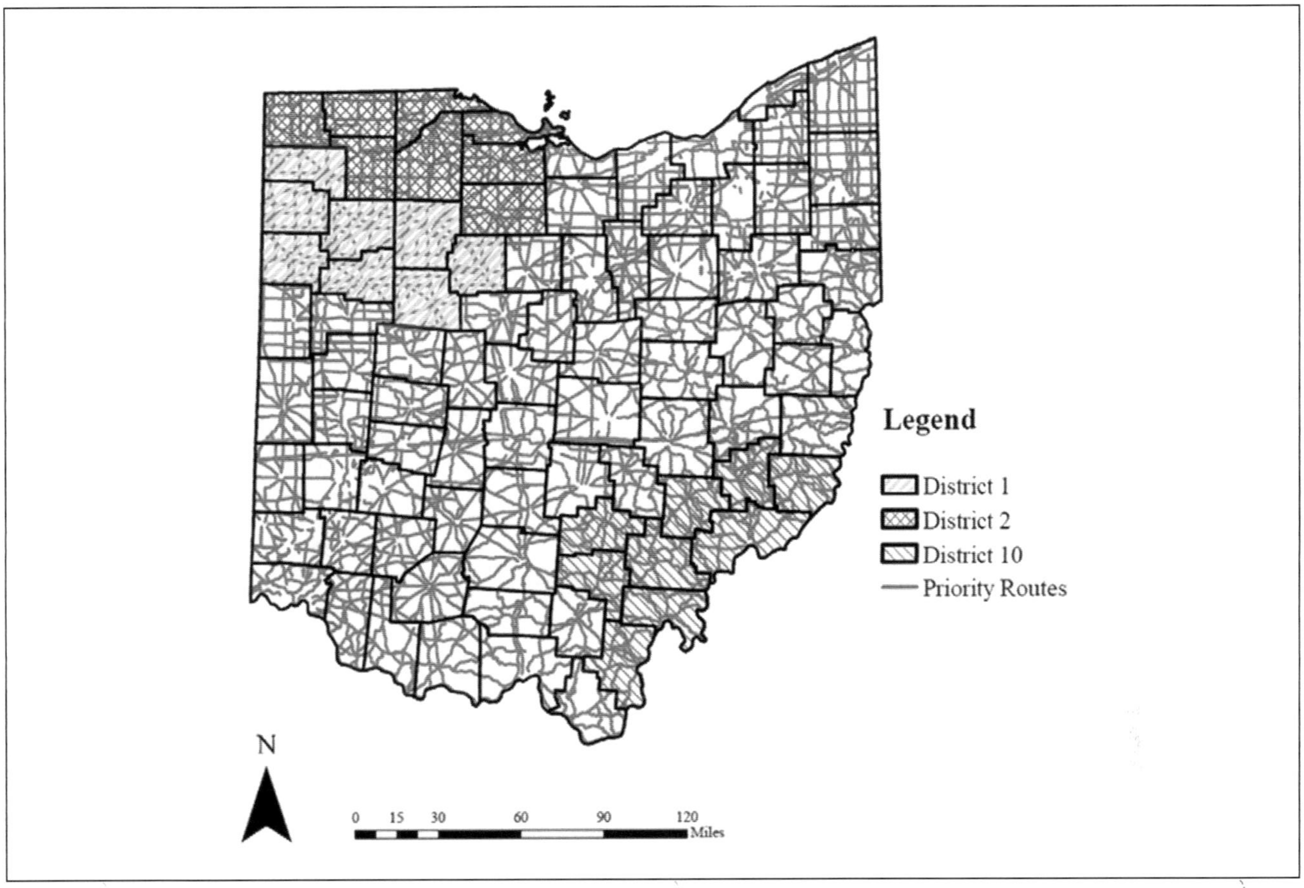

Figure 1. ODOT districts involved with the route optimization project.

individual vehicles in state departments of transportation. A summary of the findings is shown in Table 1.

Project Setting

The route optimization project was conducted in ODOT Districts 1, 2, and 10. As shown in Figure 1, these districts represent the northwestern and southeastern corners of the state.

These districts entail unique geographic and meteorological demands for winter maintenance operations. Table 2 provides a general background of the snow and ice routes as well as the operational and environmental management of the three districts that are used in this study. In winter maintenance operations, the districts are allowed to set some of their own individual performance metrics criteria. Table 3 shows the suggested level of service (LOS) and corresponding plow speeds that are commonly used during winter maintenance operations. As shown in Table 3, the performance metric requirements are divided into three levels of priority routes. Priority one routes (interstate highways, for example) typically have the highest design standards and volumes of traffic whereas priority three routes typically have lower design standards and traffic volumes. In winter maintenance operations as shown in Table 3, it is common to have more monitoring and activity on priority one segments than priority three segments.

ROM Methodology

The development of the ROM is set within Esri's ArcGIS software platform. This setting was selected due to the fact the end user currently operates within this setting and required all solutions to be in a commercially available software and that all additional computer coding was done within general accepted practices.

Base Layer

The building of the ROM base layer began with the preparation of data in the form of layers for the map of the state of Ohio, which included the roads dataset for the entire state of Ohio. Once all the roads in Ohio were uploaded, a network analyst dataset was created in order to define the roadway edges and capture elevation

Table 1. Literature Review Summary

Project goal	Methodology	Findings	Reference
Provide a sustainable optimization of winter maintenance service by maximizing the potential of the service to ensure steady traffic conditions for Kraljevica, Croatia.	Develop a model to take route optimization away from human managers and use computer models based on the Soyster Heuristic.	The analysis shows that current maintenance fleet numbers and depot locations are sufficient for optimal road maintenance of the city.	(*2*)
Develop snow and ice control operations for storm-specific routes to maximize the efficiency of the service provided in terms of man-hours and fuel.	Optimize the vehicles by garage based on the combined service time/fuel consumption metric. Evaluate the competing vehicle allocations based on the speed at which high priority roads are serviced.	Allocations based on roadway Network Robustness Index (NRI) allowed the state to optimize both man-hours and fuel consumption.	(*3*)
Implement a synchronized routing problem for the snowplowing operations	Develop a model to synchronize arc routing for winter maintenance operations.	An improvement was added to the optimizing algorithm that resulted in significant improvement to the efficiency of the plowing model.	(*4*)
Optimize the routes and fleet allocations for Missouri Department of Transportation (DOT) to provide a sufficient LOS.	Develop an integrated algorithm for Missouri DOT to determine the most efficient route plans and fleet allocations	A list of various conditions and responses as needed for the Missouri DOT.	(*5*)
Determine optimal workforce planning and shift scheduling for snow and ice removal.	Develop a methodology for deployment of available crews and equipment to maintain the most efficient implementation of resources.	Use of contract employees reduces the total cost to Missouri DOT.	(*6*)
Develop a data model to represent the transportation network of an urban area to be used for route planning.	Develop an urban transportation network using standard and customized GIS software tools.	A model was developed that incorporated a multimodal network with restrictive attributes to represent real-world scenarios.	(*7*)
Determine methods for producing optimal deployment schedules to conduct winter maintenance operations.	Develop a model to take into account a variety of road and weather conditions to aid in winter maintenance operations planning.	Provides a method to produce optimal deployment schedules and a framework to compare future research.	(*8*)
Enhance a decision support system for assisting the Maryland State Highway Administration's Office of Maintenance staff in designing snow emergency routes for Calvert County, MD.	Assign segments of the treated road network to trucks so that the number of trucks is minimized and all routes are continuous.	The Genetic Algorithm with First Fit heuristic reduces the number of trucks for Calvert County from 14 to 12 trucks (14%).	(*9*)
Decide how to use arc routing methods to determine routing.	Analyze arc routing methods and applications.	Developed a list of steps to use for conducting arc routing methods.	(*10*)
Develop a model to predict costs and benefits of winter maintenance operations in the state of Idaho.	Using historical data, develop a model for each district to accurately predict costs and benefits of winter maintenance operations.	The model assists in estimating the benefits to safety, travel time, and fuel cost.	(*11*)
Provide a description of a computer application to assist in determining the routing of street sweepers.	Use computer programs to optimize the routes for street-sweeping trucks.	Provides a list of steps to create route optimizing models.	(*12*)

Note: The findings listed in Table 1 plus articles (*13, 14*) support the idea that the optimization of ODOT's winter maintenance fleet may result in cost savings while maintaining current level of service (LOS) in the districts. In addition, the findings from the literature review assisted in developing a methodology to conduct route optimization in ODOT Districts 1, 2, and 10.

differences. In addition to elevation differences, network attributes in the road layer included road hierarchy (freeway, arterial, collector, or local road), direction of travel (two-way vs. one-way roads), cost attributes (such as travel time), and distance. The one-way road attribute is utilized to ensure that the model does not route a vehicle in the wrong direction. The directional basis of the roads, which is an especially important consideration when routing on divided highways, was built into the road layer.

Once the routes were digitized, the next step was to locate the garages, outposts, and yards within ArcGIS.

Table 2. Summary of District Snow and Ice Routes

District snow and ice routes	District operational and environmental management
 District 1	District 1 is responsible for maintaining approximately 3,200 lane miles of state and federal highways. District 1 operates from eight garages and nine outposts. District 1 is level terrain that receives 20–30 inches of snow annually with a significant amount of blowing snow.
 District 2	District 2 is responsible for maintaining approximately 3,381 lane miles of state and federal highways. District 2 operates from eight garages and two outposts. District 2 is level terrain that receives 30–40 inches of snow annually with a significant amount of blowing snow.
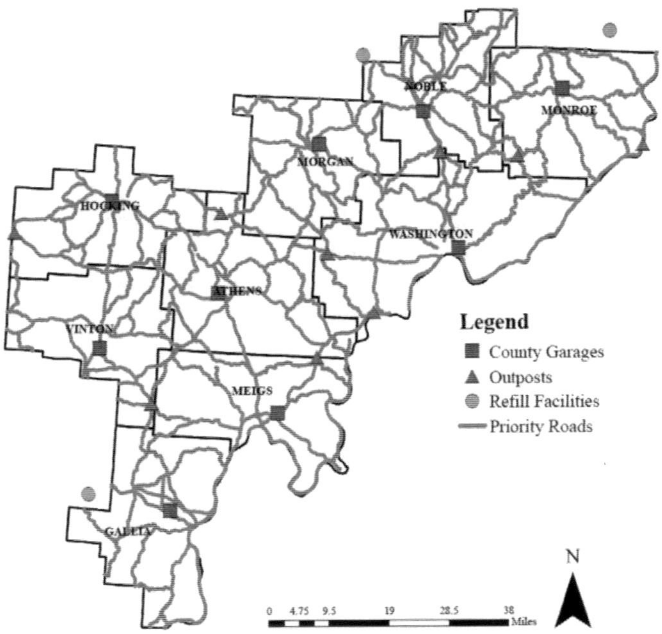 District 10	District 10 is responsible for maintaining approximately 4,000 lane miles of state and federal highways. District 10 operates from nine garages, eight outposts, and three refill stations. District 10 is hilly terrain with an elevation change of 933 feet and receives 20–30 inches of snow annually with a significant amount of blowing snow.

Note: The source of this information is provided by ODOT. The maps were created using ArcGIS.

Table 3. Winter Operations Performance Metrics for ODOT Districts 1, 2, and 10

District	Suggested LOS cycle time			Suggested plow speed		
	Priority 1 (min)	Priority 2 (min)	Priority 3 (min)	Priority 1 (mph)	Priority 2 (mph)	Priority 3 (mph)
1	60	90	120	40	30	30
2	60	90	120	30	30	30
10	90	120	150	35	25	20

Note: The LOS requirements shown above were acquired from each district's leadership.

Next, the current number of trucks stationed at each garage, outpost, and yard was inputted in the model. Then, the snow and ice routes provided by ODOT were added so that trucks could be routed from each garage along the routes.

Vehicle Routing Problem

After the necessary data collection to develop the initial inputs for the ROM, the next step was to use Esri's Vehicle Routing Problem (VRP) in the ArcGIS software platform. The VRP was utilized to create and finalize the model before route optimization for the districts. The following applications were addressed to create and finalize the ROM within each district: loading plowing locations to be used in the VRP, limiting trucks from traveling on county and township roads, assigning an associated capacity and salt application rate to each truck so that the VRP would account for when a truck will run out of material and will need to refill its hopper. Finally, LOS inputs were included into the VRP such as assigning cycle times and truck speeds (Table 3) to ensure that LOS was maintained.

For the VRP to be capable of using the previously described data, point locations or "orders" were inserted along the roadways. Within each of these orders was included the information on the allowed salt application rates and the prioritization of the routes. Both of these parameters could be modified to account for various application rates and the unique cycle time requirements within each district.

Initial Route Optimization

In order to optimize the routes in a district, the ROM was utilized with the entire truck inventory that each district maintained. By first utilizing all trucks within a district, the ROM was able to provide a baseline of the area that each facility was most capable of maintaining while best satisfying the LOS requirements within the district. The initial route optimization was accomplished by restricting the trucks from traveling on county and township roads, by normalizing driving at typical treating speeds for each district (Table 3), and by utilizing an application rate of 250 lbs/ln mile. This initial optimization of routes also took into account the removal of county border limits within the district, thus allowing the ROM to determine the most efficient routes and treating areas for each facility.

The cycle times used to determine if the LOS requirements were satisfied for each truck within the district included the deadhead time to travel to and from the treating area, the time required to treat the determined section of a road, and a refill time of ten minutes. The cycle times for the current, the initially optimized, and the fleet optimized trucks were calculated and the percent of trucks meeting the LOS requirements determined.

Another aspect that was considered when presenting the initial optimized routes was the number of cycles that could be completed before a refill was needed. ODOT has both single- and tandem-axle trucks in their respective garages. Single-axle trucks have an 8 ton capacity while tandem-axle trucks have an 11 ton capacity. These capacities and the general application rates are extremely important in determining the number of cycles a truck is able to complete without having to refill. The goal was to maximize the number of completed cycles, as it is not effective for a truck to complete only 1.4 cycles for an assigned route before it needs a refill. Once the initial model using the base number of trucks was developed, the next step was to verify the accuracy of the routes and cycle times.

Verification Process

The verification of the initial model using the VRP developed by Esri was implemented by driving all the snow and ice routes in all three districts. In all driving cases, a GPS unit was installed in all 300 plus trucks to record the actual field cycle times. The data for the driven routes were collected from GPS transponders and further analyzed, the results of which are shown for the three districts in Table 4.

In all cases the results were positive and showed that the software-predicted cycle times, and the in-field cycle

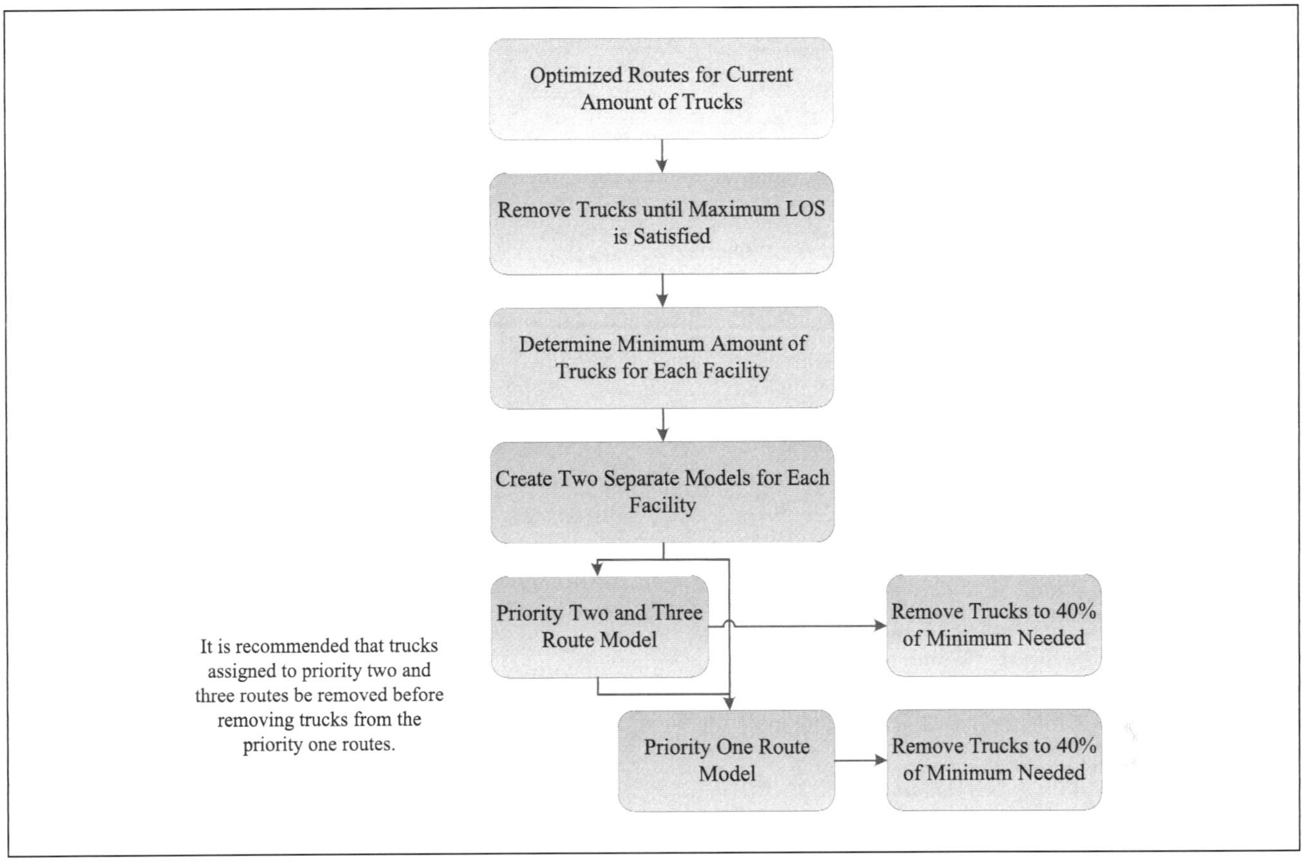

Figure 2. Fleet optimization methodology.

Table 4. ROM Initial Model Verification

District	Equation	R^2
1	Y=1.01X	0.99
2	Y=1.03X	0.98
10	Y=1.01X	0.97

Note: Y = verification time; and X = ROM time.

times were consistent and reliable. Based on this validation of the data, the next phase of this study was to optimize the fleet.

Fleet Optimization

The fleet optimization section summarizes the final results obtained from the optimized winter maintenance operations in Districts 1, 2, and 10. The optimization of the fleets consisted of removing the least efficient trucks from the initially optimized routes in order to determine the minimum number needed at each facility to maintain the current LOS in the districts. This analysis was completed by using the LOS and speed requirements (Table 3) for each district, the typical treating speeds, the removal of county borders, and utilizing an application rate of 250 lbs/ln mile (typical application) and 400 lbs/ln mile (worst-case scenario application). The range of application rates allowed the ROM to determine an accurate number of operational trucks needed to best treat the roads in each district. The flowchart in Figure 2 describes the methodology used to optimize the fleet in these districts.

Figure 2 provides a visual representation of the fleet optimization process that has been conducted in all three districts. Fleet optimization used the ROM to take stock of all trucks in a district and then remove those from the least efficient routes at each facility. After removing the trucks, the ROM was re-run and new routes were created to accommodate the missing trucks. The process of removing trucks from the least efficient routes and running the optimization model was repeated until the current LOS requirements were satisfied in the district.

It is important to note that the efficiency for fleet-optimized trucks falls within a range of values and is not a fixed number. To account for this range, the number of cycles that may be completed by each truck at an application rate of 250 lbs/ ln mile and 400 lbs/ln mile was taken into account. The lower end of the efficiency range was calculated as the number of cycles that may be completed at the higher application rate and the higher end of the efficiency utilized the lower application rate. The efficiency of each route is determined from the application rate applied and the capacity of the truck, which resulted in a range of efficiencies rather than a specific efficiency value, in which the low efficiency is calculated by determining the amount of cycles that may be completed by using 400 lbs/ln mile application rate and the high efficiency determined from the amount of cycles that may be completed at a 250 lbs/ln mile application rate. The equation used to calculate the low and high efficiencies for each route is

$$\text{Individual Truck Efficiency} = \frac{\text{TCT}}{\text{TCT} + \text{DHT}} \quad (1)$$

where

TCT = treated cycle time, and

DHT = deadhead time.

TCT is the time to drive and treat the optimized roads for each truck and DHT is the time to drive to and from the treating area for each truck and total time traveled. Not included in either equation is the calculated time to complete one cycle, which includes the time to refill the truck.

The range of the potential efficiencies for each truck was then used to calculate the overall district efficiency range. The equation used to calculate the district efficiency is

$$\text{District Efficiency} = \frac{\Sigma\ \text{IRE}}{T} \quad (2)$$

where

IRE = individual route efficiencies, and

T = number of operational trucks.

After the minimum number of routes for the district was determined, trucks were continuously removed from the ROM in order to analyze the changes in cycle times, deadhead times, and lane miles treated for each route. Trucks were continuously removed from each facility until approximately 40% of the minimum was reached to satisfy the LOS requirements in the district. The removal of trucks until reaching 40% of the minimum amount needed would yield enough data to conduct an initial analysis of how the routes change to accommodate the missing trucks and provide a valuable tool for winter maintenance managers as they determine how to allocate their limited resources to best conduct winter maintenance operations. It is important to note that trucks assigned to the priority two and three routes were removed before the trucks assigned to the priority one routes. This was done to ensure that the priority one routes in the district are effectively maintained regardless of the resources available at each facility at any given time.

Results

The primary parameters used in this paper consisted of the LOS requirements, the average speed at which a truck travels, the time required for refill, and the capacity of each truck. The LOS requirements represent the maximum time required for a road to be treated and are a primary factor in comparing the current routes with the optimized routes. A route was considered to satisfy the LOS if the truck was able to leave the facility, treat the assigned roadways, return to the facility, and then refill. If the truck was able to complete a full cycle under the LOS requirements, it was determined that the truck satisfied the LOS.

The capacity of each truck in the district directly correlates with the efficiency calculations for the individual routes, each facility, and the district as a whole. It is important to note that the efficiency is not a fixed value but rather a range of values dependent upon the application rate. The highest efficiency of a truck has the potential for numerous cycles to be completed at an application rate of 250 lbs/ln mile. The lowest efficiency occurs in the worst-case scenario of 400 lbs/ln mile with each truck being able to complete one cycle before requiring a refill.

Performance Metrics

There are three methods to compare the results for the ROM. The three methods by which this paper evaluates the before- and after-optimization are:

- Method One: Maintain the same LOS,
- Method Two: Maintain the same number of trucks, and
- Method Three: Maximize the LOS for the district.

The first method compares the number of trucks required to maintain the same LOS. In this example, the initial model, which is the current operating conditions, sets the expected LOS for the district under a set number of trucks. In this case, the district under normal conditions is able to use 100 trucks to maintain a LOS of 65%, the optimized model would hold the 65% LOS constant, and would begin to remove trucks until the model reached 65%. This method would show in theory, savings in the number of trucks used in the field. The second method holds the number of trucks constant. In this case, say, the district uses 100 trucks and achieves an overall LOS of 65%, the optimized model would hold the number of trucks equal to 100 and then would determine the new LOS. This method in theory would show an improvement in the overall LOS for the district. The third method compares the maximum LOS that is available given additional trucks to the district. This third

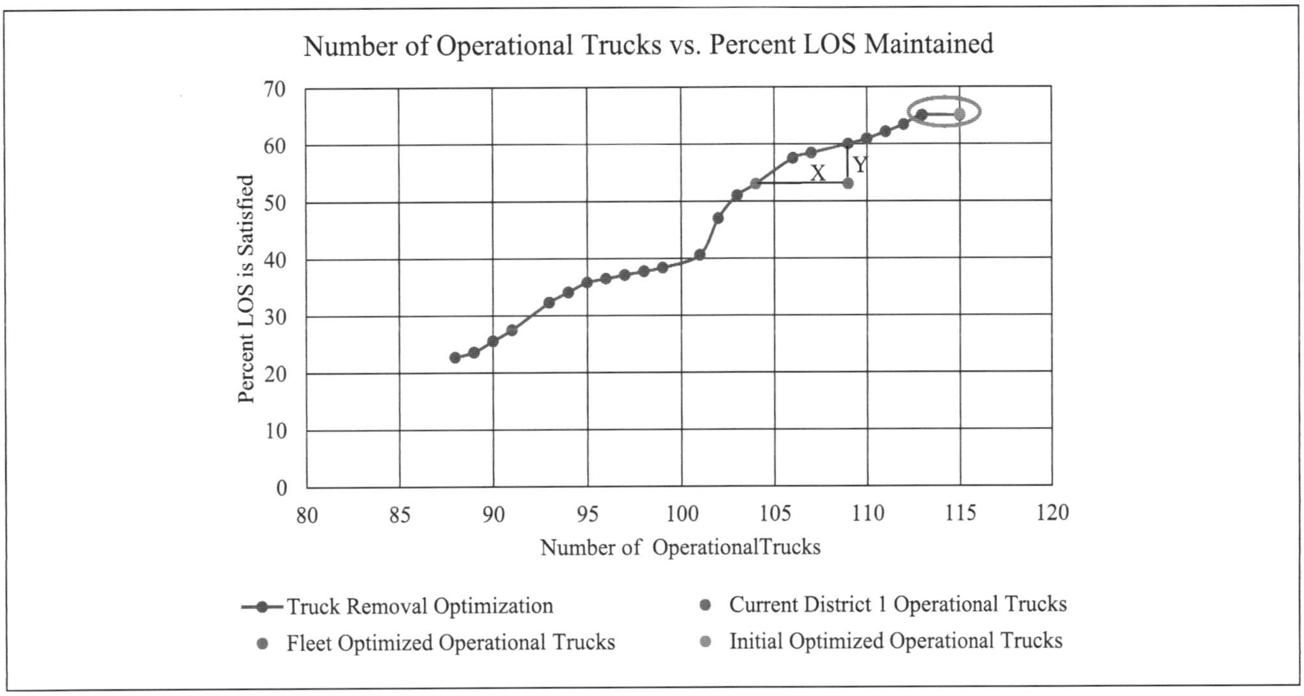

Figure 3. District 1 LOS analysis.

method is important to show that, without additional resources, such as new garages, outposts, or yards, it may be impossible to improve the LOS of a district using the criteria shown in Table 3. Simply stated, there may be too much travel time between garages and treating areas. Figure 3 shown below provides an example of the three performance metrics that may be used to evaluate the efficiency of route optimization for winter maintenance operations.

As may be observed in Figure 3, District 1 currently operates 109 operational trucks with 53% of the trucks satisfying the LOS requirements. A fleet optimization determined that the same LOS, the first method, might be obtained with 104 operational trucks. Line "X" in Figure 3 above provides a visual representation of how the number of operational trucks may be decreased from 109 to 104 at the same LOS. The second method maintains the same 109 trucks. The results show that the LOS increases from 53% to 60%. And the third method indicates that, if additional trucks, 113, are added to the snow and ice routes, the maximum LOS that may be satisfied is approximately 65%, as shown by the green circle in Figure 3. This plateau occurs after implementing operational constraints (Table 3), as well as safe turnaround locations for plow trucks and ensuring that an acceptable amount of lane miles is maintained. Ultimately, what this plateau suggests is that ODOT would need to add more facilities, most likely a seasonal yard where trucks could re-fill their salt supply, to improve the LOS to the district above 65%.

The analysis shown in Figure 3 was done for all three districts. The summary of these findings is shown in Table 5. The number of operational trucks shown in Table 5 represents the three methods currently used to evaluate the performance of the districts during winter maintenance operations.

The results for Method One analysis show savings of five trucks, eight trucks, and seven trucks respectively for Districts 1, 2 and 10. In addition to the reduction in time, the total cycle times improve by 208 min/cycle, 371 min/cycle, and 827 min/cycle. In other words, Method One improves both the fleet reduction and time. These results are important because they may represent capital savings of 3 million dollars, assuming the cost of one truck at approximately $150,000. These results may also represent 3,750 hours in saved time, assuming 40 storms per winter and 4 cycles per storm.

The results for Method Two show an improvement of 7% and 11% respectively for Districts 1 and 2. District 10 is an interesting case. The specific results for District 10 suggest that the district is unable to improve on its existing LOS without either i) modifying the evaluation criteria as indicated in Table 3, or ii) adding new yards where the trucks could re-fill during their hours of operation.

The results for the third method of analysis show that District 1 would be able to increase its fleet by four trucks if it wanted to meet its maximum LOS at 65%. District 2 would be able to increase its fleet by nine trucks to meet the district's maximum LOS at 50%. As

Table 5. Summary of Route Optimization

	Current[a]	Method one: same LOS	Method two: same number of trucks	Method three: maximum LOS
		District 1 results		
Fleet size	109	104	109	113
Total travel time (min/cycle)	7,651	7,443	7,636	7,692
Percent LOS maintained	53	53	60	65
		District 2 results		
Fleet size	114	106	114	123
Total travel time (min)	9,095[b]	8,724	8,864	9,100
Percent LOS maintained	27	27	38	50
		District 10 results		
Fleet size	116	109	–	–
Total travel time (min)	12,650	11,813	–	–
Percent LOS maintained	60	61	–	–

Note: – = adding more than 109 trucks will not increase the other performance metrics because additional outposts need to be added.
[a]This is the number of required trucks in the field. Typically garages maintain a 10% buffer to account for trucks that are mechanically unavailable.
[b]This value is a representation of typical routes performed for a base storm.

stated in the previous paragraph, District 10 would not be able to increase its LOS without adding additional facilities. In all cases, one significant finding is that there will come a threshold point where adding more trucks as a way to increase operational capacity brings no more practical benefit to the overall winter maintenance operations. This trend is shown in the plateau seen in Figure 3. When this plateau occurs, the suggested way to improve is to add additional facilities throughout the region which in turn will lower deadhead time and will improve the overall efficiency of the region.

Facility Optimization

Although the fleet optimization determined the optimal fleet size as shown in Table 5 to maintain the current high LOS in Districts 1, 2, and 10, there are other possible strategies to contemplate. When considering the number of operational trucks satisfying the LOS requirements, there is a pivotal point, the plateau shown in Figure 3, at which the number of operational trucks yields the highest LOS satisfaction possible in a district. This occurs after implementing operational constraints on turnaround locations and lane mile requirements for individual routes. The maximum LOS obtained is limited to the facility locations within a district because remotely located roadways may not be treated under the LOS requirements due to the typical speeds traveled during snow and ice events.

To prove this point, each facility in Districts 1, 2, and 10 was assigned 60 trucks to be utilized in the ROM under the average treating speeds each district specified. The assignment of 60 trucks at each facility removed fleet size concerns since each facility typically utilizes five to 20 operational trucks. The results of this analysis were used to determine the challenging areas in each district where construction of future facilities may accommodate an increase in the number of operational trucks that satisfy the LOS.

In order to show the areas that would benefit from additional facilities, the ROM that utilized the current facility locations with 60 available trucks at each facility to conduct winter maintenance operations was operated. By limiting each truck to strictly adhere to the district requirements, as per Table 3, the "demanding areas" shown in Figure 4 were determined and should be considered the starting point for locating new yards to store salt. These demanding areas were classified as such due to the greatest potential to increase the number of trucks that would satisfy the LOS requirements if facilities were constructed in these areas.

Conclusion

Through the implementation of Esri's Network Analyst Extension and VRP, a ROM was developed to optimize each district's winter maintenance fleet. The ROM was proven to be capable of producing accurate cycle times based on the speeds utilized in the model, as shown in the route verification process. In addition, the ROM was capable of providing a current route analysis for each district. This analysis may prove useful to state agencies as they assess their current winter maintenance operations and continue to pursue methods that improve safety along roadways in a cost-efficient manner.

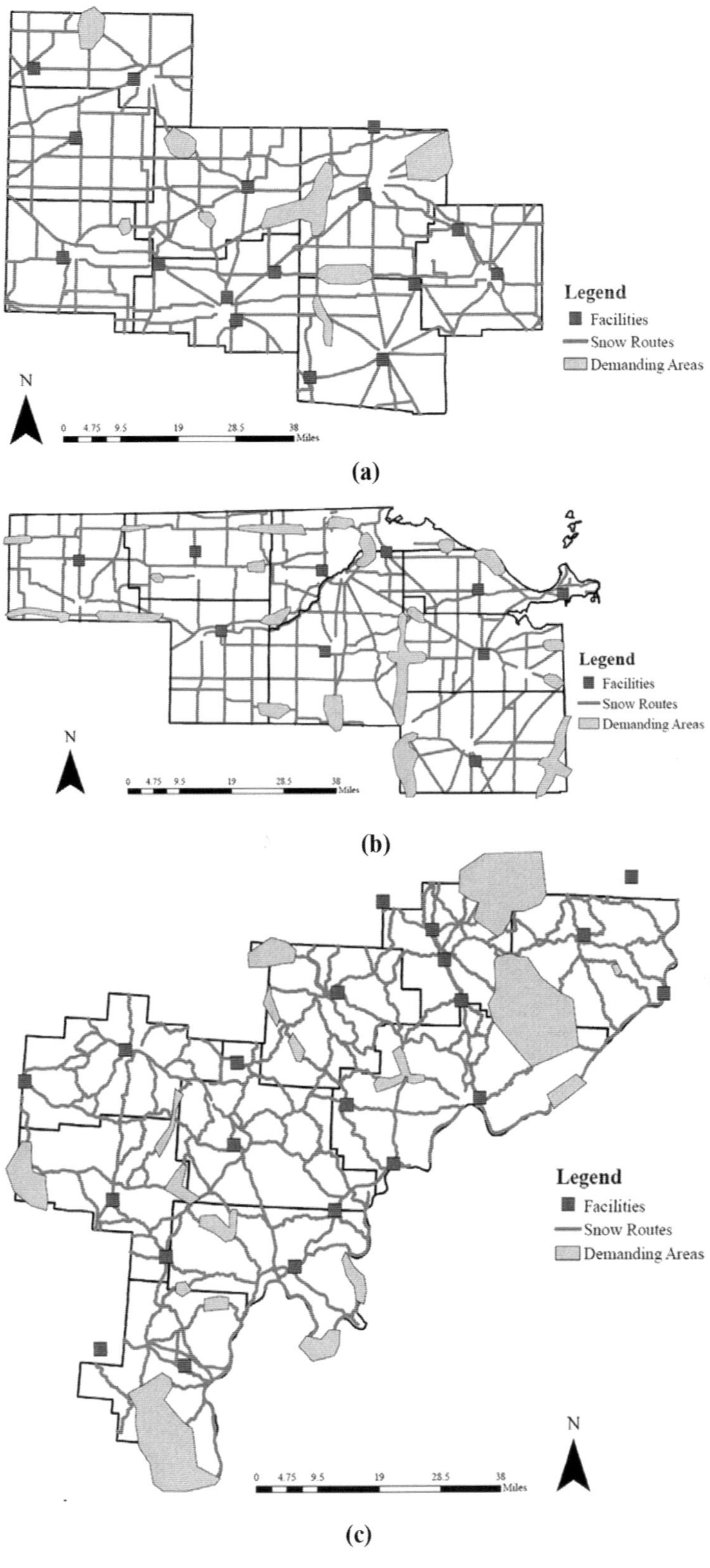

Figure 4. District-level demanding areas: (*a*) District 1 vulnerable areas; (*b*) District 2 vulnerable areas; and (*c*) District 10 vulnerable areas.

Note: Demanding areas are based on optimized vehicle routes. They are not based on land use or cost.

The ROM produced optimized fleets that were capable of maintaining the same or greater LOS currently achieved in Districts 1, 2, and 10. In addition to maintaining the same or greater LOS, each district showed time savings between 208 and 837 min/cycle that may result in safer roads for travelers and decreased costs for ODOT. As state budgets fluctuate, the potential time and cost savings from optimizing ODOT's district fleets may become essential in ensuring the safety of the state's roadways for the driving public.

The second result showed by the ROM is that the addition of new facilities may result in a higher LOS if the current fleet is maintained or increased. A lack of new facilities is the primary cause of significant deadhead times due to the inconvenience of current facility locations. Additional facilities constructed in strategic locations may improve the LOS for the districts. The ROM serves ODOT as a valuable tool in determining where future facilities are best constructed. As shown in Figure 3 of this study, there is a threshold point at which additional trucks will not increase the LOS. This finding is the result of the nature of current facility locations and typical speeds traveled during snow and ice events. By utilizing a saturated fleet that must strictly adhere to the LOS requirements within each district, the ROM was capable of producing areas that could not be maintained within the desired LOS requirements. These areas, listed as "demanding areas", require construction of new facilities in order to provide a greater LOS.

Author Contributions

The authors confirm contribution to the paper as follows: T. Miller, B. Gleichert, H. Crabtree, J. Hendershot, R. Nuveman, and W. Schneider worked on all aspects of this paper. This includes the data setup, data analysis and writing of the paper.

References

1. The Ohio Department of Transportation. *Snow and Ice Practices*. The Ohio Department of Transportation Division of Operations, Columbus, OH, 2011.
2. Gudac, I., T. Hanak, and I. Marovic. *Optimization of Winter Road Maintenance Services Under Real Time Information*. University of Rijeka, Rijecka, Croatia, 2014.
3. Dowds, J., D. Novak, D. Scott, and J. Sullivan. *Optimization of Snow Removal in Vermont*. Vermont Agency of Transportation, Montepelier, VT, 2013.
4. Salazar-Aguilar, A., A. Langevin, and G. Laporte. Synchronized Arc Routing for Snow Plowing Operations. *Computer & Operations Research*, Vol. 39, No. 7, 2012, pp. 1432–1440.
5. Jaung, W. *Optimizing Winter/Snow Removal Operations in MoDOT St. Louis District- Includes Outcome Based Evaluation of Operations Final Report*. Missouri Department of Transportation, Jefferson City, MO, 2011.
6. Gupta, D. *Optimal Workforce Planning and Shift Scheduling for Snow and Ice Removal*. Missouri Department of Transportation, Jefferson, City, MO, 2010.
7. Mandloi, D., and J.-C. Thrill. Object-Oriented Data Modeling of an Indoor/Outdoor Urban Transportation Network and Route Planning Analysis. In *Geospatial Analysis and Modelling of Urban Structure and Dynamics*, Springer, Dordrecht, the Netherlands, 2010, pp. 197–220.
8. Fu, L., M. Trudel, and V. Kim. Optimizing Winter Road Maintenance Operations Under Real-Time Information. *European Journal of Operational Research*, Vol. 196, 2009, pp. 332–341.
9. Haghani, A., and M. Hamedi. *Calvert County Snow Emergency Decision Support System Phase II*. University of Maryland, 2002.
10. Assad, A. A., and B. Golden. Network Routing. In *Handbook in Operations Research* (M. O. Ball, T. L. Magnanti, C. L. Monma, G. L. Nemhauser, eds.), Vol. 8, 1995, pg 375–483.
11. Haber, D., and U. Limage. *Benefit Cost Analysis of Winter Maintenance Levels of the Idaho Transportation Department*. Idaho Transportation Department, Boise, ID, 1990.
12. Bodin, L., and S. Kursh. A Detailed Description of a Computer System for the Routing and Scheduling of Street Sweepers. *Computers & Operations Research*, Vol. 6, 1979, pp. 131–138.
13. Tang, W. J., G. H. Zhang, and M. J. Liao. A Framework of Winter Road Maintenance Optimization. *Proc., 2015 11th International Conference on Natural Computation (ICNC)*, Zhangjiajie, China, IEEE, New York, NY, 2015.
14. Hajibabai, L., S. Nourbakhsh, Y. Ouyang, and F. Peng. Network Routing of Snowplow Trucks with Resource Replenishment and Plowing Priorities: Formulation, Algorithm, and Application. *Transportation Research Record: Journal of the Transportation Research Board*, 2014. 2440: 16–25.

The Standing Committee on Winter Maintenance (AHD65) peer-reviewed this paper (18-02887), which was selected by the Maintenance & Preservation Section (AHD00) to receive its 2018 Impact on Practice Paper Award in Winter Maintenance and Surface Transportation Weather.

Sage Publications USA
sagepub.com

ISBN 978-1-5108-8180-8

9 781510 881808

CURRAN ASSOCIATES INC.
proceedings
.com

International Conference on Design & Construction of Super & Mega Yachts 2019

Genoa, Italy
14 - 15 May 2019

The Royal Institution of
Naval Architects